Mechanical Engineering Series

Frederick F. Ling
Editor-in-Chief

Mechanical Engineering Series

(continued after index)

George Chryssolouris

Manufacturing Systems:
Theory and Practice

 Springer

George Chryssolouris
University of Patras, Greece

Editor-in-Chief
Frederick F. Ling
Earnest F. Gloyna Regents Chair Emeritus in Engineering
Department of Mechanical Engineering
The University of Texas at Austin
Austin, TX 78712-1063, USA
   and
Distinguished William Howard Hart
   Professor Emeritus
Department of Mechanical Engineering,
   Aeronautical Engineering and Mechanics
Rensselaer Polytechnic Institute
Troy, NY 12180-3590, USA

Manufacturing Systems: Theory and Practice, 2nd Ed   Printed on acid-free paper.

ISBN 978-1-4419-2067-6   e-ISBN 978-0-387-28431-6
e-ISBN 0-387-28431-1

9 8 7 6 5 4 3 2 1

springeronline.com

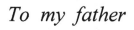

To my father

Mechanical Engineering Series

Frederick F. Ling
Editor-in-Chief

The Mechanical Engineering Series features graduate texts and research monographs to address the need for information in contemporary mechanical engineering, including areas of concentration of applied mechanics, biomechanics, computational mechanics, dynamical systems and control, energetics, mechanics of materials, processing, production systems, thermal science, and tribology.

Series Preface

Mechanical engineering, and engineering discipline born of the needs of the industrial revolution, is once again asked to do its substantial share in the call for industrial renewal. The general call is urgent as we face profound issues of productivity and competitiveness that require engineering solutions, among others. The Mechanical Engineering Series is a series featuring graduate texts and research monographs intended to address the need for information in contemporary areas of mechanical engineering.

The series is conceived as a comprehensive one that covers a broad range of concentrations important to mechanical engineering graduate education and research. We are fortunate to have a distinguished roster of consulting editors, each an expert in one of the areas of concentration. The names of the consulting editors are listed on page vi of this volume. The areas of concentration are applied mechanics, biomechanics, computational mechanics, dynamic systems and control, energetics, mechanics of materials, processing, thermal science, and tribology.

Preface for the 2nd Edition

The new edition of this book reflects a number of new developments in the world of manufacturing during the past decade.

A number of my associates in our Laboratory for Manufacturing Systems and Automation, in the University of Patras, have been involved with the revision of the manuscript; among them, Dr. D. Mourtzis, Dr. N. Papakostas and Dr. V. Karabatsou were the key persons in putting the revision of this book together. K. Alexopoulos, Dr. S. Makris, Dr. D. Mavrikios, M. Pappas, K. Salonitis, P. Stavropoulos, A. Stournaras, G. Tsoukantas, and others contributed substantially to the revised text and the necessary literature research. I would like to thank all of them. Finally, I would like to thank greatly my administrative assistant in our Laboratory, Angela Sbarouni, for her typing and editing of the final manuscript.

George Chryssolouris
Patras, Greece, May 2005

Preface

This book is the derivation of my notes from a course which I have been teaching at M.I.T. on Machine Tools and Manufacturing Systems. The book is intended for students at the undergraduate or graduate level who are interested in manufacturing, industry engineers who want an overview of the issues and tools used to address problems in manufacturing systems, and managers with a technical background who wan to become more familiar with manufacturing issues.

The six chapters of this book have been arranged according to the sequence used when creating and operating a manufacturing system. Chapter 1 of this book provides a general decision making framework for manufacturing. In Chapter 2 an overview of manufacturing processes, the "backbone" of any manufacturing system, is given. Chapter 3 is devoted to manufacturing equipment and machine tools, which are the "embodiment" of manufacturing processes. Chapter 4 provides an overview of process planning methods and techniques, and Chapter 5 deals with the design of manufacturing system. Chapter 6 covers manufacturing systems operation. At the end of each chapter a set of questions is provided for the purpose of reviewing the material covered in the chapter. An instructor's manual containing solutions to these questions, along with suggested laboratory exercises, is also available.

I would like to thank the National Science Foundation, Ford Motor Co., CAM-i, and the Leaders For Manufacturing program at M.I.T. for their financial support during my work on this book. I would also like to thank Professor David Wormley, associate Dean of Engineering at M.I.T., for his encouragement and support. Professor Nam Suh, the head of the Mechanical Engineering Department at M.I.T., has provided many challenging remarks and constructive comments on the broad subject of manufacturing over the years, and I would like to thank him for his leadership and intellectual inspiration. My colleagues Professor David Hardt, the director of M.I.T.'s Lab for Manufacturing and Productivity, and Professor Timothy Gutowski, director of M.I.T.'s Industry Composites and Polymer Processing Program, have also provided me over the years with helpful remarks and discussions which have contributed significantly to my understanding of manufacturing, and for that I thank them.

A number of my graduate and undergraduate students have been involved with this manuscript; among them, Moshin Lee was instrumental in putting this book together. He, together with Nick Anastasia, Velusamy Subramaniam, Mike Domroese, Paul Sheng, Kristian Dicke, Jeff Connors, and Don Lee contributed substantially to the text, the literature research and the overall layout of the book. Their excellent contributions and helpful, friendly attitudes were great assets during the creation if this manuscript. I feel greatly indebted to all of them. Finally, I would like to thank my former administrative assistant at M.I.T., Jennifer Gilden, for her assistance in typing and editing the manuscript and for drawing many figures, and my current administrative assistant at M.I.T., Kendra Borgmann, for her help in coordinating the production of this book.

George Chryssolouris
Cambridge, September 1991

Contents

3. Machine Tools and Manufacturing Equipment

6. The Operation of Manufacturing Systems

Acknowledgements

I would like to acknowledge the following publishers and authors for the use of a number of different figures:

Fig. 1.1 to 1.4	Reprinted form <u>EUROSTAT</u>: http://europa.eu.int/comm/eurostat/
Fig. 1.5	Reprinted from the <u>Organization for Economic Co-operation and Development – OECD,</u> www.oecd.org
Fig.1.9 & 1.10	Reprinted from the <u>Organization for Economic Co-operation and Development – OECD,</u> www.oecd.org
Fig. 2.2 & 2.3	Reprinted from <u>Handbook of Machine Tools</u> by Manfred Weck, John Wiley & Sons, New York, 1984.
Fig. 2.7	Reprinted from <u>Manufacturing Processes for Engineering Materials</u> by Serope Kalpakjian, Addison-Wesley Publishing Company Inc., Reading, MA, 1984.
Fig. 2.9	Reprinted from <u>Manufacturing Processes for Engineering Materials</u> by Serope Kalpakjian, Addison-Wesley Publishing Company Inc., Reading, MA, 1984.
Fig. 2.22 & 2.23	Reprinted from <u>Manufacturing Processes for Engineering Materials</u> by Serope Kalpakjian, Addison-Wesley Publishing Company Inc., Reading, MA, 1984.
Fig. 2.26 & 2.27	Reprinted from <u>Manufacturing Processes for Engineering Materials</u> by Serope Kalpakjian, Addison-Wesley Publishing Company Inc., Reading,

Wesley Publishing Company Inc., Reading, MA, 1984.

Fig. 2.29   Reprinted from <u>Machining Data Handbook</u>, 3rd Edition, Vol 2, Metcut Research Associates Inc., 1980.

Fig. 2.37   Reprinted from Manufacturing Engineering and Technology, by Serope Kalpakjian, Addison-Wesley Publishing Company Inc., Reading, MA, 1989.

Fig. 2.39 to 2.42   Reprinted from <u>Engineering Fundamentals</u> http://www.efunda.com/processes/rapid_prototyping/sla.cfm

Fig. 2.44   Reprinted from <u>Overview of Activities on Nanotechnology and Related Technologies</u> by Budworth, D.W, Institute for Prospective Technological Studies, Seville, 1996.

Fig. 3.1   Reprinted from <u>Handbook of Machine Tools</u> by Manfred Weck, John Wiley & Sons, New York, 1984.

Fig. 3.2   Reprinted from <u>Handbook of Machine Tools</u> by Manfred Weck, John Wiley & Sons, New York, 1984.

Fig. 3.5   Reprinted from <u>Handbook of Machine Tools</u> by Manfred Weck, John Wiley & Sons, New York, 1984.

Fig. 3.6 (a)   Reprinted from <u>Handbook of Machine Tools</u> by Manfred Weck, John Wiley & Sons, New York, 1984.

Fig. 3.6 (b)   Reprinted from <u>Forging & Industrial Equipment</u>, www.whitemachinery.com

Fig. 3.7   Reprinted from <u>Handbook of Machine Tools</u> by Manfred Weck, John Wiley & Sons, New York, 1984.

tem for Industrial Applications by Astheimer, P., W. Felger; and S. Mueller, Computers & Graphics, Vol. 17, no. 6: 671-677 , 1993.

Fig. 4.1   Reprinted from An Introduction to Automated Process Planning Systems by Tien-Chien Chang, and Richard A. Wysk, Prentice-Hall, Inc., Englewood Cliffs, New Jersey, 1985.

Fig. 4.2 & 4.3   Reprinted from Modern Manufacturing Process Engineering by Benjamin Niebel, Alan Draper, and Richard A. Wysk, McGraw-Hill, New York, 1989.

Fig. 4.5   Reprinted from An Introduction to Automated Process Planning Systems by Tien-Chien Chang, and Richard A. Wysk, Prentice-Hall, Inc., Englewood Cliffs, New Jersey, 1985.

Fig. 4.7   Reprinted from An Introduction to Automated Process Planning Systems by Tien-Chien Chang, and Richard A. Wysk, Prentice-Hall, Inc., Englewood Cliffs, New Jersey, 1985.

Fig. 5.31   Snapshot of Witness simulation software
http://www.lanner.com

Fig. 5.32   Reprinted from Technomatix,
http://www.technomatix.com

Fig. 5.33   Reprinted from Delmia , http://www.delmia.com

Fig. 6.40   Reprinted from Genetic Algorithms + Data Structures = Evolution Programs, Michalewicz Z., Springer Verlag, 3rd Edition, 1996.

1. Introduction

1.1 A Guide to this Book

The first chapter of this book attempts to provide a perspective of manufacturing issues, including an economic point of view, and to establish a framework for decision making in manufacturing. For such a framework to be implemented, two major challenges must be met: The first challenge, scientifically sound definitions of relevant manufacturing attributes must be settled upon. This will enable the trade-offs between attributes to be quantitatively assessed in the decision making process. The second challenge, which is directed in particular to the engineering community, is to develop *technoeconomical* models, which allow the decision making process to be scientifically executed. Engineering science has made progress in terms of rigorously analyzing many engineering problems, but often these analyses are not performed in the context of relevant manufacturing attributes, and as such, remain purely academic exercises, which cannot be effectively utilized in industrial manufacturing practice.

This book attempts to provide a systematic overview of manufacturing systems issues and tools for addressing these issues. Since this book follows the sequence that will be followed during the design and operation of a manufacturing system, the second chapter is devoted to manufacturing *processes*. Often, manufacturing systems are treated in a "block diagram" approach, which neglects their processing aspects. To avoid this, Chapter 2 summarizes relevant manufacturing processes. The intent of this chapter is not to make the reader an expert on the different processes, but rather to provide a descriptive overview of the most commonly used processes in discrete part manufacturing, thus enabling the reader to put manufacturing systems issues in perspective from a process point of view.

The *machines* that perform manufacturing processes are the embodiment of these processes. There is a close relationship between machines and processes, since the capabilities and limitations of a process often depend on the design and operation of the machine that performs it. Chapter

3 reviews relevant knowledge and practice regarding machine design, control and automation. The purpose of this chapter is to provide background information on the machines used in manufacturing systems, since the machines are the physical building blocks of these systems.

Once the reader has become familiar with processes and machines, it is important that he discuss *process planning* (Chapter 4), which provides the instructions necessary to manufacturing parts, and creates a bridge between design and manufacturing, which is why this chapter precedes the chapters on design and operation of manufacturing systems. Process planning requires knowledge of processes and machine capabilities; therefore, its discussion follows the chapters on processes and machines.

Chapter 5 addresses the *design and planning of a manufacturing system*. Here we relate design to process and machine selection, system layout, and capacity planning. In a broader context, the manufacturing system design is particularly concerned with establishing material and information flows, which dictate the architecture of a manufacturing system. This perspective is incorporated in Chapter 5.

From a practical point of view, the design of a system is interwoven with its operation. In order to design a system, one must have a preliminary idea of the way it will be operated, while the operation of the system depends very much on its design. However, for educational purposes it is essential that these issues be separated in order to address them systematically. Thus, Chapter 6 is concerned with scheduling in manufacturing systems, assuming that the system design is established and given. In operation and scheduling there is a surplus of academic literature that unfortunately has little impact on the practical aspects of scheduling in the manufacturing industry. Chapter 6 attempts a compromise between theory and practice in scheduling, as does the entire book in all addressed manufacturing issues.

1.2 Economic Perspective

Manufacturing defined as the transformation of materials and information into goods for the satisfaction of human needs, is one of the primary wealth-generating activities for any nation (Fig.1.1) and contributes significantly to employment (Fig. 1.2). Nevertheless, the advances of technology in the manufacturing sector, lead to gradual reduction in the workforce, employed in manufacturing industries (Fig. 1.2). This is because the technological development in manufacturing, leads worldwide to increased productivity (Fig. 1.3), which also requires higher level of investment (Fig. 1.4).

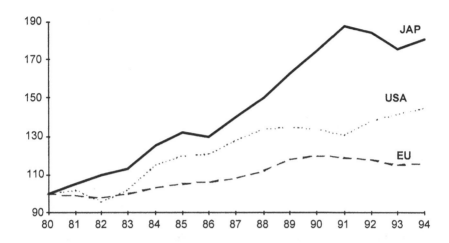

Fig. 1.1. Output in manufacturing, 1980=100 (gross value added at 1985 prices).
Source: European Commission

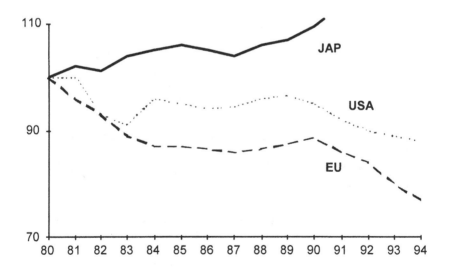

Fig. 1.2. Employment in manufacturing, 1980=100 (at 1985 prices). Source: European Commission

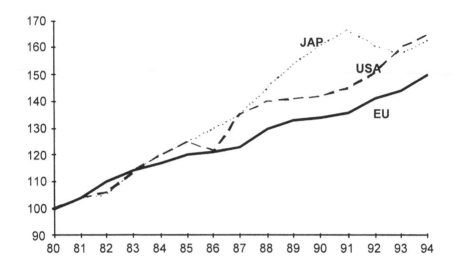

Fig. 1.3. Productivity in manufacturing, 1980=100 (gross value added at 1985 prices per person employed). Source: European Commission

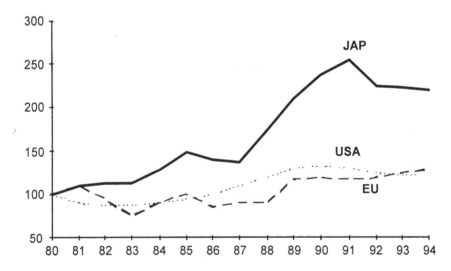

Fig. 1.4. Investment in manufacturing, 1980=100 (at 1985 prices). Source: European Commission

From a worldwide perspective, one can see that in some manufacturing sectors, such as these of computers and pharmaceuticals, we have simultaneously, a significant growth both in R&D expenditures as well as in em-

ployment (Fig. 1.5) while in some other sectors, notably in the ship building industry, both R&D expenditures and employment growth are sinking
(Fig.1.5).

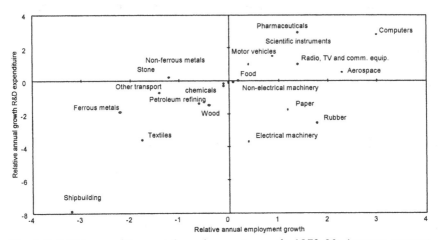

Fig. 1.5. R&D expenditures and employment growth, 1973-90. Average percentage growth rates by industry relative to total manufacturing growth for 13 OECD-
countries. Source: OECD STAN database

One of the most important indicators, reflecting to the industrial and
technological development of a country, is the growth of machine tools in
the manufacturing sector. As a supplier of technologies to all manufacturing industries, the machine tools sector plays a major role in all industrial-
ized countries. This is the reason for Germany being the major machine
tools manufacturer in Europe (Fig. 1.6)

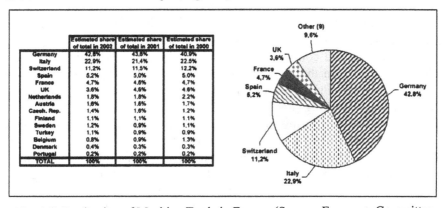

Fig. 1.6. Production of Machine Tools in Europe (Source: European Committee
for Co-operation of the Machine Tool Industries – CECIMO)

The machine tools manufacturing sector is also strongly related with that of employment. More than 150.000 people are employed in 1.474 machine tools building companies, in all 15 European countries, which participate in the European Committee for Co-operation of the Machine Tool Industries (CECIMO) (Fig. 1.7).

Employment by MT builders in 2002		Number of MT building companies (1)	
	People		In units
Austria	6.281	Austria	45
Belgium	2.060	Belgium	15
Czech Rep.	7.010	Czech Rep.	77
Denmark	450 *	Denmark	6 *
Finland	1.400	Finland	10
France	7.850	France	120
Germany	67.000	Germany	320 **
Italy	30.320	Italy	450
Netherlands	735	Netherlands	10
Portugal	700 *	Portugal	18 *
Spain	6.100	Spain	120
Sweden	1.400 *	Sweden	24 *
Switzerland	10.800	Switzerland	89
United Kingdom	8.600	United Kingdom	100
Turkey	7.300	Turkey	70
CECIMO (15)	158.006	CECIMO (15)	1.474

* 2000 data
** 2001 results

(1) The number of units (or brand names) does not take into account the numerous clusters of companies having financial links or having various commercial agreeme regarded as separate units.

Fig. 1.7. Number of people employed in 2002 by the major European Machine Tools Builders (Source: European Committee for Co-operation of the Machine Tool Industries – CECIMO)

Another important indicator reflecting to the industrial and technological development of a country is the consumption of machine tools. In figure 1.8, is presented the consumption of machine tools, taking into consideration the major machine tools builders all over the world.

Governments and companies worldwide, recognize the significance of R&D (Fig.1.10). In the manufacturing sector, industrially developed nations, such as the United States, spend significant amounts of money on R&D for manufacturing (Fig.1.9).

These observations show the importance of manufacturing within the overall economic development of a nation and also the significance of R&D for sound and productive development in the manufacturing industry. Thus, the study of issues related to manufacturing is becoming increasingly important not only to the industry but also to the academic community, which is often challenged with finding solutions to industrially relevant problems.

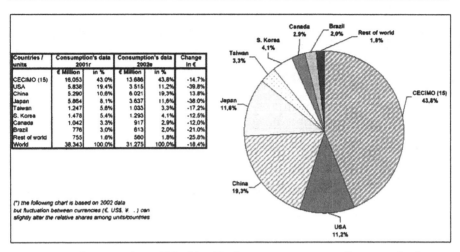

Countries / units	Consumption's data 2001r		Consumption's data 2002e		Change in €
	€ Million	in %	€ Million	in %	in %
CECIMO (15)	16.053	43.0%	13.686	43.8%	-14.7%
USA	5.838	19.4%	3 515	11.2%	-39.8%
China	5.290	10.6%	6.021	19.3%	13.8%
Japan	5.864	8.1%	3.637	11.6%	-38.0%
Taiwan	1.247	5.6%	1.033	3.3%	-17.2%
S. Korea	1.478	5.4%	1.293	4.1%	-12.5%
Canada	1.042	3.3%	917	2.9%	-12.0%
Brazil	776	3.0%	613	2.0%	-21.0%
Rest of world	755	1.6%	560	1.8%	-25.8%
World	38.343	100.0%	31.275	100.0%	-18.4%

(*) the following chart is based on 2002 data but fluctuation between currencies (€, US$, ¥ ..) can slightly alter the relative shares among units/countries

Fig. 1.8. Consumption of Machine Tools *(Source: European Committee for Co-operation of the Machine Tool Industries – CECIMO)*

MANUFACTURING SECTOR

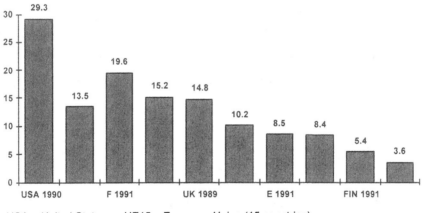

USA = United States   UE15 = European Union (15 countries)
F = France   I = Italy
UK = United Kingdom   S = Sweden
E = Spain   D = Germany
FIN = Finland   DK = Denmark

Fig. 1.9. Share of Industrial R&D Expendture Financed by the State, as %. *Source: Estimates of Commission services from OECD data & national sources*

Overview of indicators for comparing research financing (million US$ at current PPP)					
INDICATORS		E.U.	U.S.	JAPAN	YEAR
GROSS DOMESTIC EXPENDITURE ON R&D (BERD)	IN MILLION US$ *AT CURRENT PPP*	123 308	167 122	75 047	1992
	AS % OF GDP	1.96	2.81	3.00	1992
GROSS DOMESTIC EXPENDITURE ON CIVIL R&D	AS % OF GDP	1.8	2.2	3.0	1992
IN HOUSE BUSINESS EXPENDITURE ON R&D (BERD)[1]	IN MILLION US$ *AT CURRENT PPP*	77 042	122 000	49 431	1993
	AS % OF GDP	1.22	1.95	1.93	1993
	AS % OF GIPmp	1.64	2.34	2.30	1992
STATE-FINANCED BERD	AS % OF TOTAL BERD	12.2[2]	20.3	1.1	1992
	AS % OF TOTAL SECTOR				
	- MANUFACTURING	12.6	25.4	1.2[1989]	1991
	- ELECTRONICS	20.4	55.9[1990]	0.3[1989]	1991
	- AEROSPACE	48.1	90.8	9.0[1989]	1991
	- AUTOMOBILE	1.3	117.7[1985]	0.04[1989]	1991
BREAKDOWN OF CURRENT GOVERNMENT R&D EXPENDITURE AS A FUNCTION OF MARKET PROXIMITY	AS % OF BASIC R&D	25.9	15.8	12.7[1989]	1991
	AS % OF APPLIED R&D AND DEVELOPMENT	74.1	84.2	87.3[1989]	1991

Note: Some of the values for EU12 are estimated
Source: Commission services from OECD data and national sources

Fig. 1.10. Overview of indicators for comparing research financing

---

[1]   It concerns expenditure for research carried out within companies (excluding search subcontracted to an external contractor) wherever the financing comes from

[2]   If community funding is taken into account, the corresponding figure would be approximately 14%. This being the case, funding from different States should be taken into account for the United States: ultimately the difference would not be reduced.

1.3 A Decision Making Framework for Manufacturing Systems

Manufacturing can be thought of as a system (Fig. 1.11) in which product design is the initial stage, and the delivery of finished products to the market is the final output [1]. Since the field of manufacturing integrates many disciplines in engineering and management, it is useful to divide it in such a way so as to facilitate the identification of issues and to allow a scientific approach to the problems encountered. Manufacturing can be subdivided into the areas of manufacturing *processes*, which alter the form, shape and/or physical properties of a given material; manufacturing *equipment*, used to perform manufacturing processes; and manufacturing *systems*, which are the combinations of manufacturing equipment and humans, bound by a common material and information flow. *Design and manufacturing interfaces* should also be considered, since manufacturing always begins with a design such as that of a drawing, sketch, CAD file, virtual or physical prototype, or another way of communicating the features and characteristics of the product or subassembly that is to be manufactured.

Decisions regarding the design and operation of manufacturing systems require technical understanding and expertise, as well as the ability to satisfy certain business objectives. Thus, a combination of engineering and management disciplines are required to provide a decision making framework for manufacturing.

In general, there are four classes of *manufacturing attributes* to be considered when making manufacturing decisions: *cost*, *time*, *quality* and *flexibility* (Fig. 1.12). These manifest themselves, depending on the particular problem, specific objectives, goals, and criteria. An *objective* is an attribute to be minimized or maximized. A *goal* is a target value or range of values for an attribute, and a *criterion* is an attribute that is evaluated during the process of making a decision. A specific cost objective can be "minimize the cost per part;" a specific quality objective can be "minimize the deviation from the target value of the part dimension." A goal related to time can be "produce no less than 120 and no more than 123 parts per hour." As a final example, a flexibility criterion relevant to deciding on the design of a manufacturing system is the ease with which the system can be set up for the production of a new part type. Not all manufacturing attributes are relevant to every manufacturing decision, and, as the examples show, each class of attributes arises in a specific form, in individual manufacturing decisions.

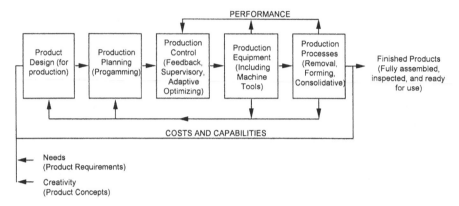

Fig. 1. 11. The System of Manufacturing

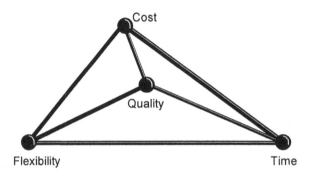

Fig. 1.12. The Manufacturing Tetrahedron

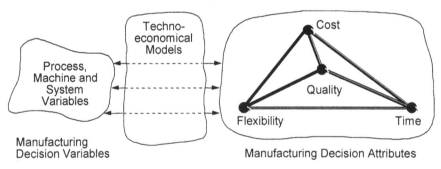

Fig. 1.13. Mapping Manufacturing Decision Variables into the Manufacturing Attributes

A manufacturing decision is basically the selection of values for certain decision variables related either to the design or to the operation of a manufacturing process, machine or system (Fig. 1.13). Such a decision making process is based on performance requirements, which specify the values of the relevant manufacturing attributes. In this light, the decision making process can be regarded as a mapping from desired attribute values onto corresponding decision variable values. In order for this mapping to be scientifically founded, it must be based on *technoeconomical models*. This mapping essentially constitutes a modeling process, in which the levels of manufacturing attributes are related to the levels of different decision variables.

The tetrahedral arrangement of manufacturing attributes emphasizes the interrelationship of these attributes. It is not possible to simultaneously optimize cost, time, flexibility and quality. The overall outcome of a manufacturing decision is rather governed by trade-offs between the different manufacturing attributes. Assessment of these trade-offs requires a means of quantitatively evaluating each attribute. The more precisely defined these are, the easier it is to trade them off and make decisions. In industry to date, most quantitative measures in manufacturing refer to cost- and time-related attributes. In the United States, cost and time (or production rate) have been emphasized by the proliferation of mass production systems through the 1960s. Consequently, the trade-off between cost and production rate has been extensively treated. In the '70s, with the entrance of Japan and Germany into the world markets, quality became a significant driving force, and it is anticipated that in the new millennium, flexibility will become a major competitive weapon for the manufacturing industry. Thus, although cost-and time-related attributes remain extremely important for any manufacturing decision, emphasis must be placed on finding quantitative definitions for quality and flexibility, so that appropriate trade-offs among the four different classes of manufacturing attributes can be established in order to achieve a comprehensive treatment of manufacturing problems.

In the following pages these four classes of attributes will be briefly discussed in general terms with the help of some simple examples.

1.3.1 Cost

Costs related to manufacturing encompass a number of different factors, which can be broadly classified into the following categories:

- *Equipment and facility costs.* These include the costs of equipment necessary for the operation of manufacturing processes, the facilities used to house the equipment, the factory infrastructure, etc.

- *Materials.* This includes the cost of raw materials for producing the product, and that of tools and auxiliary materials for the system, such as coolants and lubricants.
- *Labor.* The direct labor needed for operating both equipment and facilities.
- *Energy* required for the performance of the different processes. In some manufacturing industries, the cost of energy may be negligible, compared with other factors, while in others it contributes significantly to the financial burden of the manufacturing system.
- *Maintenance and training.* This includes labor, spare parts, etc., that are needed to maintain the equipment, facilities and systems, as well as the training necessary to accommodate new equipment and technology.
- *Overhead.* This is part of the cost that is not directly attributable to the operation of the manufacturing system, but supports its infrastructure.
- *The cost of capital*, which may not be readily available within the manufacturing firm, and therefore must be borrowed under specific terms.

This classification provides a general framework of how cost issues can be addressed in the manufacturing environment by establishing a systematic way to measure the cost performance of different solutions.

Example

To clarify the meaning of a "technoeconomical" model and its relationship to the manufacturing attributes, consider the cost of a machining (turning) operation [2]:

$$
\begin{bmatrix} \text{Cost per} \\ \text{Workpiece} \end{bmatrix} = \begin{bmatrix} \text{Machine} \\ \text{Rate} \end{bmatrix} + \begin{bmatrix} \text{Labor} \\ \text{Rate} \end{bmatrix} + \begin{bmatrix} \text{Overhead} \\ \text{Rate} \end{bmatrix} \times \underbrace{\begin{bmatrix} \text{Feeding} \\ \text{Time} \end{bmatrix} + \begin{bmatrix} \text{Rapid} \\ \text{Feeding} \\ \text{Time} \end{bmatrix} + \begin{bmatrix} \text{Portion of} \\ \text{tool insert} \\ \text{replacement} \\ \text{time per} \\ \text{workpiece} \end{bmatrix}}_{\text{Operating Time per Workpiece}} + \frac{1}{\begin{bmatrix} \text{No. of work-} \\ \text{pieces bet.} \\ \text{tool insert} \\ \text{changes} \end{bmatrix}} \times \begin{bmatrix} \text{Cost} \\ \text{per} \\ \text{tool} \\ \text{insert} \end{bmatrix}
$$

Factors contributing to the cost per workpiece include machine, labor, and overhead rates. The contribution of these factors is proportional to the operating time, which consists of feeding time for material removal, rapid movement time for tool positioning, and tool replacement time (Fig. 1.14). The tool is a replaceable insert whose cost is another factor contributing to the cost per workpiece.

This "technoeconomical" model can be derived from the mechanics and geometry of the turning process. In turning, material is removed from a workpiece rotating about an axis of symmetry by a cutting tool or insert, which travels in the axial and radial workpiece directions (Fig. 1.14).

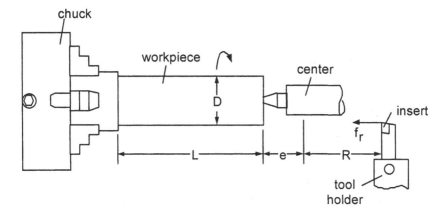

Fig. 1.14. Geometry of the Turning Process

The decision variables of the process are two parameters that are typical for machining, namely the feed rate, f_r [mm/rev], usually expressed as length per revolution, and the cutting speed, v [m/min]. These decision variables are mapped into the manufacturing attribute "cost" expressed in [$/piece] by a technoeconomical model as follows:

$$C = M[T_F + T_{RF} + T_{IR}] + \frac{1}{N_{IR}}[C_l]$$

Where:

$$T_F = \frac{D(L+e)}{318f_r v}$$

$$T_{RF} = \frac{R}{r}$$

$$T_{IR} = \frac{DLt_d}{318f_r vT}$$

$$\frac{1}{N_{IR}} = \frac{DL}{318f_r vT}$$

The meaning of each term is:

C   Cost for turning one workpiece [$]
f_r Feed per revolution [mm]
v   Cutting speed of the insert along the workpiece surface [m/min]
M   Machine + labor + overhead cost on lathe [$/min]
T_F Feeding time per workpiece [min]
T_{RF} Rapid feeding time per workpiece [min]
T_{IR} Insert replacement time per workpiece, assuming insert replacement time is distributed evenly over workpieces [min]
N_{IR} Number of workpieces between inserts changes

C_I Cost per insert [\$]
D   Workpiece diameter [mm]
L   Length of workpiece [mm]
e   Extra travel at feed rate f_r [mm]
R   Total rapid traverse distance for one part [mm]
r   Rapid traverse rate [mm/min]
t_d Insert replacement time [min]
T   Tool life [min]

For a given workpiece, insert material, and feed rate, the machining cost [\$/piece] is primarily a function [2] of the cutting speed v (Fig. 1.15). As the cutting speed increases, the material removal rate increases, thereby reducing the machining cost by allowing more pieces to be made per unit time. However, the machining cost also depends, among other things, on tool wear; inserts have to be replaced periodically, creating costs both in terms of acquiring new inserts and time lost from production. The wear behaviour of tool materials as a function of cutting time and cutting speed (Fig. 1.16) is a subject of engineering analysis; the point here is that machining cost is an *economic* attribute whose value is determined via a techoeconomical model, based on *engineering* analysis of tool wear behaviour. It is exactly this behaviour that determines the shape of the curve in Figure 1.15. As the cutting speed increases further, the wear of the tool also increases, leading to higher tooling costs and more frequent tool replacements; this in turn increases the cost per piece.

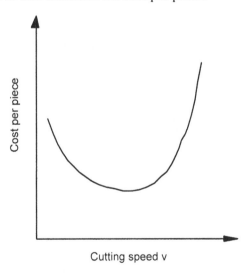

Fig. 1.15. Machining Cost as a Function of Cutting Speed

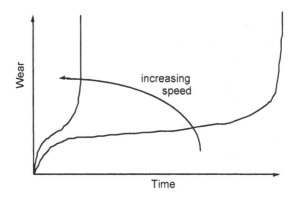

Fig. 1.16. Tool Wear as a Function of Time and Cutting Velocity

This example shows that in order to treat manufacturing problems and make decisions in their proper context, one must consider both technical and economic or business issues. Tool wear behaviour is a purely technical issue, requiring scientific analysis and technical innovation which, however, may be irrelevant if this is not performed in the context of a business objective such as that of cost. By the same token, consideration of the cost of a manufacturing operation without the proper engineering background and physical understanding is irrelevant and cannot lead to the proper business decision.

1.3.2 Time

In manufacturing systems, time attributes refer to

- How quickly a manufacturing system can respond to changes in design, volume demand, etc. The responsiveness of the system to changes will be discussed later on, in the section of flexibility.
- How quickly a product can be produced by the system – usually expressed as the *production rate* of the system.

Production rate affects, directly or indirectly, all other types of attributes. Higher production rates typically result in lower cost and possibly lower quality as well. Furthermore, to achieve high production rates, it is often necessary to resort to automation, which may have an impact on the flexibility of the system. An example of such automation is a transfer line (Fig. 1.17), which due to its fully automated part transfer mechanisms and machining operations, provides production rates on the order of one piece every 20 seconds, even for very complex parts (Fig. 1.18). However, a transfer line lacks significantly in flexibility.

The theoretical production rate of a system, or *machine cycle*, refers to pieces produced per unit of time when a machine or system is running with no interruptions or delays. This rate is typically constrained by the physics of the processes and the robustness of the machines. The actual production rate of a system is called the *process* or *system yield,* which is the number of acceptable pieces produced per unit of time, incorporating both delays, occurring regularly during production and unpredicted interruptions, such as machine breakdowns. The efficiency of the overall process, as far as production rate is concerned, is the ratio of process yield or system yield as a percent of machine cycle.

Achievable production rates in a manufacturing system are significantly affected by the reliability of the equipment and the overall structure of the system. An assessment of these factors requires the definition of a number of terms. The failure rate λ of a component is the ratio of the number of failures over a time period. The reciprocal of the failure rate is known as the mean time between failure (MTBF). The probability of a system or a component to perform its required function is called reliability. The magnitude of reliability is measured under stated conditions for a specified period of time (t) and it is given by the expression $R = e^{-\lambda t}$.

For a given failure rate λ the reliability decreases as the time period increases. For example, if the failure rate λ of a given component is 0.001 per hour, the reliability of the component for an eight-hour period is $R(8) = e^{-(0.001*8)} = 99\%$, while the reliability for a 20-hour period is $R(20) = e^{-(0.001*20)} = 90\%$. Availability (A) is defined as the mean time between failure (MTBF) divided by the sum of MTBF and mean time to repair (MTTR): $A = MTBF/(MTTR + MTBF)$.

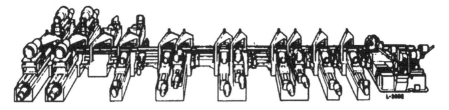

Fig. 1.17. An Example of a Transfer Line

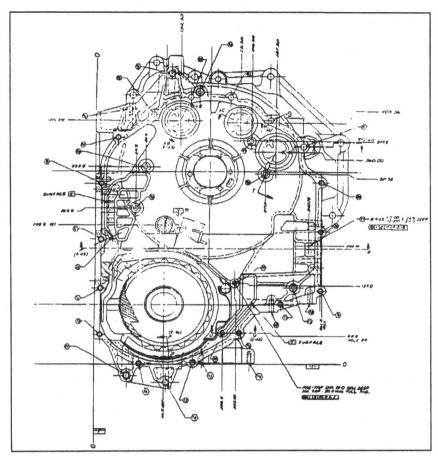

Fig. 1.18. A Component Machined on a Transfer Line

The actual production rate of a manufacturing system is given by its availability multiplied by its machine cycle. The availability of a system is a function of the system's reliability (or failure rate) and the system's mean time to repair. The system's reliability is in turn, a function of the structure and material flow of the system and that of the reliabilities of the system's individual components. For example, in a transfer line (Fig. 1.17), the material flow is purely serial, and the total system reliability is given by the product of the reliabilities of the individual machines: $R = R_1$ x R_2 x ... x R_n.

If all individual machine reliabilities R_i are taken to be 98%, then reliability of a ten-machine transfer line is $(0.98)^{10} = 82\%$. Clearly, the reliability of a system may be very different from that of an individual component of the system.

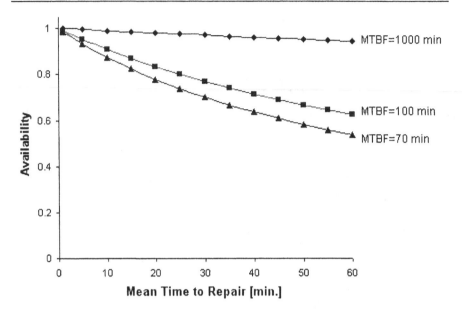

Fig. 1.19. Availability as a Function of Mean Time to Repair

Availability is often a more informative indicator of breakdown behaviour than reliability, because it incorporates the mean time to repair. The availability of a system with a long MTBF is relatively insensitive to the duration of the MTTR, while the availability of a system, with a short MTBF, is very sensitive to the duration of the MTTR (Fig. 1.19).

Example

Consider a transfer line, which is composed of a number of individual machines linked in series. This type of manufacturing system is usually used for the high volume production of a single part type. Two factors dictate transfer line design: the production rate requirement and the design of the part to be produced. The part design can be thought of as a collection of features, which must be produced by the machines of the transfer line.

If the required production rate is to be achieved, each part must be permitted to spend only a certain maximum amount of time on each machine. Within this time constraint, the general aim of a transfer line designer is to maximize the number of features produced at each machine. This minimizes the number of machines required – a desirable objective not only because it reduces the acquisition cost of the transfer line, but also because the reliability of the line is improved. For a given machine reliability, the reliability of the transfer line as a whole, increases dramatically [19] as the number of machines decreases (Fig. 1.20). Recall that the reliability of a transfer line is the product of the reliabilities of its individual machines. However, if the transfer line contains safety stocks of parts between machines (buffers), then individual machines are to a certain extent decoupled from the failures of other machines, and the overall transfer line reliability becomes less sensitive to the reliabilities of individual machines. It is thus, possible to have a transfer line that is both large and reliable, but only at the expense of increased part inventories (work-in-process).

The actual production rate of a transfer line (the system yield) can never be as great as the theoretical production rate (the machine cycle). However, one can come closer to the machine cycle if the individual machine availability is improved. The availability and hence production rate of an individual machine is a function of its operating speed (i.e. the angular velocity of the tool spindle and the feed rate of the tool into the workpiece). In machining, the production rate of an individual machine initially increases as operating speed increases, due to the increase in material removal rate. However, as the operating speed increases beyond a certain point, the increasing frequency of the tool changes due to increased tool wear acts in order to decrease the production rate (Fig. 1.21). Since production rate is the product of availability and machine cycle (a constant), the shape of the availability versus operating speed curve would be identical to that of the production rate versus operating speed curve (Fig. 1.21). In general, the production rate or availability of a transfer line machine is maximized at an operating speed, which is below the maximum operating speed.

Given the individual machine behaviour shown in Figure 1.21, the production rate behaviour of the transfer line (assuming no safety stocks) can be predicted. Should λ, R and A be the total transfer line failure rate, reliability and availability respectively, and λ_i, R_i and A_i be the individual machine failure rates, reliabilities and availabilities respectively in a transfer line with n machines, then we know:

$$R = R_1 \times R_2 \times \cdots \times R_n$$
$$e^{\lambda t} = e^{\lambda_1 t} \times e^{\lambda_2 t} \times \cdots \times e^{\lambda_n t}$$
$$\therefore \lambda = \lambda_1 + \lambda_2 + \cdots + \lambda_n$$

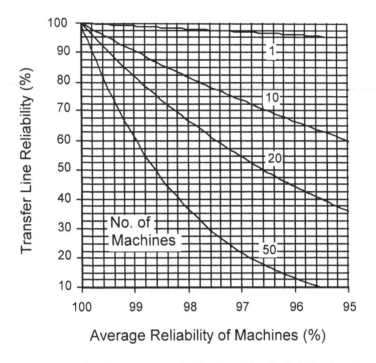

Fig. 1.20. System Reliability versus Individual Machine Reliability in a Transfer Line

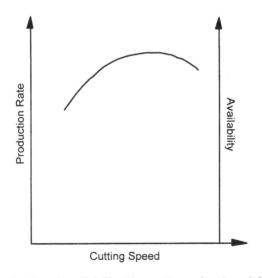

Fig. 1.21. Production Rate/Availability Versus Operating Speed for an Individual Machine

We also know that, for a given MTTR,

$$A = \frac{MTBF}{MTBF + MTTR}$$

$$= \frac{1/\lambda}{1/\lambda + MTTR}$$

$$= \frac{\dfrac{1}{\left(\lambda_1 + \lambda_2 + \cdots + \lambda_n\right)}}{\dfrac{1}{\left(\lambda_1 + \lambda_2 + \cdots + \lambda_n\right)} + MTTR}$$

$$= \frac{1}{1 + MTTR\left(\lambda_1 + \lambda_2 + \cdots + \lambda_n\right)}$$

$$\cong \left(\frac{1/\lambda_1}{1/\lambda_1 + MTTR}\right) \times \left(\frac{1/\lambda_2}{1/\lambda_2 + MTTR}\right) \times \cdots \times \left(\frac{1/\lambda_n}{1/\lambda_n + MTTR}\right)$$

$$= A_1 \times A_2 \times \cdots \times A_n$$

This relationship shows how the individual machine behaviour of Figure 1.21 can be correlated with the availability and hence the production rate of the transfer line as a whole. The overall transfer line production rate, as calculated using the above relationship, can be drawn as a function of the number of machines and the operating speed (Fig. 1.22). In the figure, all operating speeds have been divided by the maximum allowable operating speed; the resulting normalized value is called the relative speed. A relative speed of 0.6 means that each machine of the transfer line is operated at 60 percent of its maximum speed. While this may yield a lower theoretical production rate than if operated at 100 per-cent of maximum speed, in actuality, the availability of each machine may be improved when operating at this speed (Fig. 1.21) and thus, the overall production rate of the transfer line may be improved as well.

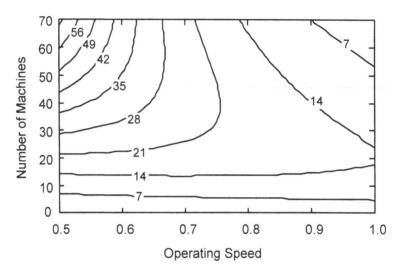

Fig. 1.22. Contours of Constant Transfer Line Production Rate (PPH)

Each line in Figure 1.22 is a contour of constant production rate. For example, following the 28 PPH contour from left to right, we see that a transfer line of 29 machines operating at a relative speed of 0.5 can produce as much (28 PPH) as a transfer line of 70 machines operating at a relative speed of 0.67. The advantage of using 70 machines, lies in the greater flexibility of the system with respect to production rate. It is possible to expand the production rate to 56 PPH with 70 machines (a horizontal line drawn at number of machines = 70 intersects the 56 PPH contour), but it is not possible to do so with 29 machines. This example emphasizes the importance of a quantitative understanding of manufacturing attributes, in making manufacturing decisions (in this case, deciding on the number of machines and their operating speeds).

1.3.3 Flexibility

For many decades, cost and production rates were the most important performance criteria in manufacturing, and manufacturers relied on dedicated mass production systems in order to achieve economies of scale. However, as living standards improve, it is increasingly evident that the era of mass production is being replaced by the era of market niches. The key to creating products that can meet the demands of a diversified customer base, is a short development cycle yielding low cost, high quality goods in sufficient quantity to meet demand. This makes flexibility an increasingly important attribute to manufacturing.

Flexibility, however, cannot be properly considered in the decision making process if it is not properly defined in a quantitative fashion. The quantification of flexibility has been the focus of academic work, but industrial applications have been meagre. Academic research has focused on one-of-a-kind or small lot size production systems, such as those found in the aerospace industry, but not on mass production systems. This is because the flexibility debate has concentrated on the ability of a manufacturing system to produce a spectrum of products both quickly and economically. However, other aspects of flexibility are equally important. A detailed classification of the different aspects or types of flexibility [3] includes:

- *Machine flexibility:* the ease of making changes required to produce a given set of part types.
- *Process flexibility:* the ability to produce a given set of part types, possibly using different materials, in different ways.
- *Product flexibility:* the ability to change over to produce new (set of) products very economically and quickly.
- *Routing flexibility:* the ability to handle breakdowns and to continue producing a given set of part types.
- *Volume flexibility:* the ability to operate profitably at different production volumes.
- *Expansion flexibility:* the ability to expand the system easily and in a modular fashion.
- *Operation flexibility:* the ability to interchange the ordering of several operations for each part type.
- *Production flexibility:* the universe of part types that the manufacturing system can produce.

A different way of viewing flexibility [4] considers two types of flexibility: the ability of the system to cope with external change (e.g. jobs to be processed) and the ability to cope with internal change (e.g. machine breakdowns). For the first type, the proposed measure is the probability of a job occurring and can be processed by the system; for the second type, the proposed measure is the ratio of the expected production rate with disturbances (e.g. breakdowns) to the expected production rate without disturbances.

Furthermore, measures such as the number of units produced per downtime hour, have been proposed for machine flexibility, and a plant-wide ratio of setup times to cycle times has been proposed for operation flexibility [5].

Simulation of Petri nets [6] can provide estimates of the times required for a system to adapt to various random disturbances, times which can be used as flexibility measures.

A more mathematical way to treat flexibility [7, 8] is based on the premise that flexibility should be a function of the number of available options, and the relative freedoms at which these options can be selected. This leads to a measure, which is identical to the expression for entropy, both in thermodynamics and information theory. Its calculation is as follows:

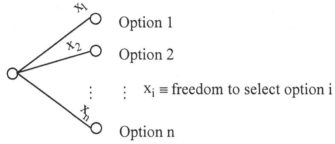

flexibility index $FI = - (x_1 \ln x_1 + x_2 \ln x_2 + \dots + x_n \ln x_n)$

This flexibility index allows, assuming n available options and the freedom x_i to select the option i, the calculation of a measure that characterizes the flexibility of a machine, a process, or a system. However, it may only be applicable to situations when preferences among various choices may be quantified.

An alternative mathematical approach [9] takes the viewpoint that the flexibility of a system or machine should be measured relatively to a reference task set. The proposed measure is the weighted average efficiency with which the system or machine can perform the tasks in the reference task set (each task is weighted according to its relative importance):

$$E = \frac{\displaystyle\int_{\tau \in T} e(A,\tau)\, w(\tau)\, d\tau}{W(T)} \qquad (1.1)$$

where:

E	$\equiv$	weighted average efficiency;
τ	$\equiv$	an individual task;
A	$\equiv$	the system or machine;
T	$\equiv$	the reference task set;
$w(\tau)$	$\equiv$	the weight or importance of task τ;

$W(T)$ ≡ the sum of the weights of the tasks in T;

$e(A,\tau)$ ≡ an efficiency rating between 0 and 1 which indicates how well A performs τ.

This measure accounts only for the flexibility that is actually needed, through the weights w(t) and the selection of the reference task set T (irrelevant tasks can be excluded from T).

"Versatility indices" for the selection of assembly systems have also been proposed [10]. The versatility index of an assembly *process* (V_p) is defined to be the number of design changes per year, in the model to be assembled, while the versatility index of an assembly *system* (V_s) is defined as the number of times per year that the system can be reconfigured in order to assemble a new model. Assembly system selection occurs by matching V_p and V_s.

Finally, a number of approaches have been attempted to quantify the flexibility of manufacturing systems by using discounted cash flow (DCF) techniques. One strategy is to estimate the cash flows, associated with the intangible benefits of flexible systems (such as shorter lead times) [11, 12]; a second strategy is to combine the results of a standard DCF analysis with a separate evaluation of intangible benefits [13].

The measures of flexibility reviewed so far reflect two distinct viewpoints about flexibility. From the first viewpoint, flexibility is an *intrinsic* attribute of a manufacturing system itself, in the same way that length and width are intrinsic attributes of a rectangle. Measures, which ascribe to this viewpoint, compute flexibility as a function solely of the characteristics of the manufacturing system. A good example is the entropic flexibility index FI, which evaluates a system on the basis of the number of choices (of, say, part types to process) that the system can accommodate. The second viewpoint is that flexibility is a *relative* attribute, which depends not only on the manufacturing system itself, but also on the external *demands* placed upon the manufacturing system. To make the point concrete: it is not the total number of part types produced that makes a system flexible; it is rather the fraction produced from the *demanded* part types. The various discounted cash flow flexibility measures [11-13] are based upon this second viewpoint, since they require, for the calculation of cost and revenue cash flows, forecasts of demand for the products of the manufacturing system.

Both viewpoints of flexibility have merit. By considering flexibility as an attribute, intrinsic to a manufacturing system, the need for forecasts or estimates of external demands is obviated. This may make a measure of flexibility easier to apply. On the other hand, any flexibility calculation which omits external demand runs the risk of omitting relevance as well. The number deriving at the end may be meaningless, unless some idea as to how much flexibility is actually needed is known. Here, measures, which consider external demand, have the advantage; estimation beats omission.

This fact is a basis for a generic measure of flexibility. Of the flexibility measures reviewed so far, some lack rigorous justification, while others have a firmer theoretical basis but are more difficult to apply, and perhaps pertain only to certain manufacturing situations. A more generic measure that is nonetheless relatively easy to apply to realistic manufacturing situations, is based on the premise that the *flexibility of a manufacturing system is determined by its sensitivity to change*. The lower the sensitivity is, the higher the flexibility.

High flexibility or low sensitivity to change brings three principal advantages to a manufacturing system. It is convenient to think of these advantages as arising from the various types of flexibility, which can be summarized in three main categories.

- *Product flexibility* enables a manufacturing system to make a variety of part types with the same equipment. Over the short term, this means that the system has the capability of economically using small lot sizes to adapt to changing demands for various products. Over the long term, this means that the system's equipment can be used across multiple product life cycles, increasing *investment efficiency*.
- *Operation flexibility* refers to the ability to produce a set of products using different machines, materials, operations, and sequences of operations. It results from the flexibility of individual processes and machines, the flexibility of product designs, and the flexibility of the structure of the manufacturing system itself. It provides *breakdown tolerance* – the ability to maintain a sufficient production level even when machines break down or humans are absent. This affects mass production in particular, where a large number of identical or very similar parts must be produced, and production quantity is often the most important indicator of manufacturing success.
- *Capacity flexibility* allows a manufacturing system to *vary the production volumes of different products to accommodate changes in demand*, while remaining profitable. It reflects the ability of the manufacturing system to contract or expand easily. It has traditionally been seen as critical for make-to-order systems, but is also very important in mass production, especially of high-value products such as automobiles.

Since flexibility is inversely related to the sensitivity to change, a measure of flexibility must quantify the *penalty of change* (POC). If change can be implemented without penalty, then the system has maximum flexibility, and POC is 0. If, on the other hand, change results in a large penalty, then the system is very inflexible, and POC should be high.

An important, practical question concerning manufacturing flexibility is, "How flexible should a manufacturing system be if I were to acquire it *now* in order to accommodate changes in the *future*?" This question addresses future demand for changes, which cannot be predicted with certainty. Demand for change, stated in probabilistic terms to deal with prediction uncertainty, should therefore be accounted for in the POC. A system that can only accommodate changes that would never occur is not very useful and should not be considered flexible.

The above consideration leads to the conclusion that a measure of flexibility (POC) should account for the *penalty for change* and the *probability of change*. A definition of POC can then take the general form:

$$POC = Penalty \times Probability$$

The lower POC is, the higher the flexibility. If the penalty for change is low, then POC will be low, indicating high flexibility. If the probability of change is low, then POC will again be low, even if the penalty for change is relatively high.   This reflects the fact that a system should not be considered *in*flexible when it has a high penalty for changes, which have little probability of occurring. By the same token, a system should not be considered much more flexible than another system having a minimal penalty for changes, when those changes have little probability of occurring.

The value of POC is based on two inputs: penalty for potential change and probability of potential change, where change is a transition from one "state" to another. The nature of a state depends on the type of flexibility considered: for product flexibility, a state may be the type of product manufactured by the system; for operational flexibility, it may be the operational status of the system (e.g. "fully operational" or "partially operational"); for capacity flexibility it may be the demanded production rate. Both penalty and probability can be viewed as functions of a discrete variable X, which represents potential change.

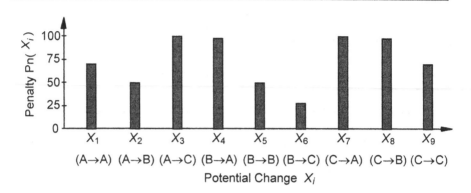

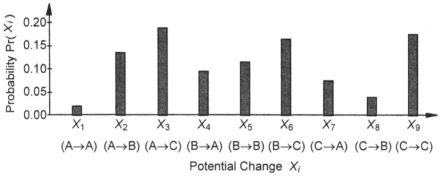

Fig. 1.23. Penalty and Probability as Discrete Functions of Potential Change

The ith value of X is denoted X_i. For example, with three states, A, B, and C, the possible values of X are X_1 (A→A), X_2 (A→B), X_3 (A→C), X_4 (B→A), X_5 (B→B), X_6 (B→C), X_7 (C→A), X_8 (C→B), and X_9 (C→C), where (A→A) means that the system stays in the same state A, (A→B) means that the system changes from State A to State B, etc. (Fig. 1.23).

The Penalty of Change (POC) can be defined as:

$$POC = \sum_{i=i}^{D} Pn(X_i)\,Pr(X_i) \qquad (1.2)$$

where:

D	$\equiv$	the number of potential changes
X_i	$\equiv$	the ith potential change
$Pn(X_i)$	$\equiv$	the penalty of the ith potential change
$Pr(X_i)$	$\equiv$	the probability of the ith potential change

and can be interpreted as the expected value of the penalty to be incurred for potential changes.

The calculation of POC can be viewed as an application of single-attribute decision making under uncertainty, where X_i are possible future scenarios, $\text{Pn}(X_i)$ are the attribute values for each future scenario, and $\text{Pr}(X_i)$ are the probabilities of occurrence of the future scenarios. The quantity POC would then be the expected attribute value, or utility, of the manufacturing system. In a comparison of a number of manufacturing system alternatives, the decision making process selects the alternative with the best (in this case, lowest) utility. Decision theory will be described in more detail in Chapter 6.

Returning to the definition of POC, we note that in the most general case, one may think of a system as not having a finite set of discrete states but a *continuous* range of states. In this case, the number of possible state transitions is unlimited, and the variable of potential changes X can be considered continuous. Then the penalty for potential change can be represented by a continuous distribution $\text{Pn}(X)$, while the probability of potential change can be represented by a continuous probability distribution or probability density function $\text{Pr}(X)$. The product $\text{Pn}(X)\,\text{Pr}(X)$ is a "normalized" penalty distribution whose integral is the expected value of the penalty (Fig. 1.24), and which provides a quantitative measure of the system's flexibility:

$$\text{POC} = \int_{X_1}^{X_2} \text{Pn}(X)\,\text{Pr}(X)\,\mathrm{d}X \qquad (1.3)$$

where:

X_1	$\equiv$	the lowest value of the potential change variable X
X_2	$\equiv$	the highest value of the potential change variable X
$\text{Pn}(X)$	$\equiv$	the distribution of the penalty of potential change
$\text{Pr}(X)$	$\equiv$	the distribution of the probability of potential change

The measurement of flexibility in terms of Penalty of Change (POC) has an easily understandable meaning, and, through the $\text{Pr}(X_i)$ or $\text{Pr}(X)$ term, it accounts for the demand for flexibility instead of blindly counting flexibility which may not be needed. POC can be used as a quantitative attribute in manufacturing decision making.

Example

Penalty of Change can be applied to the evaluation of all three types of manufacturing flexibility: product flexibility, operation flexibility, and capacity flexibility. Consider the evaluation of product flexibility for two mass production manufacturing systems, A and B. Product flexibility reflects the ability of the system to make a variety of products with the same equipment. In order to quantify it with an appropriate POC value, the probability and penalty of change must be established. In this case, the relevant change is a potential change in the product to be manufactured. Let us assume that there is a 70% probability that the next product to be manufactured will be product 1, and a 30% probability that it will be product 2. Let us also assume that product 1 is more similar to the currently manufactured product, and therefore can be accommodated on system A with only $20 million in modifications, as opposed to $50 million in modifications for Product 2. System B is a dedicated system which must be completely replaced (at a cost of $80 million) in order to accommodate any product change. Evaluating flexibility as the product of penalty and probability, we get POC_A = $20 million x 70% + $50 million x 30% = $29 million for system A, and POC_B = $80 million x 70% + $80 million x 30% = $80 million for system B. POC is much lower for system A than for system B, which means that system A has much more product flexibility than system B.

The same methodology may be applied to the calculation of other types of flexibility. Operation flexibility reflects the ability of the system to provide alternative routings for parts and thereby they withstand breakdowns. The penalty of change can be expressed as the average decrease in production rate, caused by an operational change. Let us assume that the decrease in production rate is 10 parts/hour for system A, and 20 parts/hour for system B. Without considering the probability for change, one would conclude that system A has more operational flexibility than system B, since the penalty is less for system A than it is for system B. However, considering the probability of change, namely the probability of system failure (20% for system A and 5% for system B), we get POC_A = 10 parts/hour x 20% = 2 parts per hour for system A, and POC_B = 20 parts/hour x 5% = 1 part per hour for system B. System B therefore, has more operational flexibility than system A.

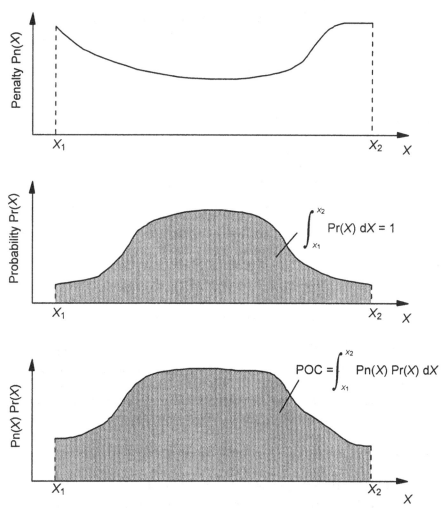

Fig. 1.24. POC as the Integral of the Product of Penalty and Probability

Capacity flexibility reflects the ability of the system to expand or contract in order to accommodate changes in product demand. Say, that product demand is forecast to increase from 100 parts/hour to 115 parts/hour with a 50% probability, to 150 parts/hour with a 30% probability, and to 200 parts/hour with a 20% probability. System A can accommodate these potential changes with penalties of 0, $50 million, and $100 million respectively, where the monetary values represent the acquisition costs of additional equipment, necessary to meet the increased demands. System B has the following penalties for change: $5 million, $40 million, and $120 million respectively. We then calculate POC_A = $0 x 50% + $50 million x

30% + $100 million x 20% = $35 million for system A and POC_B = $5 million x 50% + $40 million x 30% + $120 million x 20% = $38.5 million for system B. System A therefore, has slightly more capacity flexibility than system B.

Determining the flexibility of a manufacturing system by a mechanical analogy method

Modern manufacturing (MFG) systems have to respond to a dynamic demand, namely, a demand that changes over time. One way to characterize the behaviour of such a system is by considering its inputs and outputs. One can view MFG systems, having as input a "demand", typically expressed on some quantity of some products that are due at a particular time, and as output the "delivery" of industrial products. The orders that constitute the demand on the system may satisfy the needs of particular customers (make-to-order-production), or they may go to the inventory of the company (make-to-inventory-production).

Flexibility of an MFG system can be thought of as the ability, and the rapidness by which the system responds to the dynamic demand. This resembles the behaviour of a mechanical (MEC) system under the excitation of a force that fluctuates over time (Fig. 1.25). Such considerations about MFG systems have led to the establishment of a quantifiable analogy between MFG and MEC systems [22, 24]. This analogy between MFG and MEC systems is based on the fact that a structure responds to a particular input - typically, a force varying over time - in a way characterised by its transfer function, which determines the behaviour of the structure. Similarly, the "transfer function" of an MFG system characterises the way the latter responds to changes in the input - i.e., to various orders of different products and in different volumes. The dimensionless damping factor ζ in the case of a MEC system, represents the system's ability to respond to changes in the excitation, i.e, it characterises the MEC system's flexibility. In a similar manner, this dimensionless damping factor ζ represents the MFG system's ability to respond to changes in the input, i.e., it constitutes a quantifiable flexibility measure for the MFG system under consideration.

The analogy between MEC and MFG systems has led to a method, the ζ analogy method [23], for estimating the transfer function of the MFG system. As inputs are taken the incoming orders for the MFG system, and as output the delivery of these orders, expressed in flow times of the orders within the MFG system (Fig. 1.22). This modelling method makes use of the established ζ analogy between an MFG and a MEC system [24].

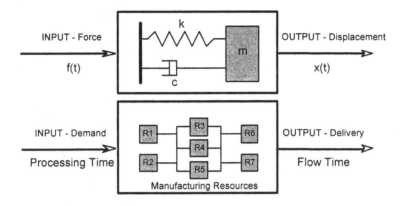

Fig. 1.25. The analogy between an MFG and a MEC system

As it has been stated above, ζ is the dimensionless damping factor, which, in the case of a single degree of freedom MEC system, is defined by the relation:

$$\zeta = \frac{1}{2}\sqrt{\frac{c^2}{km}} \qquad (1.4)$$

where c, k, and m denote the MEC system's well-defined lumped parameters of damping, stiffness, and inertia, respectively. This damping factor can be estimated also from the frequency spectrum of the system as:

$$\zeta \approx \frac{1}{2Q} \qquad (1.5)$$

where Q is the amplitude in the frequency spectrum that corresponds to the fundamental natural frequency. Equation (1.5) is valid, provided the damping factor is in the range $0 < \zeta < 2^{-1/2}$ otherwise, the system is over-damped and the peak in the amplitude disappears. The higher the damping factor ζ is in a MEC system, the less sensitive is the response of the system to changes in the input/excitation, while the lower the value of ζ, the more sensitive is the response of the system to changes in the input. Thus, the damping factor ζ constitutes a measure of the MEC system's ability to respond to changes in its input.

In the case of an MFG system, interrelations among the system's elements are very difficult to be mathematically described in their totality, since the dynamic behaviour of an MFG system is in reality non-linear, due to the system's complexity [25, 26], complexity, which is sometimes necessary to produce products that modern markets require. Therefore, it has not been yet possible to establish relations - in the form of equation (1.4) - which could describe the dynamic behaviour of an MFG system. Estimations - in the form of the approximate equation (1.5) - which can describe the dynamic behaviour of an MFG system are possible, though. The latter statement establishes in reality the ζ- analogy between an MFG and a MEC system, and offers the possibility of using some tools - typically applied to analysing the dynamic behaviour of MEC systems - in the analysis of MFG systems.

In order for the approximate equation (1.5) to be used in the analysis of MFG systems, a frequency spectrum is required for the determination of the amplitudes Q_i that correspond to the natural frequencies $\hat{}_i$ of the MFG system under consideration. This can be achieved through the implementation of Discrete Fourier Transform (DFT) techniques, which estimate the transfer function of a system, given its inputs and outputs. The input/excitation to an MFG system can be the demand on the system, namely, the sequence of orders and their Processing Times *PT*. The response of the system, which can be expressed in Flow Times *FT*, is the delivery of the products, for the different orders, similar to the resulting displacement in a MEC system. Since the input to and the output of an MFG system have both time units, the amplitudes Q_i in the frequency spectrum are dimensionless numbers, and so is the flexibility measure, calculated from equation (1.5). By processing Time is meant, in the context of this analysis, the time required to process a particular order through an MFG system, having considered the system as empty and working only on this particular order. On the other hand, the response of the real system, which is not empty and has to manage different orders, under real conditions, results in time "delays". This means that the orders spend typically, more time in the system than the absolute required processing time, namely, the orders "flow" through the system for periods called Flow Times. An "ideally flexible" MFG system will then have zero delays in responding to incoming orders, and the Processing and Flow Times will be identical. In general, the smaller the discrepancy between Processing Times/input and Flow Times/output is, the better an MFG system behaves and the more it is able to respond to the incoming demand.

Example

A simple simulation model [23], comprising one machine preceded by an input buffer, is used to describe the MFG system under trial. The ζ analogy method, described above, is implemented in the study of this simulation model. A set of assumptions has been made, regarding the operation of the simulation model, such as: set-up times for the orders do not depend on the sequence of incoming orders; the system's machine is always available and it is never kept idle, when work is waiting; and, once processing on an order begins, it continues without interruption until completion. The flow time for an individual order is given by the relation:

$FT = WT + ST + PT$

where:

FT	=	Flow Time
WT	=	Waiting Time
ST	=	Set-Up Time
PT	=	Processing Time

Seven sets of simulation experiments have been designated in order to test the validity of the proposed ζ analogy method in the dynamic modelling of MFG systems. Table 1.1 shows design details of the individual sets of simulation experiments, regarding the input, output and variable(s) used.

Set of Experiments	Inputs	Variables	Outputs
1	Inter-arrival time, Processing times	Set-up times	Flow times
2	Inter-arrival time, Set-up times	Processing times	Flow times
3	Inter-arrival time, Set-up times, Processing times	Sequencing of orders (FIFO)	Flow times
4	Inter-arrival time, Set-up times, Processing times	Sequencing of orders (LIFO)	Flow times
5	Inter-arrival time, Set-up times, Processing times	Sequencing of orders (SPT)	Flow times
6	Inter-arrival time, Set-up times, Processing times	Sequencing of orders (LPT)	Flow times
7	Arrival time, Set-up times, Processing times, Due times	Sequencing of orders (EDD)	Flow times

Table 1.1. Design of simulation experiments

A random number generator is employed to produce the required simulation data sets. A tenfold repetition - while using different "seeds" to initialise the generator - for each individual experiment, in all sets of the designed simulation experiments, is performed. Inter-arrival time data sets, for incoming orders, follow an exponential distribution ($\lambda = 30$). A normal distribution is followed for the generation of both processing time data sets ($\grave{I}=30$, $\hat{U}=5$) and set-up time data sets ($\grave{I}=10$, $\hat{U}=1$), respectively. Exceptions are made in the first set of simulation experiments - in the set-up time data sets - and in the second set of simulation experiments - in the processing time data sets - where the mean value of distribution varies.

In the first and second sets of simulation experiments, the "First In - First Out" control shop-floor policy is considered. The impact of variable in-the-mean-value set-up times against stable in-the-mean-value processing times, and vice versa, on the behaviour of the MFG system's flexibility measure ζ is, thus, investigated. Similarly, the impact on the consistency and sensitivity-to-changes of the system's transfer function is also considered. In the remaining sets of simulation experiments, the impact of different shop-floor policies on the MFG system's transfer function is studied. Shop-floor policies such as "First In-First Out" (FIFO), "Last In-First Out" (LIFO), "Shortest Processing Time" (SPT), "Longest Processing Time" (LPT) and "Earliest Due Date" (EDD) are employed at this part of the investigation study.

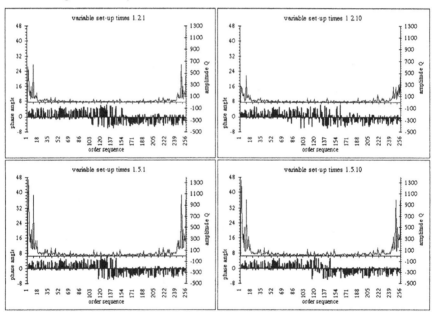

Fig. 1.26. Example of Transfer functions calculated for the variable set-up time experiments

Regarding the first set of simulation experiments, it can be concluded that similar inputs/excitations - following the same distribution (normal) with the same mean and standard deviation values - produce similar Fourier transfer-function spectra. The variation, though, of set-up times affects the MFG system's response, i.e., the flow times, resulting in quantifiable differences among the Fourier transfer-function spectra, produced through experiments 1.1. to 1.10 (Fig. 1.26).

The behaviour, the flexibility measure ζ exhibits, is shown both in table and diagram form (Table 1.2 and Fig. 1.27, respectively). It can be easily seen from the diagram in Fig. 1.27, that for different experiments, the variation of mean ζ inversely follows the variation of set-up times - i.e., mean ζ decreases with increasing mean set-up time values among the processed sets of orders, and vice versa. The corresponding variation of mean ζ, though, in the tenfold repetition series - within each designated experiment - is minimal, i.e., the mean ζ-value remains stable on the average.

1st Set of Experiments	Number of Repetitions	Average ζ	Standard deviation of ζ
1.1	10	0,093489813	0,016540874
1.2	10	0,062891590	0,005945673
1.3	10	0,047820171	0,004647306
1.4	10	0,037983476	0,002367296
1.5	10	0,031747400	0,001925188
1.6	10	0,026242502	0,001132202
1.7	10	0,023189019	0,000821031
1.8	10	0,020815637	0,001114379
1.9	10	0,018648438	0,000942566
1.10	10	0,016616332	0,000657407

Table 1.2. Average ζ calculated for the variable set-up time experiments

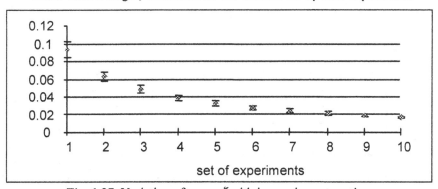

Fig. 1.27. Variation of mean ζ with increasing set-up times

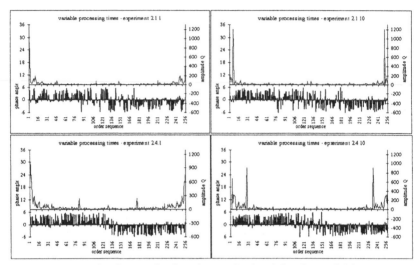

Fig. 1.28. Example of Transfer functions calculated for the variable processing time experiments

Regarding the second set of simulation experiments, it can be concluded that similar inputs/excitations - following the same distribution (normal) with the same mean and standard deviation values - produce similar Fourier transfer - function spectra. The variation, though, of processing times affects the MFG system's response, resulting in quantifiable differences among the Fourier transfer-function spectra, produced through experiments 2.1 to 2.10 (Fig. 1.28).

2nd Set of Experiments	Number of Repetitions	Average ζ	Standard deviation of ζ
2.1	10	0.073547679	0.018534621
2.2	10	0.045317974	0.006284231
2.3	10	0.036082030	0.004081434
2.4	10	0.030498370	0.004478283
2.5	10	0.025445020	0.003073631
2.6	10	0.021567257	0.002635345
2.7	10	0.019022359	0.002165440
2.8	10	0.015342111	0.002105147
2.9	10	0.014610654	0.002318876
2.10	10	0.013535416	0.000909943

Table 1.3. Average ζ calculated for the variable processing time experiments

The behaviour, the flexibility measure exhibits, is shown both in table and diagram form (Table 1.3 and Fig. 1.29, respectively). It can be easily seen, from the diagram in Fig. 1.29 that for different experiments the variation of mean ζ inversely follows the variation of processing times - i.e., mean ζ decreases with increasing mean set-up time values among the processed sets of orders, and vice versa. The corresponding variation of mean, though, in the tenfold repetition series - within each designated experiment - is minimal, i.e., the mean ζ-value remains stable on the average.

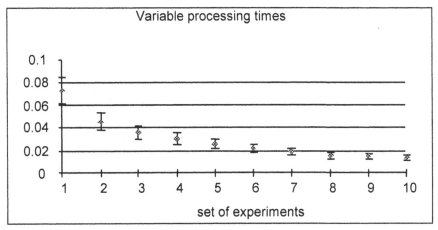

Fig. 1.29. Variation of mean with increasing processing times

Results obtained verify expectations made, concerning the outcome of this set of simulation experiments. Comparing these results with those obtained from the first set of experiments, it is concluded that both processing and set-up times affect the behaviour of the MFG system under consideration, in a similar manner.

Sets 3 to 7 of the simulation are considered, here. Different control shop-floor policies have been studied. Inter-arrival times and processing times follow an exponential ($\ddot{I}$=30) and a normal ($\grave{I}$=30, $\hat{U}$=5) distribution, respectively. Two experiments have been considered in each set, with a tenfold repetition series each. Different "seeds" have been used to initialise the random number generator to produce the inter-arrival and processing time data sets, respectively, within the repetition series. The obtained results are shown in Fig. 1.30.

It is clear, from Fig. 1.30 that the MFG system's transfer function changes both in form and magnitude, when passing from a control shop-floor policy to another. On the other hand, the stable character of the MFG system's transfer function remains - within a predetermined control shop-floor policy and a tenfold repetition series (Table 1.4 and Fig. 1.31, respectively). A result is also extracted from the first and second sets of experiments.

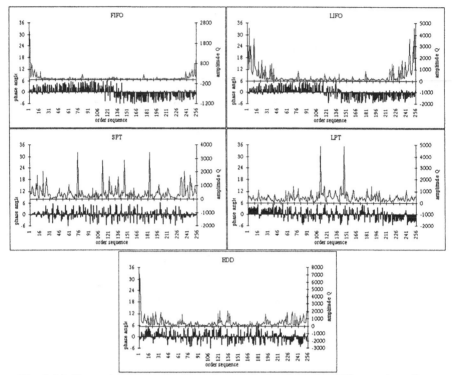

Fig. 1.30. Example of Transfer Functions calculated for the different shop-floor
policy experiments

Set of Experiments	Number of Repetitions	Average ζ	Standard deviation of ζ
FIFO 3.1	10	0,021447998	0,002865465
FIFO 3.2	10	0,012909885	0,001195650
LIFO 4.1	10	0,004213074	0,000491931
LIFO 4.2	10	0,004427646	0,000630286
SPT 5.1	10	0,002517886	0,000444466
SPT 5.2	10	0,002716877	0,000439027
LPT 6.1	10	0,001877170	0,000348230
LPT 6.2	10	0,001818793	0,000330677
EDD 7.1	10	0,003035999	0,000778058
EDD 7.2	10	0,003042973	0,000482010

Table 1.4. Average ζ calculated for different shop-floor policies

That the transfer of an MFG system changes, when considering different
control shop-floor policies, suggests that the policy itself is a characteristic
of the system under consideration.

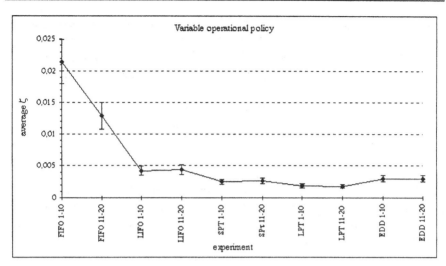

Fig. 1.31. Variation of the mean, while considering different shop-floor policies

The results obtained from the simulation experiments with the employment of the ζ-analogy method have revealed a number of interesting issues, concerning the character of the Fourier-produced transfer function of an MFG system. Firstly, processing times, set-up and waiting times as well as control shop-floor policy are constituent elements or properties of an MFG system. Secondly, the transfer function of an MFG system, with a given configuration, is consistent and its sensitivity to changes in the input, is characterised by the behaviour of the system's flexibility measure (ζ). Finally, the behaviour of the MFG system's flexibility measure is consistent with the behaviour exhibited by a real MFG system.

1.3.4 Quality

The quality of a product, broadly related to customer satisfaction, is often difficult to be defined in quantitative terms, since customer satisfaction not only does it depend on the actual features of a product, but also on its feasibility, maintainability, and on a host of other factors that are often subjective and thus, difficult to quantify. However, customer satisfaction can be traced to two major factors at the origin of a product: its *design* and *manufacture*.

In manufacturing, quality typically refers to how well the production process meets design specifications, related to the different features and properties of a product. The quality of a product is an aggregate of the quality of individual features and properties, which can be broadly divided into two major groups: the geometric characteristics, and the physical and/or chemical properties of the materials, which constitute the product.

Quality, pertaining to manufacturing systems, has many different aspects. When identical parts or products are mass-produced, the repeatability of the manufacturing process becomes important, so as for every item produced to conform to predefined specifications. This aspect of quality has been the subject of extensive studies in the area of statistical quality control. However, as manufacturing systems move towards low volume or even one-of-a-kind production, the tools based on statistical methods may have to be adjusted, since the population of items, typically used for deriving statistically meaningful measures for quality, may not be available.

Since manufacturing quality reflects the meeting by the manufacturing process of established expectations within tolerances, it is important to note that such tolerances may be over-or under-estimated. An overestimation of tolerances leads to unnecessary cost, during product manufacture, and unnecessary pressure on the manufacturing system; while an underestimation of tolerances may lead to a malfunctioning product.

The functional properties of technical products are determined, to a considerable extent, by the geometrical properties of their individual components. Tolerances and variations affect the ability to assemble the final product, and also the production cost, the process selection, tooling, and set-up cost, required operator skills, inspections and gauging, scrap and rework. Product performance and robustness of the design are directly affected by the variations - constrained or bounded by the tolerances. The goal is to ensure functioning of the product during assembly and usage, since poorly performing products will eventually lose out in the market place. However, current tolerancing standards do not provide the ground for a theoretic tolerancing model, consistent with 3D geometry (solid) models. This is due to the fact that existing standards are not *assembly-oriented* but *part-oriented*. Tolerance specifications are a critical link between engineering and manufacturing, i.e., a common meeting ground where competing requirements may be solved. The complexity of the subject is obvious as it combines tolerancing and measurement of a dimension, form, location and orientation, deviation as well as surface roughness and waviness.

There is a need to develop new IT-based tools for TQM (Total Quality Management) capable of handling problems in flexible manufacturing and complex assembly with small batches, a situation commonly encountered in the automotive and aerospace industries. Recent research work in this field is for example in the QUETA (a European Community funded project), comprising three modules (tools): The first one, covering the area of quality, focused on analysis of assembly with the ability to identify and predict critical components, dimensions, and characteristics; the second one, covering the area of error analysis of components, subassemblies and assembly with the ability to identify quality critical manufacturing and assembly processes; and the third one, covering the area of generation of cost-effective action plans, based on the identified criticality with the ability to ensure, and improve quality.

These tools provide a number of feedback loops to different areas in manufacturing, such as design, production planning, processes, assembly, management, etc., enabling the fast response of QA (Quality Assurance) systems to changes in manufacturing processes and business requirements.

Measuring quality is critical for manufacturing, as it reflects the performance of the production process as a whole, and facilitates the establishment of trade-offs between quality and other manufacturing attributes. Quality is measured from the most aggregate level, in terms of acceptance or rejection of a product, down to the elementary characteristics of a component. In general, the more aggregate the quality consideration is, the less quantifiable it is and the more it relies on customer reports and subjective judgment. More elementary characteristics, such as dimensions and physical properties, such as hardness and strength, are easier to quantify, and therefore, provide an easier measuring task for testing and inspection, which is usually employed during the production process or immediately thereafter.

As the manufactured parts tend to become larger, more complex, and require higher accuracy, both the dimensional and surface inspections become costly and time consuming processes [27]. Co-ordinate Measurement Machines (CMMs) have been used in manufacturing, mainly for off-line control, to measure parts with contours and irregular surfaces and to provide geometric validation of 3D surface features. CMMs, which are contact measurement systems, are relatively slow, non-portable and unable to measure accurately parts with large dimensions. These limitations have driven the manufacturing industry to the development and use of non-contact measurement systems, which allow fast and accurate checking of large structures, e.g. aircraft wings, assembly jigs, etc., and provide accurate estimates of the positions of 3D surface features. These measurement systems are typically theodolite systems, laser tracking systems, camera systems, close-range and real time photogrammetry or triangulation systems.

Two common aggregate measures of quality are *percent defective*, which reflects the percentage of parts produced outside the required specifications, and *warranty costs*, which reflect the costs that occur during operation, due to failure of products. Both measures are relatively easy to determine and provide some trade-off, particularly with cost attributes. On the other hand, they do not provide insight into the production process, but they rather provide a global assessment of the capabilities of a manufacturing system. The *process capability index* C_p is a more process-oriented indicator of quality, and is defined as:

$$C_p = \frac{\text{tolerance}}{6 \times \text{standard deviation}} \tag{1.6}$$

The *tolerance* is defined as the range of values within which the particular dimensional characteristic of the product is acceptable and the *standard deviation* is derived from the distribution of the particular characteristic or dimension, as produced by the process, which is being evaluated. While this measure is very process-oriented and provides a good base for comparison among processes, it is difficult to interpret it physically, and to identify any trade-off between it and the other manufacturing attributes. The *quality loss function* [13], which will be discussed below with the help of an example, is another metric, which measures the loss that occurs, when a product's quality characteristic, deviates from the target value, *within* or *outside of* pre-specified tolerances.

Measures of quality such as warranty cost, or even the process capability index or loss function, have an aggregate character and do not necessarily "connect" to the physics of the process. Surface quality, on the other hand, often relates more to process physics. It is particularly relevant for the metalworking industry and discrete parts manufacturing, and is measured by using electromechanical devices. In machining processes, the achieved surface quality is directly related [2] to the feed rate (Fig. 1.32), a typical process decision variable, which must be selected before beginning the process. This is because the surface generation mechanism in machining processes is such that the surface quality [2] is dictated by the rate of relative motion (Fig. 1.33) of the tool and the workpiece (see Chapter 2). It is interesting to note, however, that the feed rate not only does it influence the surface quality but also the production rate and cost of the process. Higher feed rates will produce more parts per unit of time, with lower quality. Furthermore, higher feed rates may cause greater tool wear and thus, increase the operational cost of the process. So, we can see that with machining processes, a trade-off among different manufacturing attributes allows the process to be adjusted in order to meet particular objectives.

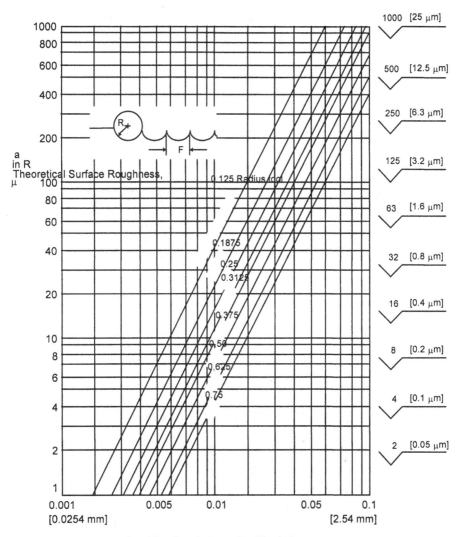

Feed Per Revolution or Per Tooth, in

Fig. 1.32. Theoretical Surface Quality as a Function of Feed Rate for Face Milling or Turning Tools with Round Cutting Edges

Example

Figure 1.34 [14] shows two different processes, 1 and 2, which produce the same product, but with different distributions of a particular quality characteristic.

Assuming that the tolerance is 10, and that the standard deviations of the quality characteristic for the two processes are 2.89 and 1.67 respectively, the C_p's for these two processes can be calculated as follows:

$$C_{p1} = \frac{10}{6 \times 2.89} = 0.58 \qquad \text{for Process 1}$$

$$C_{p2} = \frac{10}{6 \times 1.67} = 1.00 \qquad \text{for Process 2}$$

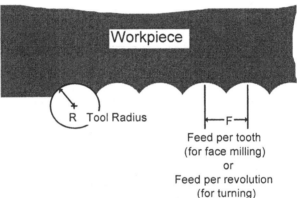

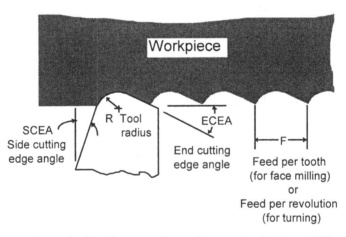

Fig. 1.33. Theoretical Surface Roughness Geometries for Face Milling and Turning

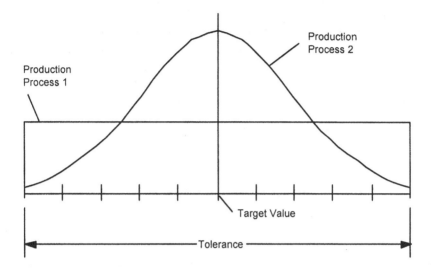

Fig. 1.34. Quality Characteristic Distributions for Processes 1 and 2

Alternatively, one can introduce a quality loss function L(y), which measures the loss occurring when a product's quality characteristic, y, deviates from a target value m (Fig. 1.35).

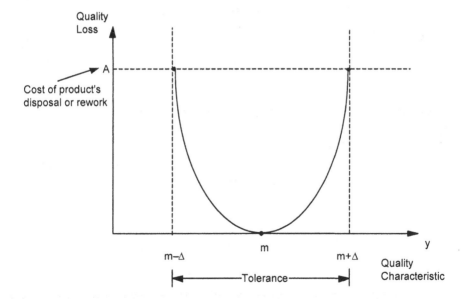

Fig. 1.35. Quality Loss Function

A, is the loss, due to a defective part and represents the monetary value of reproducing or repairing that part [14]. If a product is produced and the value of the quality characteristic y is exactly at the target value m, then there is no loss for that product. One can assume that the quality loss increases exponentially (Fig. 1.35) as the quality characteristic of the product moves away from the target value m. Based on these observations, one can write the quality loss function as a Taylor series:

$$L(y) = L(m) + \frac{L'(m)}{1!}(y-m) + \frac{L''(m)}{2!}(y-m)^2 + \cdots \qquad (1.7)$$

or, if we neglect terms of higher order,

$$L(y) = L(m) + \frac{L'(m)}{1!}(y-m) + \frac{L''(m)}{2!}(y-m)^2 \qquad (1.8)$$

Since there is no loss at the target value m, L(m) = 0. Furthermore, since the loss at y = m-δ and at y = m+δ (where δ is a small positive number) should be small and positive, we postulate that L′(m) = 0. Therefore,

$$L(y) = \frac{L''(m)}{2!}(y-m)^2 = K(y-m)^2 \qquad (1.9)$$

When the quality characteristic y is at the tolerance limit Δ, away from the target value, say y = m+Δ, then the value of the loss function is A:

$$L(m+\Delta) = K(m+\Delta-m)^2 = K\Delta^2 = A \qquad (1.10)$$

Therefore, $K = A/\Delta^2$, and

$$L(y) = \frac{A}{\Delta^2}(y-m)^2 \qquad (1.11)$$

The following numerical example illustrates this definition of the quality loss function. Let the cost of failure be A = $2, and the tolerance be 10 (which implies that Δ equals 5), and let us also consider two processes, 1 and 2, with variances from target $(y-m)^2$ of 2.89^2 and 1.67^2 respectively. Then, the quality loss function for these two processes can be calculated as:

for Process 1

$$L_1 = \frac{2}{5^2} \times 2.89^{\,2} = \$0.69/\text{piece}$$

for Process 2

$$L_2 = \frac{2}{5^2} \times 1.67^{\,2} = \$0.22/\text{piece}$$

Further Reading

For those interested in reading more about manufacturing productivity, the report of the MIT Commission on Industrial Productivity, entitled *Made in America* and written by Dertouzos, Lester and Solow [15], provides an analysis of eight industrial manufacturing sectors in the United States and contrasts them with the same sectors in other nations, such as Germany and Japan. These eight industries include semiconductors, computers and copying machines, commercial aircraft, consumer electronics, steel, chemicals, textiles, automobiles, and machine tools.

For a more general treatment of economic development and its relationship to industrial productivity, one can read two U.N. reports: *World Economic Survey 1990: Current Trends and Policies in the World Economy* [16]; and *Economic Survey of Europe in 1989-1990* [17].

For the cost and production rates of a number of processes, we recommend the *Tool and Manufacturing Engineers Handbook* published by SME [18]. A thorough treatment of the subject of reliability, which is becoming increasingly important in the manufacturing environment, is provided in *An Introduction to Machinery Reliability Assessment* by Bloch and Geitner [19].

Total Quality Control by Feigenbaum [20] offers an excellent look at statistical approaches to quality. Other useful books on this subject include *Quality Engineering in Production Systems*, by Taguchi, El-Sayed and Hsiang [14], and *Quality Engineering Using Robust Design* by Phadke [21].

In the area of flexibility the literature is rather restricted. *Flexible Manufacturing Systems, Decision Support for Design and Operation* by Tempelmeier, H. and Kuhn [28], is a useful book on this subject. Furthermore, the *International Journal of Flexible Manufacturing Systems*, published by Springer – Verlag, offers a number of papers on flexibility and the measurement of flexibility.

References

1. *NSF Report on Research Priorities for Proposed NSF Strategic Manufacturing Research Institute.* Dr. E.M. Merchant, Workshop Chairman. March 11-12, 1987.

2. *Machining Data Handbook, 3rd Edition, Vol. 2,* Metcut Research Associates Inc., 1980.

3. Browne, J., K. Rathmill, S.P. Sethi and K.E. Stecke, "Classification of Flexible Manufacturing Systems," *The FMS Magazine* (April 1984), pp. 114-117.

4. Buzacott, J.A., "The Fundamental Principles of Flexibility in Flexible Manufacturing Systems," *Proceedings of the First International Conference on Flexible Manufacturing Systems* (1982), Amsterdam, N. Holland, pp. 13-22.

5. Falkner, C.H., "Flexibility in Manufacturing Plants," *Proceedings of the Second ORSA/TIMS Conference of Flexible Manufacturing Systems: Operations Research Models and Applications* (1986). K.E. Stecke and R. Suri, Editors, pp. 95-106.

6. Barad, M. and D. Sipper, "Flexibility in Manufacturing Systems: Definitions and Petri Net Modelling," *International Journal of Production Research* (Vol. 26, No. 2, 1988), pp. 237-248.

7. Kumar, V., "Entropic Measures of Manufacturing Flexibility," *International Journal of Production Research* (Vol. 25, No. 7, 1987), pp. 957-966.

8. Kumar, V., "On Measurement of Flexibility in Flexible Manufacturing Systems: An Information–Theoretic Approach," *Proceedings of the Second ORSA/TIMS Conference of Flexible Manufacturing Systems: Operations Research Models and Applications* (1986). K.E. Stecke and R. Suri, Editors. pp. 131-143.

9. Brill, P.H. and M. Mandelbaum, "On Measures of Flexibility in Manufacturing Systems," *International Journal of Production Re-*

search (Vol. 27, No. 5, 1989), pp. 747-756.

10. Makino, H., "Versatility Index – An Indicator for Assembly System Selection," *Annals of the CIRP* (Vol. 39, No. 1, 1990), pp. 15-18.

11. Primrose, P.L. and R. Leonard, "Conditions Under Which Flexible Manufacturing is Financially Viable," *Proceedings of the Third International Conference on Flexible Manufacturing Systems* (1984), pp. 121-132.

12. Son, Y.K. and S.P. Chan, "Quantifying Opportunity Costs Associated With Adding Manufacturing Flexibility," *International Journal of Production Research* (Vol. 28, No. 7, 1990), pp. 1183-1194.

13. Srinivasan, V. and R.A. Millen, "Evaluating Flexible Manufacturing Systems as a Strategic Investment," *Proceedings of the Second ORSA/TIMS Conference on Flexible Manufacturing Systems: Operations Research Models and Applications* (1986). K.E. Stecke and R. Suri, Editors, pp. 83-93.

14. Taguchi, G., E.A. Elsayed and T. Hsiang, *Quality Engineering in Production Systems*, McGraw-Hill Publishing Company, New York, 1989.

15. Dertouzos, M.L., R.K. Lester, R.M. Solow and The MIT Commission on Industrial Policy, *Made in America: Regaining the Productive Edge*, The MIT Press, Cambridge, MA, 1989.

16. *World Economic Survey 1990, Current Trends and Policies in the World Economy*, United Nations, New York, 1990.

17. *Economic Survey of Europe in 1989-1990*, United Nations, New York, 1990.

18. *Tool and Manufacturing Engineers Handbook, Fourth Edition*, Society of Manufacturing Engineers. Charles Wick, Editor-in-Chief. 1987.

19. Bloch, H.P. and F.K. Geitner, *An Introduction to Machinery Reliability Assessment*, Van Nostrand Reinhold, New York, 1990.

20. Feigenbaum, A.V., *Total Quality Control*, McGraw-Hill Publishing

Company, New York, 1983.

21.   Phadke, M.S., *Quality Engineering Using Robust Design*, Prentice Hall, Englewood Cliffs, New Jersey, 1989.

22.   Chyssolouris, G., "Flexibility and Its Measurement", *Proceedings of the CIRP Annals*, (Vol. 45, No. 2, 1996), pp. 581-587.

23.   Bechrakis, K., S. Karagiannis, and G. Chryssolouris, "The -Analogy Method For the Modelling Of Manufacturing Systems", *Proceedings of the Second World Congress on Intelligent Manufacturing Processes & Systems*, Budapest, Hungary (June 1997). pp. 166-169.

24.   Chryssolouris, G., N. Anifantis and S. Karagiannis "An Approach to the Dynamic Modelling of Manufacturing Systems", *IJPR*, (Vol. 36, No. 2, 1998), pp. 475-483.

25.   Wiendahl, H-P et al, "Modelling and Simulation of Assembly Systems", *Keynote Paper*, *Annals of CIRP*, (Vol.40, No.2, 1991), pp. 577-585.

26.   Wiendahl, H-P et-al, "Management and Control of Complexity in Manufacturing", *Keynote Paper*, *Annals of CIRP*, (Vol.43 No.2, 1994), pp. 533-540.

27.   Chryssolouris, G., H. Vasileiou and D. Mourtzis "Critical review of non-contact measuring methods for large part inspection", *15th Int. Conference on Production Research*, University of Limerick Ireland (August, 1999).

28.   Tempelmeier, H. and H. Kuhn, *Flexible manufacturing systems: Decision support for design and operation*, Wiley: New York, 1993.

Review Questions

1.   Define manufacturing.

2.   Name five economic indicators of a nation's manufacturing performance, and discuss the advantages and disadvantages of each.

3.   What is an attribute? What are the four major classes of attributes in manufacturing?

4.   What is an objective? What is a goal? How is an objective different from a goal?

5.   What is a technoeconomical model?   For the technoeconomical model on Page 9, describe the inputs and outputs. Describe the inputs and outputs of two other technoeconomical models that you think would be useful in manufacturing.

6.   In setting up and operating a metal-cutting job shop, what specific costs would be associated with each of the major categories of manufacturing cost? Provide a rough estimate of each specific cost. Make reasonable assumptions for any required information (e.g. the number and types of machines in the job shop, the labor cost rate, etc.).

7.   Why do machining operations have an optimum (non-zero, non-infinite) cutting speed?

8.   Provide 10 specific reasons why system yield is lower than machine cycle in a manufacturing system.

9.   What is the reliability over a one year period of a machine which fails on average once every two months? Over what period will the reliability of this machine be 50%?

10.   If the reliability of each transfer line machine for a specified period is 90%, what is the longest transfer line for which the overall transfer line reliability over the specified period will be at least 80% (assuming no buffers)?

11.   If the availability of a machine is 90%, what is the ratio of its mean time to repair (MTTR) to its mean time between failures (MTBF)?

12.   In order to increase availability, is it better to: *a*) increase the MTBF by *x*%, or *b*) decrease the MTTR by *x*%? If your answer depends on *x*, find the value of *x* for which the two options will yield the same availability.

13.   What are the advantages of manufacturing flexibility?

14.   A manufacturing system currently manufactures Product A. It is to be converted to the production of a second-generation product, which will be either Product B (with a 90% probability), or Product C (with a 10% probability). The cost of converting the system to manufacture Product B is $10 million, while the cost of converting the system to manufacture Product C is $40 million. What is the product flexibility of the system as measured by the Penalty of Change (POC)?

15.   What characteristics attributable to *design* contribute to the quality of a manufactured product? What characteristics attributable to *manufacturing* contribute to the quality of a manufactured product?

16.   What are the advantages and disadvantages of percent defective as a measure of quality?

17.   What is the optimal value of the quality loss function L(y)? What is the optimal value of the process capability index C_p?

18.   If the value of C_p for a process is 0.9, what percentage of process outputs fall within the specified tolerance? What assumptions are required in arriving at this answer?

2. Overview of Manufacturing Processes

2.1 Introduction

A manufacturing process is defined as the use of one or more physical mechanisms to transform the shape of a material's shape and/or form and/or properties. Manufacturing processes can be further divided into *discrete parts* processes and *continuous* processes. The metalworking industry, where many single items are produced, uses discrete parts manufacturing. Chemical processing, used, for example, in the film- or fiber-making industries, uses continuous processing. In this chapter, we will concentrate on discrete parts manufacturing.

Manufacturing processes can be classified into five categories:

1. *Forming* or *primary forming processes* – processes in which an original shape is created from a molten or gaseous state, or from solid particles of an undefined shape. During primary forming processes, cohesion is normally created among particles.
2. *Deforming processes* – processes that convert the original shape of a solid to another shape without changing its mass or material composition. During this process, cohesion is maintained among particles.
3. *Removing processes* – processes during which material removal occurs; cohesion among particles is destroyed.
4. *Joining processes* – processes that unite individual workpieces to make subassemblies or final products. These include additive processes, such as filling and impregnating of workpieces; cohesion among particles is increased.
5. *Material properties modification processes* – processes that purposely change the material properties of a workpiece in order to achieve desirable characteristics without changing its shape.

These categories are applied to a variety of engineering materials, which can be classified into metals, ceramics, polymers and composites. Table 2.1 indicates the application of different processes to different materials.

The selection as to which process should apply to a particular material, is influenced by a number of factors, which affect cost, production rate, flexibility and part quality. Probably the two most important factors (Fig. 2.1) are:

1. *The lot size of the parts to be made.* Small lot sizes require flexible processes, such as material removal processes, capable of accommodating different geometric features, etc. Large lot sizes allow the use of primary forming or deforming processes; since the tooling cost of such processes is high, a large lot size is required to amortize the cost.

2. *Physical properties of the material (i.e. melting point).* Metals have relatively high melting temperatures, and therefore are usually processed in a solid form, using material removal and deforming processes. Polymers and composites have a much lower melting point, allowing the use of primary forming processes, where the material is often in a liquid state; however, secondary operations, usually grinding, are often needed to achieve the required dimensional accuracy and surface quality. Because these materials often contain abrasive fibers or fillers, they often exhibit abrasive behavior when conventionally machined. Ceramics are usually brittle, a property which makes it difficult to be processed in a solid form with conventional machining techniques. With ceramics, primary forming processes are used to create the basic shape of the workpiece, and secondary operations (usually machining) are used to create the final shape and surface quality.

		PROCESSES				
		Forming	Deforming	Removing	Joining	Modifying
MATERIALS	Metals	XX	XX	XX	XX	XX
	Ceramics	XX	--	X	--	--
	Polymers	XX	X	X	X	--
	Composites	XX	--	X	X	--

XX: Widely used
X : Seldom used
-- : Not used

Table 2.1. Overview of Materials and Processes

2.2 Primary Forming Processes

Primary forming processes produce a net shape through either solidification of a liquid or vapor or through cohesion of solid particles. Forming processes can be used effectively on all major classes of engineering materials. Among the primary forming processes, *casting* has most of the applications in manufacturing. Casting describes a family of processes during which the molten metal is introduced into a mold and then becomes solidified. Metal casting processes are used when large parts, possibly having intricate shapes and internal cavities, are produced with mechanical properties such as high strength, brittleness or hardness, which properties make these parts difficult to be produced through other methods. Casting processes are limited in terms of surface quality, porosity, and, consequently, the strength of the parts produced is also limited. Generally, casting processes are used in production with relatively large lot sizes so that the high capital cost, required to establish a foundry, can be justified. It should be noted, however, that some casting processes are not capital-and investment-intensive.

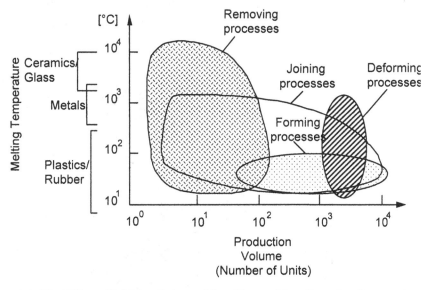

Fig. 2.1. The Effect of Melting Point and Lot Size on Manufacturing Process Application

Die casting is a widely used casting process during which molten material is forced into a die cavity, at pressures ranging from 0.7 to 700 MPa (0.1 to 100 ksi). Among the parts that may be produced by die casting are carburetors, motors, appliance components, hand tools, and toys [1]. The weight of most castings ranges from 90g to approximately 25 kg. Die casting machines may be horizontal or vertical. They require a high clamping force to be exerted during the process, in order to keep the dies closed. Machines are rated according to the clamping force they can provide, which ranges from 25 to 3000 tons. Die casting can be divided into hot chamber and cold chamber processes. The hot chamber process involves the use of a piston [2], which traps a certain volume of molten metal and forces it into the die cavity (Fig. 2.2). The metal is held under pressure until it solidifies in the die, which is cooled by circulating water or oil. Depending upon the size of the part, the production rate may reach as high as 18,000 shots per hour. Low melting-point alloys such as zinc, tin, and lead are commonly cast by this process, as they are not harmful to the machine structure, made of steel and cast iron. When brass, magnesium, or aluminum materials are cast, cold chamber processes and a cold chamber machine are used, due to the higher melting point of these materials [2] and their corrosive effect on iron and steel. In a cold chamber machine (Fig. 2.3), a ladle is used to pour molten material into an injection cylinder. The injection cylinder is not heated, and the metal is forced into the die cavity at pressures ranging from 20 to 70 MPa (3 to 10 ksi).

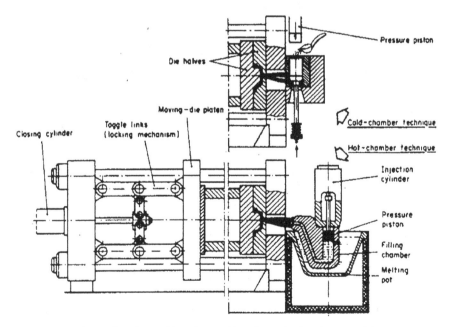

Fig. 2.2. Schematic of a Cold and Hot Chamber Die Casting Machine [2]

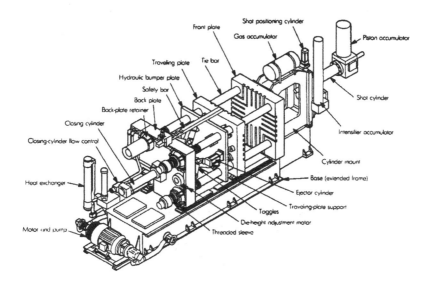

Fig. 2.3. Cold Chamber Die Casting Machine [2]

Centrifugal casting is another very important metal casting process in which hollow cylindrical parts, such as pipes, gun barrels, and street lamp posts are produced [2]. An illustration of this process is shown in Figure 2.4. Molten metal is poured into a rotating mold made of steel, iron, or graphite, which is usually coated with a refractory lining to increase its lifespan. The axis of rotation may be [2] vertical (Fig. 2.4) or horizontal (Fig. 2.5). Castings of good quality and dimensional accuracy can be obtained by using this process. The parts are cylindrical, ranging from 13 mm to 3 m in diameter, with wall thicknesses ranging from 6 to 125 mm.

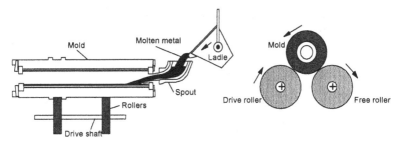

Fig. 2.4. Schematic of Centrifugal Casting

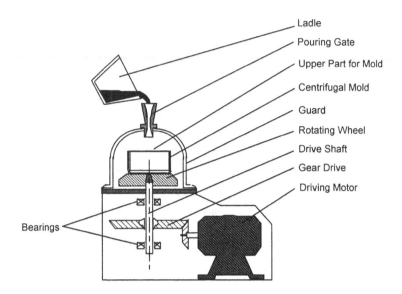

Fig. 2.5. Horizontal Centrifugal Casting Machine

Dies for parts with highly intricate features or large sizes are often difficult to design and produce. In these cases, *sand casting* can be used to produce metallic parts, ranging from cast iron bathtubs to engine blocks (Fig. 2.6). In the sand casting process, a pattern of the part to be produced is machined from a block of foam. This pattern is placed in a container and surrounded by fine compacted sand. When molten metal is poured into the pattern, the foam dissolves, leaving the cavity to be filled by the metal. When the sand is removed, the metallic part remains. With sand casting, internal geometric features, which cannot be made through die casting, can be produced. However, a separate pattern must be created for each part formed.

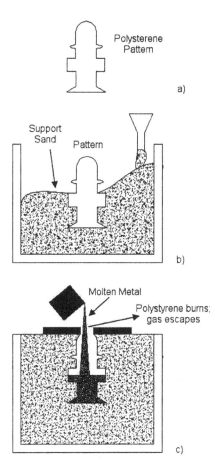

Fig. 2.6. Steps in the Sand Casting (or "lost foam") process the Polystyrene Pattern (a) is buried in sand (b) and vaporized by molten metal (c)

A similar process to sand casting is *investment casting*, (Fig. 2.7) where ceramic slurry is coated around a wax pattern. Once the desired coating thickness is achieved, heat is introduced to melt away the wax pattern, leaving a ceramic shell around the cavity. Molten metal is poured into the cavity and cooled. Then the ceramic shell is broken and a metallic part is formed. Investment casting has been extensively used for forming turbine blades, made from high-temperature alloys.

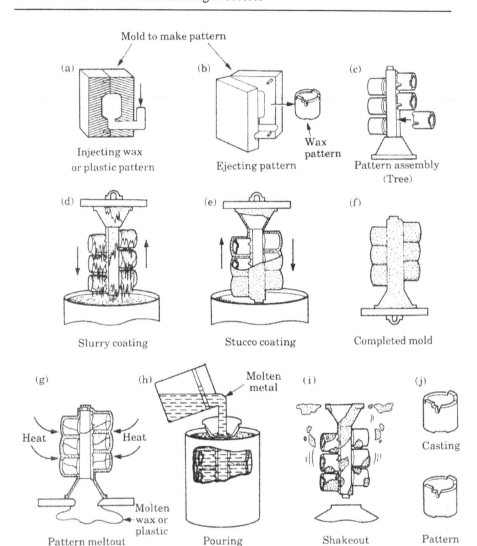

Fig. 2.7. Sequence of steps in the basic investment casting process

In metal casting, the raw material is in a molten form, and the material takes the shape of the part being produced during the solidification process. *In powder metallurgy (PM)* processes, which also belong to the primary forming category, the raw material is a powder, which is compacted in suitable dies and then sintered (heated without being molten). During sintering, the solid metallic grains will bond together through adhesion without melting. Powder metallurgy can be used to produce parts, such as gears, cams, bushings, cutting tools, and filters [1]. Figure 2.8 shows the sequence of the processes involved in making parts with powder metallurgy. Figure 2.9 shows a schematic illustration of a typical PM process.

Primary forming processes have found wide use in *polymers* and *composite* materials. Almost all processing techniques for these materials belong to the primary forming category and have a counterpart in processing metals and ceramics. In general, plastics can be molded, cast, formed, machined, and joined; and since they can be processed at much lower temperatures than metals, primary forming processes are very attractive for making components of plastics and composites.

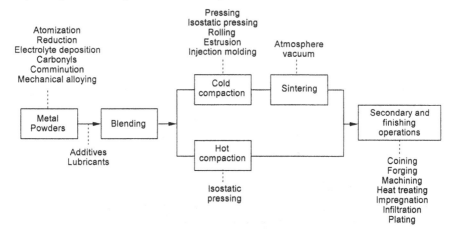

Fig. 2.8. Operations Sequence in Powder Metallurgy

The manufacturing process for plastics and composites often begins with raw materials in the form of pellets or powders, which are usually melted before the actual forming process begins. Plastic raw materials are also available as sheets, plates, rods, and tubing. Plastics, often used in producing composite materials are initially in a liquid form. Figure 2.10 outlines the different forming processes used for thermoplastics, thermosets, and elastomers [1, 3, 4].

Primary forming processes for plastics and composites belong to one of two categories: *casting* and *melt processing*. The term casting, describes the process of filling a mold by gravity, while the term *melt processing,* describes the techniques, which use applied pressure to process polymers that are too viscous, even at high temperatures, so as to flow under the force of gravity [1, 3, 4].

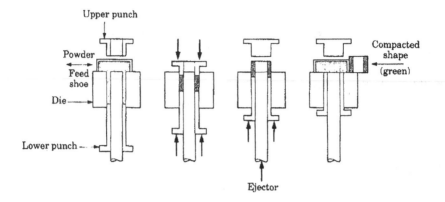

Fig. 2.9. Basic powder metallurgy process

All polymer melt processing techniques have in common some general characteristics:

- The raw material is usually granular, cut up, strained, diced, sheet, or, in the case of recycled material, chopped up and sometimes compacted scrap.
- Heating is partially external and partially internal.
- Internal heat is produced from the work of viscous shearing.
- Generally, overheating can cause permanent damage.
- There is a substantial volume change in cooling which reflects the rearrangement of molecules and the establishment of secondary bonds. Because melt processing is a time-dependent process, shrinkage increases with slower cooling, lower injection pressure and shorter injection time.
- Production rates are governed by the solidification time.

Like thermoplastics, thermosets may be granular and can be treated as powders, or they may become thermoplastics upon heating. The major difference is that when thermoplastics are cooled, their shapes are fixed, while thermosets must be held in a heated mold long enough for polymerization and cross-linking to occur, in order for their shapes to be fixed. Some polymers can be removed from a mold as soon as their shape is fixed, and then full cross-linking is obtained during cooling or holding in a separate oven. At other times, cross-linking begins immediately upon heating. Then the prepolymer must be introduced into a cold mold, and the mold must be taken into a cycle of heating and cooling for each part, resulting in a very long cycle time.

In plastics terminology, a *molding process* refers to melting processes, which occur with the aid of a mold, which is equivalent to a casting process in metals. The most widely used molding processes are *compression molding, transfer molding, injection molding, extrusion, reaction injection molding, rotational molding, calendering, and melt spinning. Particulate techniques*, in which polymers are processed from particles (powder, pellets, etc.) as in powder metallurgy, are also used [1, 4].

Reinforced plastics or *composites* are extremely important since they can be engineered to meet specific design requirements, such as high stress-weight and stiffness-weight ratios. Because of their unique structure, composites require special methods in order to be shaped into useful products. Reinforcement in composite materials can consist of loose fibers, woven fabric or mat, roving (slightly twisted fiber), or continuous fibers. In order to obtain good bonding between the reinforcing fibers and the polymer matrix, it is necessary to *impregnate* and *coat* the reinforcement with a polymer. When impregnation is made as a separate step, the resulting partially, dry sheets are called *prepregs, bulk molding compound (BMC)*, or *sheet molding compound (SMC). Compression molding* is a primary forming process, often used in composites manufacturing, in which material is placed between two molds and pressure is applied. The material may be in bulk form with a viscous, sticky mixture of polymers, fibers, and additives. Fiber lengths generally range from 3 to 50 μm. Sheet molding compounds can also be used in compression molding.

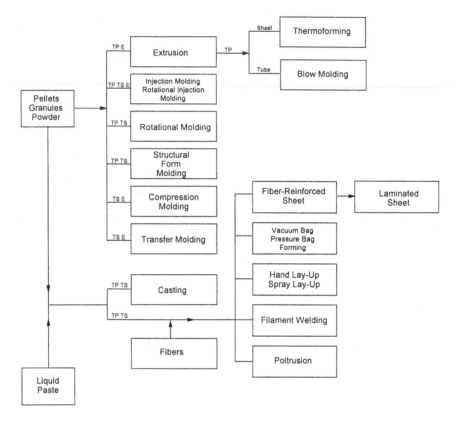

TP Thermoplastic, TS Thermoset, E Elastomer

Fig. 2.10. Forming Processes for Plastics, Elastomers, and Composite Materials

2.2.1 Characteristics of Primary Forming Processes

Cost

The applicability of forming processes is often constrained by the high cost and the relative inflexibility of tooling. The cost of forming dies and molds lies in the order of $10,000 to $100,000 and often requires weeks to months of lead-time to design and produce. However, the cost of the die can be distributed over the cost of producing a batch of parts; therefore, primary forming processes are economical when the lot size is large (in the order of 1,000 parts and up). Also, forming processes do not often require highly-skilled workers to operate them, since the tool and workpiece motions do not need to be programmed, and thus labor costs can be reduced relatively to other processes.

Production Rate

Forming processes can achieve high production rates, with cycle times from seconds to minutes, depending on part thickness. In general, casting and molding processes, which use permanent dies machined from metals, have lower cycle times than those processes which use expendable foam, ceramic or wax molds. However, expendable molds can produce intricate shapes and internal part features, not available with permanent molds. Setup times for forming are in the order of eight hours, which makes casting and forming prohibitive for low-volume production.

Part Quality

Parts produced through permanent mold or die casting can achieve surface finishes of $1\mu m$ to $2\mu m$ R_a, although typical cast parts require secondary removal processes, such as grinding or EDM to improve surface finish. Sand casting processes can produce parts with up to $25\mu m$ R_a surface roughness. Besides surface roughness, porosity in the cast or molded part, due to entrainment of gas bubbles in the flow of molten material and dimensional accuracy, are problems associated with parts produced by primary forming. The part quality can be influenced by process parameters such as die temperature, cooling time and cooling rate, as well as the design of die or mold features, such as sprues and conduits for molten material flow and mold cooling lines.

Flexibility

The flexibility of forming processes is limited. Only one part geometry can be produced for a die geometry and the part geometry cannot be changed through workpiece-tool motions. However, forming processes have the potential to produce parts with very intricate geometric features, especially internal features, and workpiece thicknesses, which can vary from under 1mm to over 100mm.

	Cost	Production Rate	Quality	Flexibility
Forming Processes	High Tooling/ Low Labor Cost	High	Medium to Low	Low

Table 2.2. Characteristics of Primary Forming Processes

2.3 Deforming Processes

Deforming processes have a variety of applications in the industrial world. Components for automobiles, industrial equipment, machine tools as well as for hand tools, such as hammers, pliers, and screwdrivers are some industrial applications of deforming processes. Containers such as metal boxes and cans, and fasteners such as screws, nuts, bolts, and rivets are other components created by using deforming processes.

The most commonly used materials in deforming processes are steel, carbon alloyed steel, stainless steel, and heat-resistant steel; non-ferrous heavy and light alloys, such as aluminum, zinc, and copper; and titanium and heat-resistant nickel alloys. The significance of deforming processes can be best understood by considering that a medium-sized automobile has somewhere between 110 and 160 pounds of steel in cold formed parts.

Generally speaking, deforming processes, used in the metal working industry, can be divided into *bulk* or *massive deforming* methods [5], such as extrusion, upsetting, or forging, and *sheet metal forming* techniques, such as deep drawing and bending (Fig. 2.11). Irrespective of the type of deforming method, the stresses required for deformation are quite high, ranging between 50 and 2500 N/mm2 (7-360 ksi) depending on the method and the material. Because the entire workpiece, or a major part thereof, is deformed, the loads, which result from a deforming process are also very high; for example, in a forging press, loads may be up to 750 MN, or 85 Ktons. Comparing this with the heavier equipment for material removal processes, such as large planing machines, the load is a few orders of magnitude (20 kN or 2.3 tons) lower than that occurring in deforming processes.

Industrially applied deforming processes can be classified in a number of ways [5]. The purpose of these classifications is to facilitate the training of people for the different processes, to provide uniform, reliable, speedy communication regarding issues of deforming processes, and also to provide some precise definitions of the different processes for use in process planning.

One way to classify these processes is to examine the *temperature* present, while the forming occurs. Following this criterion, one can distinguish between *hot forming* processes, which occur after the workpiece material has been heated, and *cold forming* processes, which occur without additional heating. *Stress* is another classification criterion, referring to the amount of loading on the material when the deforming process occurs. Following this criterion, deforming processes can be divided into five major categories:

1. *Compressive forming* – plastic deformation of the solid body is achieved by uni-axial or multi-axial compressive loading.
2. *Tensile forming* – plastic deformation of the solid body is achieved through uni-axial or multi-axial tensile stresses.
3. *Combined tensile and compressive forming* – plastic deformation of the solid body is achieved by combined both tensile and compressive loading.
4. *Bending* – plastic deformation of the solid body is achieved by means of a bending load.
5. *Forming by shearing* – plastic deformation of the solid body is achieved through a shearing load.

Within this group of processes, further subdivisions can be made, based on the kinematic considerations of the tool and the workpiece. The constraints of energy, force and travel on forming machines, are discussed in detail in Chapter 3.

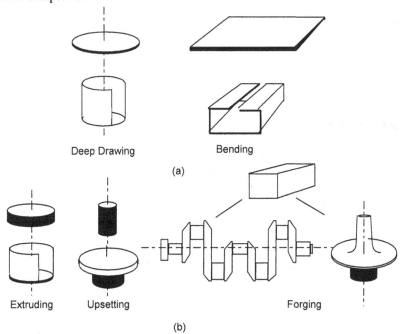

Deep Drawing   Bending

(a)

Extruding   Upsetting   Forging

(b)

Fig. 2.11. Sheet (a) and Bulk (b) Deforming Techniques

2.3.1 Characteristics of Deforming Processes

Cost

Deforming processes generally have relatively high tooling costs, similar to those for forming processes, due to the complicated die geometries required. Since the deforming loads are so high (up to 750MN), the tools used are very large, heavy, and expensive. Deforming methods are directly related to machining and metal removing methods, because the tools used for metal deformation are usually produced by machining processes, due to the small lot size (one-of-a-kind in most cases) and the high precision required. Because the cost of tooling (up to $250,000 per die) and machinery (up to $200,000 per machine) is high, deforming processes are usually applied to lot sizes large enough to economically justify the high cost of the tools and machinery required. Since a deforming tool usually has a life of several million parts, the high tooling cost can be distributed over a large number of parts, and consequently the total manufacturing cost per part will be low for large batch sizes. Finally, highly-skilled workers are usually not needed to operate deforming processes, so labor costs are also relatively low, compared with other manufacturing processes.

Production Rate

For large lot sizes, deforming processes have production rates up to 5000 parts per hour. Similar to primary forming processes, deforming dies require setup times up to several hours, thus, making deforming processes impractical for small lot size production.

Part Quality

The range of surface finishes, achievable through deforming processes, is similar to that of primary forming processes, with surface roughness down to 0.8μm Ra for extrusion and cold rolling. However, hot forming processes, such as hot rolling, during which the workpiece is preheated to an annealing temperature, generally yield much higher surface roughness (up to 50μm R_a) and require secondary finishing operations. One benefit of parts, produced through deformation, is that the deformation process produces work hardening, which increases the mechanical strength of the part. However, excessive material deformation may lead to crack and overlap formation in the workpiece.

Flexibility

Deforming processes have a relatively low degree of flexibility compared with other manufacturing processes, since the kinematics of deforming machines are constrained by either motion, force or energy (discussed in Chapter 3). The geometry of the part is governed solely by the tool geometry, as in the case of primary forming. However, since deforming dies must move relatively to the workpiece, the geometric features that are producible, are limited.

	Cost	Production Rate	Quality	Flexibility
Deforming Processes	High Tooling/ Low Labor Cost	High	Medium to Low	Low

Table 2.3. Characteristics of Deforming Processes

2.4 Removing Processes

Mechanical material removal processes are the backbone of industrial manufacturing practice. These processes provide a great deal of flexibility, since the shape and the kinematics of the tool and workpiece define the geometry of the part. The material removal mechanism is a very important aspect of removing processes. There are four types of material removal mechanisms:

- *Mechanical* – the mechanical stresses induced by a tool surpass the strength of the material
- *Thermal* – thermal energy provided by a heat source melts and/or vaporizes the volume of the material to be removed
- *Electrochemical* – electrochemical reactions induced by an electrical field destroy the atomic bonds of the material to be removed
- *Chemical* – chemical reactions destroy the atomic bonds of the material to be removed

Mechanical

Removing processes based on a mechanical material removal mechanism typically have greater flexibility, in terms of the lot size, workpiece geometry and shape, and part quality achievable compared with other manufacturing processes.

Among material removal processes, *turning* processes are the most widely used (Fig. 2.12). In turning processes, a workpiece rotates, providing the main cutting motion, while the tool is translated along the workpiece, providing the feed motion. Based on the direction of feed motion, one can differentiate among *longitudinal turning*, *facing*, and *form turning* [2]. In the first case there is an axial feed, in the second case a radial feed, and in the third case a simultaneous axial and radial feed motion. Modern machine tools, performing these processes, are usually automated to different degrees, including the implementation of computer numerical control of various kinds. Aspects of machine tool automation are discussed in Chapter 3.

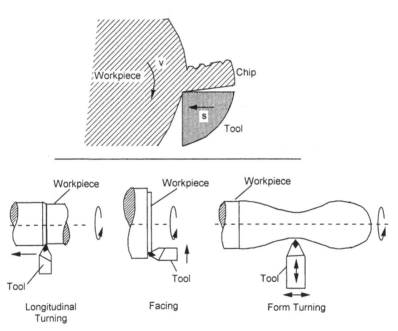

Fig. 2.12. Turning Processes

Cutting operations have been investigated considerably from a *cost* point of view because of their widespread use in industrial practice. Figure 2.13 illustrates the relationships among the cutting speed, production rate, and cost per piece for a typical machining process [7].

Drilling is a major category of processes in which the cutting action is not induced by rotational motion of the workpiece, but rather by the rotational motion of the tool (Fig. 2.14). Drilling includes processes, such as twist drilling, boring, counter-sinking, reaming, and tapping [2]. In all of these cases, the feed motion in the direction of the rotating axis is either performed by the workpiece or by the tool. Due to the nature of the drilling process, in which chip formation occurs in a closed space and chips must be periodically removed from the hole, drilling is slow and has a relatively low production rate. To overcome these disadvantages, machinery with a multiple drill head is commonly used to substantially increase the production rate, without increasing the actual cutting speed of each individual drill.

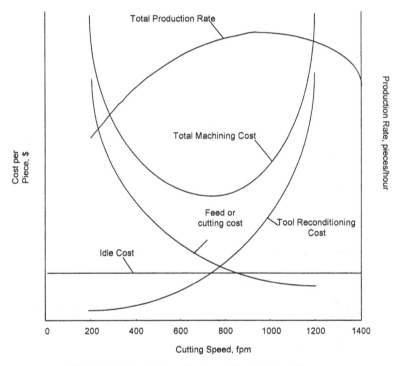

Fig. 2.13. Machining Cost and Production Rate

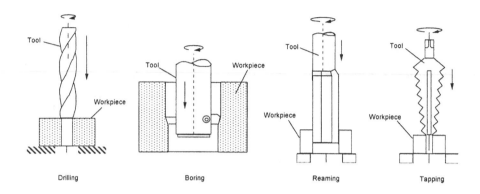

Fig. 2.14. Drilling Processes

"Deep hole drilling" is an important aspect of the drilling process (Fig. 2.15) in which the aspect ratio of the holes can be greater than 1:200. This process requires special tool design to allow for internal coolant channels, which facilitate the ejection of chips from the deep holes. Machinery for this process must take into consideration the slender tools that can create stability problems.

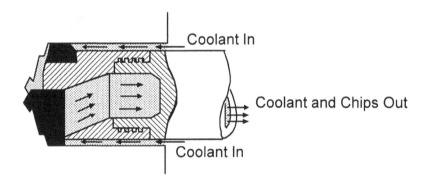

Fig. 2.15. Deep Hole Drilling (BTA method)

Milling is similar to drilling in that the main rotary cutting movement is produced by the tool. However, in milling, the feed motion is not in the axial direction of the cutting tool, as it was in drilling, but rather is orthogonal to the main axis of the tool. Figure 2.16 depicts the most commonly used milling processes. *Roller milling* and *end milling* are general-purpose processes, widely used for producing flat surfaces in any type of configuration. *Hobbing* is a major process for gear making. *Form end milling* is an important process in the machine tool industry for producing slides and other machine tool parts. *Die sinking* is used to make dies for deforming processes, and *gang milling* is a process in which a variety of tools is mounted on the same rotating axis in order to make a complex shape.

Cutting with geometrically defined single-cutting-edge tools can produce a large variety of shapes, ranging from rotationally symmetrical to prismatic parts. The workpieces produced by these processes, however, do not always satisfy the requirements for surface quality and dimensional accuracy. In order to satisfy these requirements, another set of processes, in which the cutting tool has a large number of cutting edges, is applied. In these processes, surface quality and dimensional accuracy of the workpiece are emphasized over material removal rate. Moreover, material can be removed from very hard workpiece materials. Most processes in this category are *grinding* processes. Figure 2.17 shows the contact relationship between the workpiece and the grinding wheel as well as the cutting geometry for one abrasive cutting edge or *grit*.

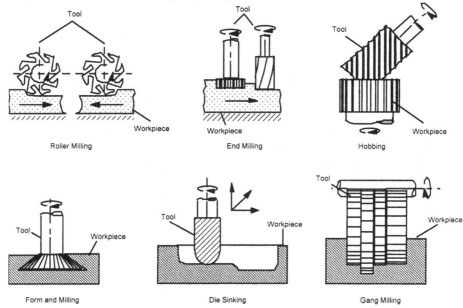

Fig. 2.16. Milling Processes

Two other high-precision families of processes have the same general characteristics as grinding: *honing* and *lapping*. With honing and lapping [2], it is the surface finish and dimensional accuracy that can be improved, but not the form of the workpiece, since the tools are not guided as they are applied to the workpiece. Figure 2.18 presents an overview of the various grinding, honing, and lapping techniques, together with their process kinematics.

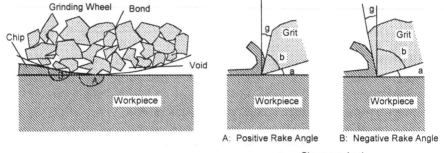

A: Positive Rake Angle   B: Negative Rake Angle

a   Clearance Angle
b   Wedge Angle
g   Rake Angle

Contact Between Workpiece and Grinding Wheel   Cutting Geometry for One Cutting Grit

Fig. 2.17. Basic Mechanism of the Grinding Process

External cylindrical grinding is perhaps one of the most widely used techniques for rotationally symmetrical parts. Another grinding process is *centerless cylindrical grinding*, which is vital to the automotive industry [2]. Figure 2.19 illustrates the various techniques, used in centerless grinding in order to produce feed motions. Figure 2.20 shows how centerless cylindrical grinding occurs in actual industrial practice.

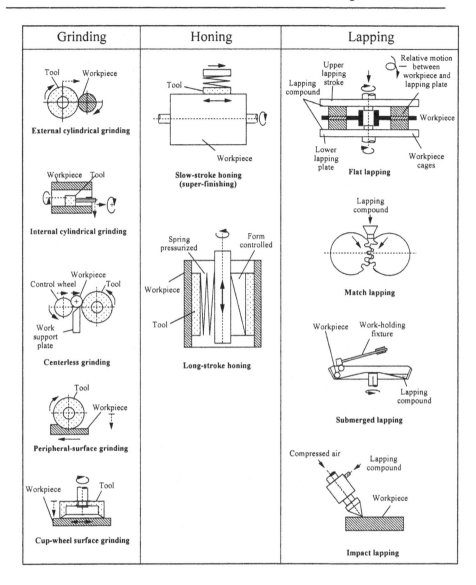

Fig. 2.18. Grinding, Honing and Lapping Processes

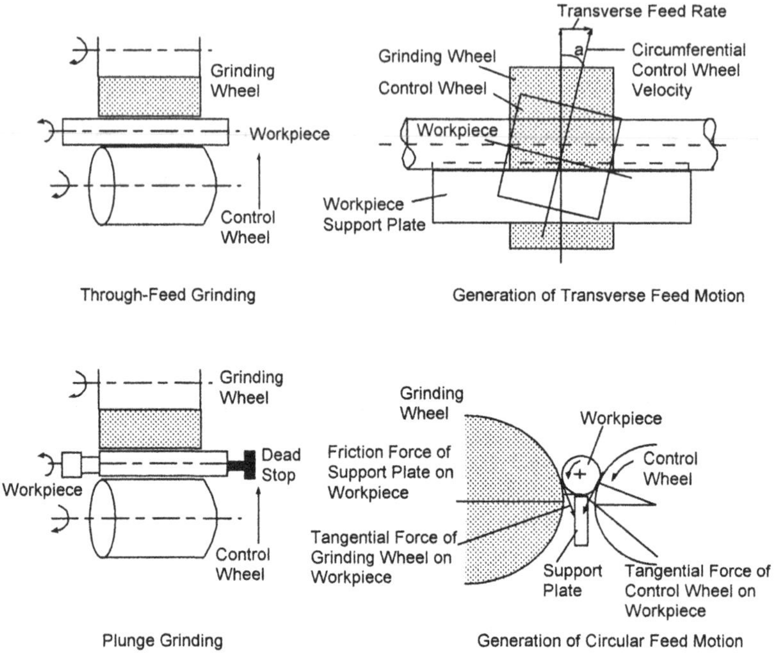

Grinding Wheel
Workpiece
Control Wheel
Through-Feed Grinding
Transverse Feed Rate
Grinding Wheel
Control Wheel
a
Circumferential Control Wheel Velocity
Workpiece
Workpiece Support Plate
Generation of Transverse Feed Motion
Grinding Wheel
Dead Stop
Workpiece
Control Wheel
Plunge Grinding
Grinding Wheel
Workpiece
Friction Force of Support Plate on Workpiece
Control Wheel
Tangential Force of Grinding Wheel on Workpiece
Support Plate
Tangential Force of Control Wheel on Workpiece
Generation of Circular Feed Motion

Fig. 2.19. Centerless Grinding Processes

Grinding wheel
Control wheel
Workpiece
Workpiece support plate

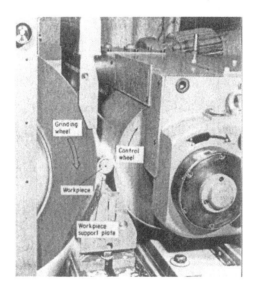

Fig. 2.20. Centerless Grinding [2]

Surface grinding techniques are also of great importance to producing good dimensional accuracy and surface quality on flat surfaces [2]. Figure 2.21 illustrates various surface grinding techniques.

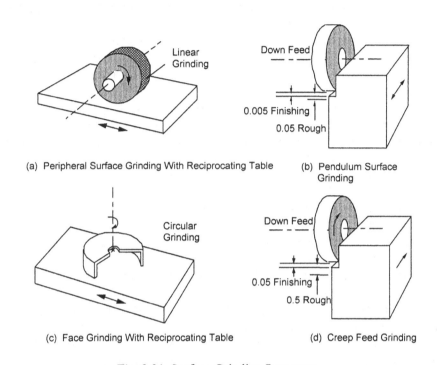

(a) Peripheral Surface Grinding With Reciprocating Table   (b) Pendulum Surface Grinding

(c) Face Grinding With Reciprocating Table   (d) Creep Feed Grinding

Fig. 2.21. Surface Grinding Processes

The term *High Speed Machining* has been used to describe the end milling at high rotational speeds, typically 10.000 to 100.000 rpm. Initially, the process was applied by the aerospace industry to the machining of light alloys, however in recent years, are being used by ,the automotive, mould and die industries for producing components from a variety of materials, including hardened steels. Major benefits of this process are the high material removal rates achieved, along with a minimum workpiece distortion and surface roughness, which can descend to 0.1 μm R_a. Its drawbacks include high tool wear, expensive tooling balance, low spindle life (5.000 – 10.000 h at maximum rotational speed) and costly control systems.

High Speed Machining is applied to the processing of a variety of materials, such as Aluminium alloys, Copper and graphite electrodes for Electrical Discharge Machining (EDM), Magnesium alloys, cast Iron, Titanium alloy turbine blades, Nickel based alloys, brasses and wood for household furniture.

Since the cutting process is performed in high rpms, high wear resistant cutting tools are required, especially when hardened steels are processed. The tool material options include nitride coated cemented tungsten carbides, cermet, conventional ceramics and Cubic Boron Nitride (CBN). The application of Polycrystalline Diamond (PCD) tooling is quite common for the High Speed Machining of non-ferrous metals and non-metallic materials, since diamond reacts with iron and turns to graphite above 750 °C [19]. We know that when we place our hand across a jet of water or air, we feel a considerable concentrated force acting on it. This force results from the momentum change of the stream and in fact, it is the principle, which the operation of water or gas turbines is based on. In water-jet machining (WJM) (Fig.2.22), also called hydrodynamic machining, this force is utilized in both cutting and deburring operations [3].

The water jet acts like a saw and cuts a narrow groove in the material. A pressure level of about 400 MPa (60 ksi) is generally used for efficient operation. Jet-nozzle diameters usually range between 0.05 mm and 1 mm (0.002 in. and 0.040 in.). Typically, the water jet contains abrasive particles, such as silicon carbide or aluminum oxide, thus, increasing the material-removal rate.

A variety of materials including plastics, fabrics, rubber, wood products, paper, leather, insulating materials, brick, and composite materials can be cut by the water jet machining technique. Thicknesses range up to 25 mm (1 in.) and higher. Vinyl and foam materials, used for automobile coverings in dashboards, are cut by using multiple-axis, and robot-guided water-jet machining equipment. Since the Water Jet Machining (WJM) is an efficient and clean operation, compared with other cutting processes, it is also used in the food processing industry for cutting and slicing products.

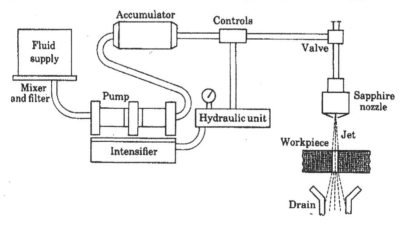

Fig. 2.22. Schematic illustration of water-jet machining [3]

In *abrasive-jet machining* (AJM) (Fig.2.23), a high-velocity jet of dry air, nitrogen, or carbon dioxide, containing abrasive particles, aims at the workpiece surface under controlled conditions. The impact of the particles, develops sufficient concentrated forces to perform operations, such as (1) cutting small holes, slots, or intricate patterns, in very hard or brittle metallic and nonmetallic materials, (2) deburring or removing small flash from parts, (3) trimming and beveling, (4) removing oxides and other surface films, and (5) general cleaning of components with irregular surfaces [3].

The gas supply pressure is in the order of 850 kPa (125 psi) and the abrasive-Jet velocity can be as high as 300 m/s (100 ft/s), controlled by a valve. The hand-held nozzles are usually made of tungsten carbide or sapphire. The abrasive size is in the range of 10-50 gm (400-2000 gin.).

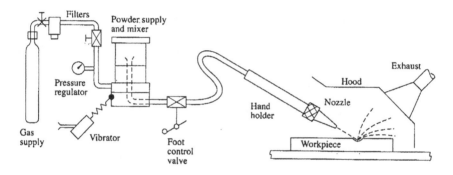

Fig. 2.23. Schematic illustration of the abrasive-jet process [3]

Thermal

Laser beams can be used in many industrial processes including machining. *Laser machining* (LM) constitutes an alternative to traditional manufacturing and can be used for machining a variety of materials, such as metals, ceramics, glass, plastics and composites. LM is characterized by a number of advantages, such as absence of tool wear, tool breakage, chatter, machine deflection and mechanically induced material damage, as well as phenomena, which are typically associated with traditional machining processes. Laser technology offers different methods of beam generation with a continuous wave or pulse operation, and wavelengths ranging from tenths to fractions of a tenth of the µm. Beam power intensity may also vary from 109 W/cm2 to 1015 W/cm2. LM is an extremely localized thermal process. During beam-material interaction, a certain portion of the incident light energy is absorbed by the material (photon absorption) and high temperature is developed in the vicinity of the beam spot, resulting in material softening, local yielding, melting, burning or evaporation.

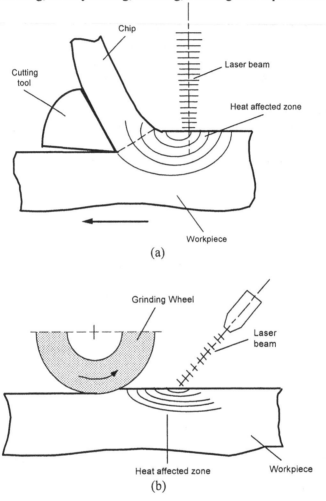

Fig. 2.24. Laser-Assisted Machining: (a) turning; (b) grinding [6]

An alternative to grinding and diamond machining is the *laser assisted machining* (LAM) process. In LAM, the material is heated by a laser, prior to the material removal location, and subsequently the material is plastically machined (Figure 2.24) [6].

Laser drilling is usually performed with a pulsed laser, which produces a higher intensity and evaporates the material easily. Pulsed laser beams create a high temperature gradient inside the material, because of their high intensity and short interaction time. In laser drilling, the hole's depth is scaled through molten material removal. *Laser cutting* is the most widely-used of all laser machining processes. In laser cutting, a laser beam penetrates the entire thickness of the workpiece and advances parallel to the surface of the workpiece. In *laser grooving*, the laser beam does not cut through the entire workpiece. (Fig. 2.25) [6].

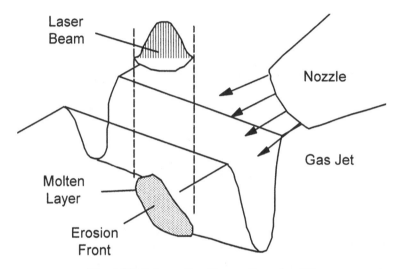

Fig. 2.25. Schematic of Laser Grooving [6]

Electrical discharge machining (EDM) is based on the erosion of metals by spark discharges. The EDM system (Fig.2.26) consists of a shaped tool (electrode) connected to a DC-power supply and placed in a dielectric fluid. When the potential difference between the electrode and the workpiece reaches a sufficient level, a transient charge is discharged through the fluid, removing a portion of metal from the surface of the workpiece.
Those discharges are reiterated at rates between 50Khz and 500Khz, with voltages ranging between 50V and 380V and currents from 0.1 A to 500A. The EDM process can be used on any material, due to its being an electrical conductor. The parameters influencing the volume of metal removed per discharge are the melting point and the latent heat of the metal. As these values increase, the volume removed decreases. Since the process does not involve mechanical energy, the hardness, strength and toughness of the workpiece do not influence the removal rate. The rate and surface roughness increase as the current density increases and the frequency of sparks decreases. Electrodes for EDM are usually made of graphite, although brass, copper, or copper-tungsten alloys may be used. The tools

are shaped by forming, casting, powder-metallurgy or machining. Typical applications for EDM are the production of die cavities for large automotive-body components, small diameter deep holes, narrow slots, turbine blades and various intricate shapes [3].

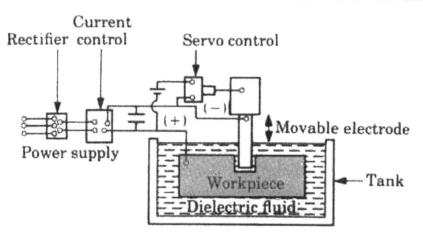

Fig. 2.26. Schematic illustration of the EDM process [3]

In *electrochemical machining* (ECM), an electrolyte (Fig. 2.27) acts as the current carrier, and the high rate of electrolyte movement in the tool-workpiece gap, washes metal-ions away from the workpiece (anode) before they have a chance to be deposited onto the tool. The cavity produced is the female mating image of the tool. The tool is made of brass, copper, bronze or stainless steel. The electrolyte is a highly conductive inorganic salt solution, such as sodium chloride, mixed in water or sodium nitrate and it is pumped at a high rate through the passages of the tool. A DC-power supply of 5V to 25V maintains the current densities, which usually range from 1.5A/mm2 to 8A/mm2. Machines with capacities as high as 40000A or as low as 5A, are available. Since the metal removal rate is only a function of the ion-exchange rate, it is not affected by the strength, hardness or toughness of the workpiece, which must be electrically conductive. Electrochemical machining is generally used for machining complex cavities in high strength material. The ECM process leaves a burr free surface and furthermore, as there is no tool wear, the process is capable of producing complex shapes [3].

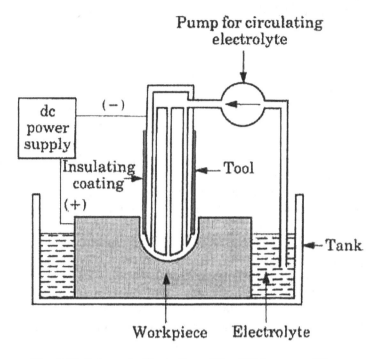

Fig. 2.27. Schematic illustration of the ECM process [3]

Chemical

In *chemical machining* (CM), the material is removed from the surface by chemical dissolution, using chemical agents or etchants, such as acids and alkaline solutions. In chemical milling, shallow cavities are produced on plates, sheets, forgings and extrusions for overall reduction of weight. Chemical milling has been used on a variety of metals, with depths of removal up to as much as 12mm (0.5 in.). Selective attack by the chemical agent on different areas of the workpiece surfaces is controlled by partial immersion in the reagent or by removable layers of material, called maskings.

Some surface damage may result from chemical milling, because of preferential etching and intergranular attack, which adversely affect surface properties [3].

2.4.1 Characteristics of Material Removal Processes

Cost

The machine and tooling costs, associated with mechanical material removal processes, are low compared with those of primary forming and deforming processes. Machines typically cost from $10,000 for manual equipment to $80,000 for CNC equipment, while tooling cost ranges from $1 to $100 per tool. However, the skill level involved for programming or manually setting the tool and the workpiece kinematics is relatively high, thus, labor costs for operating material removal processes are also correspondingly high. Machining processes are therefore, better suited for low to medium volume production with lot sizes, ranging from one to several hundred parts.

Concerning thermal material removal processes, a typical laser machine with CNC control, costs about $175,000 while an EDM machine's cost varies from $35,000 to $130,000, depending on its power and size.. Moreover, a standard-size ECM machine costs about $140,000. In spite of the increased machine costs, in mechanical material removal processes, these processes offer unique advantages, in cases that special shapes and materials are involved.

Production Rate

The production rates for machined parts are much lower than those for casting or deforming processes, since it is the tool that is required to make multiple passes over the workpiece surface in order to produce the final shape. The material removal rate is dependent on the surface quality desired, the workpiece material, the cutting tool material and the cutting fluid used. For steels, the material removal rates may range from less than 1mm3/min, for a grinding process, up to the order of 10cm3/min, for a milling process. The EDM and ECM processes have 0.3 and 2.2cm3/min production rates respectively. However, time required for workpiece setup and fixturing (in the order of several minutes to one hour) is generally much less than that for primary forming and deforming processes.

Part Quality

Surface quality and surface technology are clearly very important aspects of material removal processes. Figure 2.28 gives some of the different factors that may affect surface texture and surface integrity [7]. Surface *texture* can be described in terms of surface roughness, macro effects such as imperfections, and geometric considerations such as tolerances. On the other hand, surface *integrity* refers more to micro-structured effects such as micro-cracks and residual stresses. Surface effects are caused both by the process itself and the workpiece material properties. These effects have a direct influence on the mechanical characteristics of the workpiece and eventually on the reliability of the component. Figure 2.29 defines some of the surface quality characteristics, used in industrial standards internationally [7]. Table 2.4 summarizes the achievable surface roughness of the different processes, including commonly used removing processes, such as drilling, milling, and grinding as well as primary forming processes, such as investment casting and die casting. Table 2.4 shows that processes, based on thermal removal mechanisms, such as flame cutting, usually provide an inferior surface quality (with surface roughness up to 25μm Ra) compared with processes based on mechanical removal mechanisms (with surface roughness down to 0.05μm Ra). Figure 2.30 schematically shows the surface generation mechanisms in milling and turning [7]. This figure not only does it illustrate the way the surface is generated, but it also provides some guidance as to how a better surface quality can be achieved, by adjusting process parameters, such as the feed per tooth or feed per revolution.

Flexibility

Material removal processes are among the most flexible of the manufacturing processes. Since the geometry of the finished part is defined by the geometry and the kinematics of the tool and workpiece, material removal processes can produce parts with a wide range of sizes, shapes and surface quality. Table 2.4 shows that processes, such as milling, turning, and grinding span a large segment of achievable surface quality, exhibiting their flexibility and popularity in industrial practice [7]. The surface generation mechanism in milling and turning (Fig. 2.30) explains the great flexibility of these processes, in terms of creating a variety of surface qualities through the selection of process parameters.

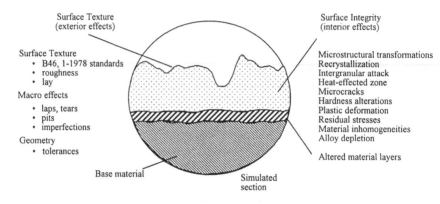

Surface technology effects

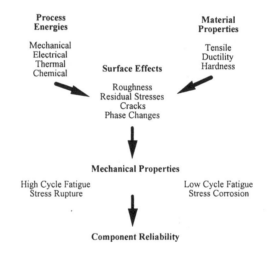

Fig. 2.28. Machined Surface Characteristics

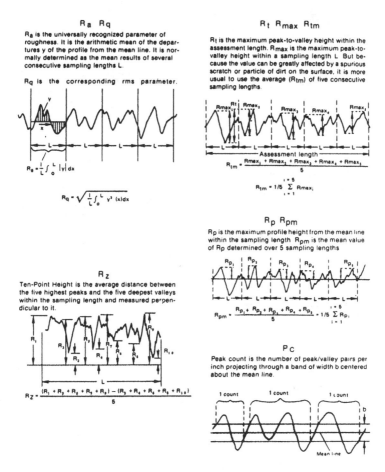

Fig. 2.29. Surface Quality Characteristics [7]

Another reason for the flexibility of material removal processes is the number of degrees of freedom, incorporated into material removal machines. Current numerically-controlled machining centers have up to nine translational and rotational degrees of freedom for a given tool-workpiece combination. Even in drilling processes, which are limited to one degree of freedom, a range of hole depths can be produced without changing tools.

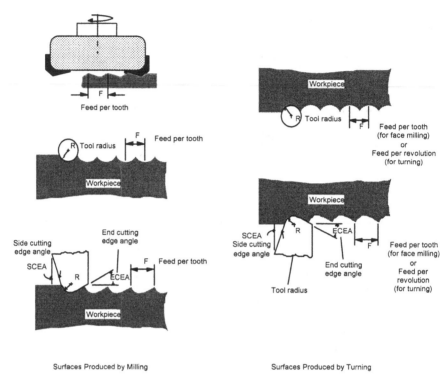

Surfaces Produced by Milling   Surfaces Produced by Turning

Fig. 2.30. Surface Generation in Milling and Turning

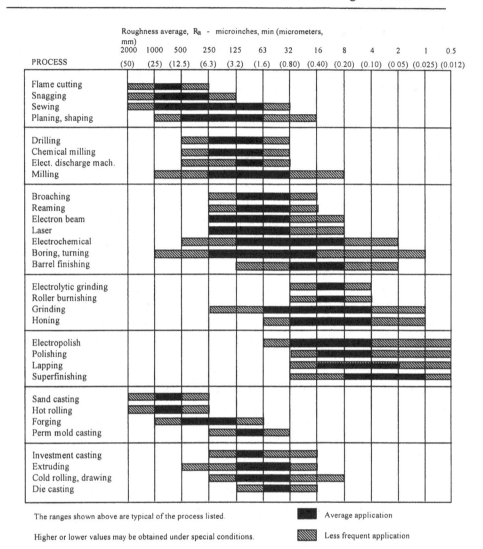

Table 2.4. Surface Quality for Different Machining Processes

	Cost	Production Rate	Quality	Flexibility
Removing Processes	Medium Tooling/High Labor Cost	Medium (Milling) to Low (Grinding)	High	High

Table 2.5. Characteristics of Material Removal Processes

2.5 Joining Processes

Joining includes a number of processes such as *welding, brazing, soldering, adhesive bonding,* and *mechanical joining* [1, 4, 8]. Each of these is an integral part of modern manufacturing operations, as most products are impossible to be made as a single piece. Indeed, joining processes enable the manufacturing of a product in individual components and then combine them into a single product, which may be easier and less expensive to manufacture than the whole product at once. Many products must be taken apart for maintenance, repair, or service, therefore, it is beneficial to join the components so that they can be easily dismantled and reassembled. Joining processes also allow the inclusion of features and properties in a particular product, which may differ from the majority of the components used in the product.

Many joining processes use *mechanical fasteners* such as bolts, nuts, and screws. Joints made with mechanical fasteners are usually non-permanent and can be used on products and machines that must be taken apart for maintenance and repair. With *liquid state* processes, such as *welding*, partial melting of the materials is necessary for joining. This requires the application of heat and/or pressure in order to permanently fuse or bond two pieces. The choice of one process over another depends on factors, such as joint design, materials size, shape, and thickness of the components to be joined.

Mechanical joining may be used because of the cost-effectiveness of making the product as components and then by joining them mechanically. In some cases, moveable joints such as hinges, sliding mechanisms for drawers and doors, and adjustable components and fixtures are required. *Fastening* is the most common method of mechanical joining. A variety of fasteners is commonly used, including bolts, nuts, screws and pins. *Mechanical assembly* is used to join the different components. When fasteners are used for joining, holes must be drilled or created, through which the fasteners can be inserted. The joints may be subjected to shear and tensile stresses and should be designed to resist these forces. Hole-making, a typical material removal process, can influence the quality of the joint and the ease of performing the assembly process. Holes may also be produced through non-material removal processes, such as casting, forging, extrusion, or powder metallurgy. Bolts and screws can be secured with nuts, or they may be self-tapping, in which case, the screw either casts or forms the thread into the part to be fastened. Self tapping is particularly effective and economical in plastic components, for which the need for pre-tapped holes or nuts can often be eliminated. Several rules of thumb can gener-ally be applied for using mechanical fasteners [1]:

- Using fewer but larger fasteners is generally less costly than when using a large number of small ones.
- The fit between parts to be joined should be as loose as possible in order to keep costs low and to facilitate the assembly process.
- Standard-sized fasteners should be used whenever possible.
- Holes for fastening should not be made too close to the edges or corners of a component in order to avoid tearing of the material when the part is subjected to forces.

The most common method of permanent or semi-permanent mechanical joining is *riveting*. Hundreds of thousands of rivets are used in the con-struction and assembly of large commercial aircraft. Installing a rivet pre-supposes having to place a rivet in a hole and deform the end of its shank by upsetting. Riveting may be done either at room temperature or with heat, using special tools or explosives in the rivet cavity. Figure 2.31 illus-trates some rivet examples and provides some general design guides for riveting.

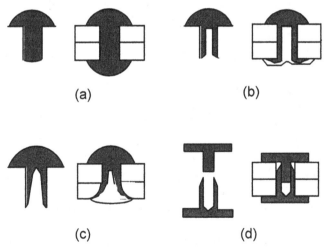

(a)   (b)

(c)   (d)

Fig. 2.31. Examples of Rivets: Solid (a) Tubular (b) Split (or Bifurcated) (c) and Compression (d)

There are other ways of mechanically assembling components without using fasteners, such as *shrink fitting* and *press fitting*. Shrink fitting is based on the different thermal expansion and contraction of two compo-nents; an outer part with a high coefficient of thermal expansion and an inner part with a low coefficient of thermal expansion. Both parts are heated and assembled. As the assembly is cooled at room temperature, the outer part will tighten around the inner part. Typical applications include mounting gears and cams on shafts. In press fitting, one component is forced over another, usually resulting in high joint strength.

Welding has a wide range of industrial applications [1,8]. The most widely used welding processes are the fusion welding processes, such as *arc welding* with consumable or nonconsumable electrodes, *oxyfuel gas welding*, and *resistance welding*. Furthermore, there are a host of solid-state welding processes, such as *cold welding, ultrasonic welding, friction welding, electron-beam welding*, and *laser beam welding*, during which joining occurs without the fusion of the workpieces.

Fusion welding processes involve a weld joint [1,8] that has three zones, as shown in Figure 2.32. These three zones are the base metal, namely the metal that is to be welded; the heat-affected zone (HAZ); and the weld metal, which is the region that has melted during welding. The metallurgy and properties of the second and third zones strongly depend on the metals joined, the welding process, the filler metals used (if any), and the process variables. A joint, produced without a filler metal, is referred to as being autogenous; the weld zone is composed of the molten, resolidified base metal. A joint, made with a filler metal, has a central zone called weld metal, which is composed of a mixture of base and weld metals. After the application of heat and the introduction of filler metal (if any) to the weld area, the molten weld joint is allowed to cool naturally at ambient temperature.

The solidification process in fusion welding is similar to that of casting. The weld metal is basically a cast structure, which, due to its having cooled slowly, has coarse grains. Consequently, the structure has low strength, toughness and ductility. However, the proper selection of filler metal composition or heat treatments following welding, can improve the joint's mechanical properties. The heat-affected zone is within the base metal itself. It has a microstructure different from that of the base metal before welding, because it has been subjected to elevated temperatures for a period of time during welding. The portions of the base metal that are at a distance from the heat source do not undergo any structural changes during welding. The strength and hardness of the heat-affected zone parts depend on how the original strength and hardness of the particular alloy was developed prior to welding.

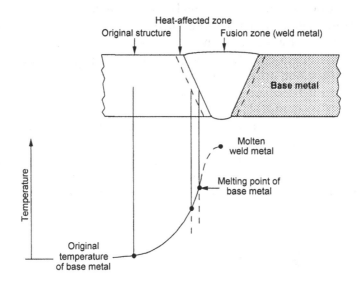

Fig. 2.32. Characteristics of a Typical Fusion Weld Zone in Oxyfuel Gas and Arc Welding

The quality of a welded joint can be affected by a number of factors including porosity, a critical factor, caused by trapped gases released during (re)solidification of the weld area, chemical reactions during welding, or contaminants [1,8]. Most welded joints contain some porosity, which generally takes the shape of spheres or elongated pockets. Porosity in welds can be reduced by proper selection of electrodes and filler materials, by better welding techniques, such as by preheating the weld area or increasing the rate of heat input, and by proper cleaning and prevention of contaminants from entering the weld zone.

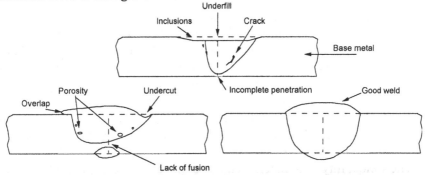

Fig. 2.33. Examples of Various Defects in Fusion Welds

Slag inclusions, namely compounds such as oxides, fluxes, and elec-trode-coating materials that become trapped in the weld zone, also influ-ence the quality of a welded joint. Welded joints of poor quality can be produced even when complete weld penetration occurs. The weld profile is important because it has an effect on the strength and appearance of the weld, and can indicate incomplete fusion, slag inclusions, etc. Figure 2.33 shows several examples [1, 8] of defective fusion weld profiles.

Cracks are a major quality problem in welding. They may occur in various locations and directions in the weld area, and are generally the re-sult of a combination of temperature gradients posing thermal stresses in the weld zone, variations in the composition of the weld zone, etc. Figure 2.34 illustrates types of cracks that may occur in welded joints.

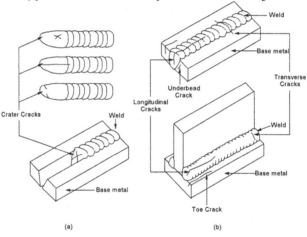

Fig. 2.34. Types of Cracks in Welded Joints Caused by Thermal Stresses that De-velop During Solidification and Contraction of the Weld Bead and the Welded Structure. Crater cracks (a) Various Types of Cracks in Butt and T Joints (b)

Residual stresses may occur in welded parts because of the expansion and contraction of the weld area, caused by localized heating and cooling. Figure 2.35 shows some of the distortions that may occur due to residual stresses. Problems of this kind can be avoided by preheating the base metal or the parts to be welded, thus, reducing both the cooling rate and the level of thermal stresses. Stress relief is very often applied in order to reduce residual stresses, with the temperature and the time required for the stress relief process dependent upon the type of material and the magnitude of the residual stresses developed.

The weldability of a material can be defined as its capacity to be welded into a specific structure of certain properties and characteristics, which will meet the functional requirements of the part. Since weldability requires a large number of variables, it is very difficult to make generalizations. The following list [1, 8] refers to the weldability of some metals, although these may vary, in case special welding techniques are used:

1. *Plain carbon steels*: excellent for low-carbon steels, fair to good for medium-carbon steels, poor for high-carbon steels
2. *Low alloy steels*: great dependence upon composition
3. *High alloy steels*: generally good under well-controlled conditions
4. *Stainless steels and nickel alloys*: weldable by various processes
5. *Aluminum and copper alloys*: weldable at a high rate of heat input
6. *Magnesium alloys*: weldable with the use of protective shielding gas and fluxes
7. *Titanium alloys and Tantalum*: weldable with the proper use of shielding gases
8. *Tungsten and Molybdenum*: weldable under well- controlled conditions
9. *Columbium* good
10.*Beryllium*: weldable under well- controlled conditions

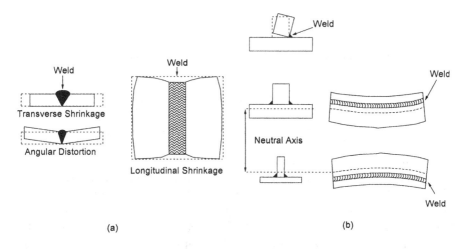

(a)   (b)

Fig. 2.35. Distortion of Parts after Welding: Butt Joints (a) and Fillet Welds (b). Distortion is caused by differential thermal expansion and contraction of different parts of the welded assembly. Warping can be reduced or eliminated by proper fixturing of the parts prior to welding

Solid-state welding, as mentioned earlier, includes a number of processes, such as cold welding and ultrasonic welding. A widely-used process belonging to this category is the *laser beam welding*, which uses as its source of heat, a focused high-power coherent monochromatic light beam. Due to the beam having a high energy density, it has deep penetrating power, and can be directed, shaped, and focused precisely onto the work-piece [1, 8]. Laser beam welding can be used successfully on a variety of materials with thickness of up to 25 mm, and it is particularly effective on thin workpieces. Welding speeds range from 40 mm/sec to as high as 1.3 m/sec for thin metals. Owing to the nature of the process, welding can be done in otherwise inaccessible locations; the welds produced are of good quality with minimum shrinkage and distortion and with depth-to-width aspect ratios as high as 30:1.

Like laser beam welding, the *electron beam welding* process uses heat, generated by a high-velocity narrow beam of electrons. The kinetic energy of the electrons is converted into heat as they strike the workpiece. Unlike laser beam welding, however, this process requires specialized equipment to focus the beam on the workpiece in a vacuum. The beam penetration and the depth-to-width ratio are both dependent on the processing environment; the higher the vacuum, the greater the beam penetration and aspect ratio.

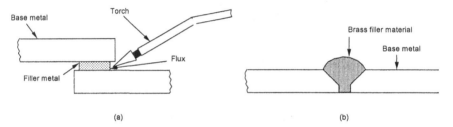

Fig. 2.36. Brazing (a) and Braze Welding (b) Operations

Brazing is a joining process in which a filler metal is placed at or between the facing surfaces to be joined. The temperature, typically above 450°C, is raised to melt only the filler metal and not the workpieces. The molten metal fills the close-fitting space through the capillary action. Upon cooling and solidifying the filler metal, a strong joint is obtained. The strength of the brazed joint depends upon the joint design and the adhesion at the interfaces of the workpiece and filler metal [1, 8]. Consequently, the surfaces to be brazed should be chemically or mechanically cleaned to ensure full capillary action. Several filler metals are available with a range of brazing temperatures and a variety of shapes, such as wire, rings, and fillings. Figure 2.36 shows the brazing process, and Figure 2.37 presents some of the joint designs commonly used in brazing operations. In general, brazing is used for dissimilar metals, which must be assembled with good joint strength, for example, carbide drill bits or inserts on steel shanks.

Soldering is a process similar to brazing except for the filler metal, called *solder*, which typically melts below 450°C. As in brazing, the solder fills a joint by capillary action between close fitting or closely placed components. The heat source for soldering is usually a soldering iron, torch, or oven. Soldering is used extensively in the electronics industry and for making air-tight joints in containers for liquids. Unlike brazed joints, a soldered joint has very limited use at elevated temperatures. Moreover, because soldered joints do not demonstrate adequate mechanical strength, they are not used for load-bearing structural members. Copper and precious metals, such as silver and gold are easy to solder, as it is also easy to solder tin plates for food containers.

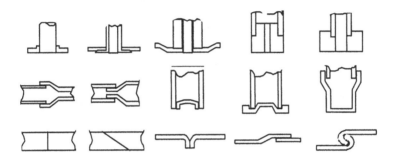

Fig. 2.37. Joint Designs Commonly Used in Brazing Operations [1]

Unlike in welding, in brazing and in soldering, the materials to be joined do not have to be heated at elevated temperatures, so it is possible to work with delicate or intricate parts, or parts made of two or more materials with different characteristics, properties, thickness, and cross-sections.

In industrial practice, components can often be joined using adhesives rather than fusion methods, especially in the assembly of plastic and composite components. Adhesive bonding is used extensively in the aerospace, automotive and appliance industries as well as for building products. Figure 2.38 illustrates a number of joints created by adhesive bonding, and the difference between a poor and a good design of the joint [1].

There are three major types of adhesives: *natural adhesives*, such as starch; *organic adhesives*, such as sodium silicate and magnesium oxide chloride; and *synthetic organic adhesives*, such as thermoplastics or thermosetting polymers. Adhesives are available in various forms, such as liquids, pastes, solutions, emulsions, powder, tape, and film. A wide variety of similar and dissimilar metallic and non-metallic materials, as well as components with different shapes, sizes, and thicknesses, can be bonded to each other with adhesives. Adhesive bonding can also be combined with mechanical joining to further improve the strength of the bond. Very thin and fragile components can be bonded with adhesives without significantly increasing their weight [1]. Porous materials and materials of different properties and sizes can also be joined. Bonding is usually carried out at temperatures between room temperature and 200°C so as to avoid causing a significant distortion to the components or a change in their original properties. Adhesives provide a joint, which distributes the load on the interface of the two parts, thus, eliminating localized stresses and providing structural integrity of the sections that are bonded.

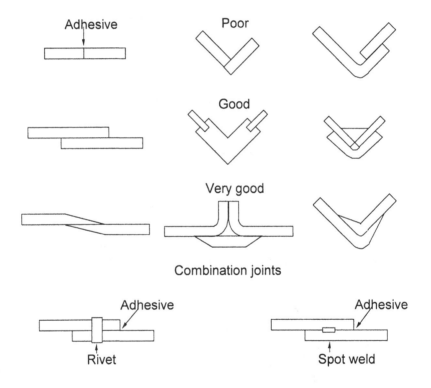

Fig. 2.38. Various Joint Designs in Adhesive Bonding (Note that good design requires large contact areas between the members to be joined.)

Rapid Prototyping

Rapid Prototyping (RP) is a term, which embraces a range of new technologies, producing parts directly from CAD models in a few hours. It has been claimed that RP can cut down new product costs by up to 70% and time to market by 90% [17].

All RP processes require input from a 3D solid CAD model, usually in the form of layers, which are sent to the RP machine for the production of the physical part by adding one layer on top of the other. By convention, the data slices are in the X-Y plane, whilst the part being built, is in the Z direction.

Since 1988 ,at the onset of RP technologies, the market has grown on an average rate of 58% world-widely. By 2000, approximately 6521 RP machines had been sold world-wide, 44.2% of which were in the U.S.A, 19.1% in Japan, followed by Germany, China, UK, Italy, France and Korea. Over 20 different RP techniques have been developed up to date [20].

Stereolithography is probably the most popular among currently available RP techniques. It uses a photosensitive monomer resin, which forms a polymer and solidifies when exposed to UV light. A stereolithography machine mainly consists of a built platform, which is immersed in a bath of liquid resin and a He-Cd (25-40mW) or ion Argon laser, including the appropriate hardware and software for control (Fig. 2.39).

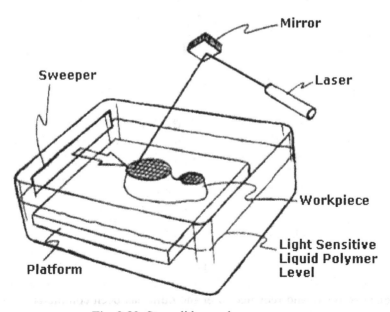

Fig. 2.39. Stereolithography process

A layer of the part is being scanned on the resin surface by the laser, according to the slice data of the CAD model. Once the contour of the layer has been scanned, the interior is crosshatched and hence solidified, the platform is being submerged into the vat, one layer below. A blade sweeps the surface to ensure flatness and the next layer is built, whilst simultaneously it is attached to the previous one.

SLS (*Selective Laser Sintering*) uses a fine powder, which is heated with a CO_2 laser (25-50 mW) in such a way so as to allow the grains to fuse together. Before the powder is sintered the entire bed is heated just below the melting point of the material in order to minimize thermal distortion and facilitate fusion in the previous layer. After each layer has been built, the bed is lowered and new powder is applied. A rotating roller is then used to spread the powder evenly (Fig. 2.40).

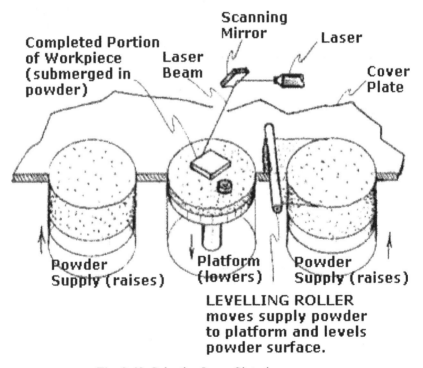

Fig. 2.40. Selective Laser Sintering process

The sintered material forms the part, while the unsintered material powder remains in place to support the structure. The unsintered material may be cleaned away and recycled after the built has been completed. Materials, such as nylon, nylon composites, sand, wax, metals and polycarbonates can be used.

The material used *in Laminated Object Manufacturing* (LOM) is a special paper having a heat sensitive adhesive applied to one of its sides. The paper is supplied from a roll and is bonded to the previous layer with the use of a heated roller, which activates the paper's adhesive. The contour of the layer is cut by a CO_2 laser, carefully modulated to penetrate into a depth of exactly one layer (paper) thickness. Surplus waste material is trimmed to rectangles to facilitate its removal, but remains in place during build in order to be used as support (Fig. 2.41).

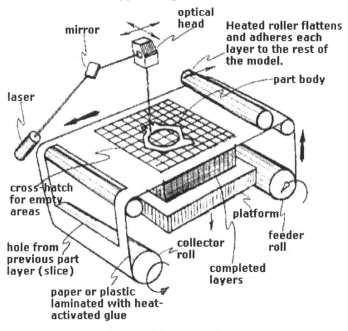

Fig. 2.41. Laminated Object Manufacturing process

The sheet of material used, is wider than the building area, so that, when the part layer has been cut, the edges of the sheet to remain intact in order to be pulled by a take-up roll and thus to continuously provide material for the next layer.

The *Fused Deposition Modeling* (FDM) technique uses a movable head, which deposits a thread of molten thermoplastic material onto a substrate. The material is heated up to 1°C above its melting point, so that it solidifies right after extrusion and subsequently welds to the previous layers (Fig. 2.42). The FDM system head includes two nozzles, one for the part material and one for the support material.

The support material breaks away from the part easily, without impairing the part's surface. An advantage of the system is that it can be viewed as a desktop prototyping facility, since it uses cheap, non toxic, odorless materials, in a variety of colors and types, such as Acrylonitrile-Butadiene-Styrene (ABS), medical ABS, investment casting wax and elastomers.

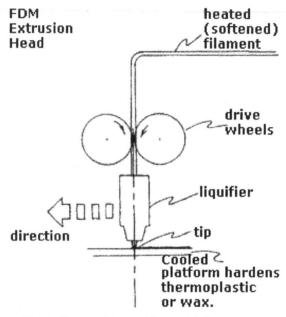

Fig. 2.42. Fused Deposition Modeling process

Other RP techniques are:

- Three Dimensional Printing (3DP)
- Solid Ground Curing
- Multi Jet Modeling (MJM)
- Ballistic Particle Manufacturing (BPM)
- Beam Interference Solidification
- Holographic Interference Solidification (HIS)
- Liquid Thermal Polymerization (LTP)
- Solid Foil Polymerization (SFP)
- Hot Plot

2.5.1 Characteristics of Joining Processes

Cost

The capital and tooling costs, associated with most joining operations are relatively low compared with those of other processes, since most joining equipment is inexpensive. A typical welding machine ranges in price from $5,000 to $50,000. However, the skill level of operators is high, especially in operations where parts with complex geometries are joined; therefore, the labor costs associated with joining processes can be high. Other costs, such as those of tooling (welding nozzles) and material (adhesive, filler, and operating gas) are relatively low.

Production Rate

Joining processes are very labor-intensive, especially in case of adhesive bonding or joining parts with complex geometries. Most joining processes require pre-processing of the joining surfaces in order to minimize surface roughness and a period of time after joining, for curing the bond or cooling the weld. These factors result in low production rates for joining processes compared with those for forming, deforming or removing processes.

Part Quality

The quality of joints is limited by the lack of effective methods for non-intrusive joint evaluation. In welding, the weld pool formation is accompanied by the propagation of a heat-affected zone, in which the microstructure of the material is changed due to heating. Other types of defects, common in joints, include porosity, entrapment of contaminants in the joint, incomplete fusion or penetration, crack formation, surface damage and residual stresses. However, effective use of joining techniques can produce joints with mechanical strength exceeding that of its joining members.

Flexibility

Joining processes also have a high degree of flexibility in part geometry and lot size. Since most welding tools have several degrees of freedom, a wide range of sizes and shapes of parts, which may be joined, can be accommodated.

	Cost	Production Rate	Quality	Flexibility
Removing Processes	Low Capital/ High Labor Cost	Medium (Welding) to Low (Adhesives)	Medium to Low	High

Table 2.6. Characteristics of Joining Processes

2.6 Modifying Material Properties Processes

In the above sections, the manufacturing processes introduced, are primarily concerned with changing the shape of a workpiece. However, since all manufacturing processes involve interactions among mechanical, thermal, acoustic, electromagnetic and other phenomena, there are some unintentional results, such as alteration of material structure and composition. In this section, the processes introduced have the intended purpose of changing the physical properties of the material, while maintaining the basic geometry of the part.

The practicality of many processes often depends on the physical properties of the workpiece materials. *Heat and surface treatment* methods can substantially influence the properties of these materials. These treatments are particularly useful to metals. Heat treatment permits the modification of the mechanical properties of metals, within certain limits, and is typically used to create suitable conditions for subsequent manufacturing operations and for the final use of the component.

Surface treatment may be necessary in order to:

- Improve resistance to wear, erosion, and indentation (guideways in machine tools, wear surfaces of machinery, and shafts, rolls, cams, and gears).
- Control friction (sliding surfaces on tools, dies, bearings, and machine ways).
- Reduce adhesion (electrical contacts).
- Improve lubrication (surface modification to retain lubricants).
- Improve corrosion and oxidation resistance (sheet metals for automotive or other outdoor uses, gas-turbine components, and medical devices).
- Improve stiffness and fatigue resistance (bearings and multiple-diameter shafts with fillets).
- Rebuild surfaces on worn components (worn tools, dies, and machine components).

- Improve surface roughness (appearance, dimensional accuracy, and frictional characteristics).
- Impact decorative features, color, or special surface texture.

In *shot peening*, [3] the workpiece surface is hit repeatedly with a large number of cast steel, glass, or ceramic shot (small balls), making overlapping indentations on the surface. This action causes plastic deformation of surfaces, to depths up to 1.25 mm (0.05 in.), using shot sizes ranging from 0.125 mm to 5 mm (0.005 in. to 0.2 in.) in diameter. Since plastic deformation is not uniform throughout a part's thickness, shot peening impacts compressive residual stresses on the surface, thus, improving the fatigue life of the component. This process is used extensively on shafts, gears, springs, oil-well drilling equipment, and jet-engine parts.

In *mechanical plating* (also called *mechanical coating, impact plating, or peen plating*), fine metal particles are compacted over the workpiece surface by impacting them with spherical glass, ceramic or porcelain beads. The beads are propelled by rotary means. The process is used on hardened-steel parts of automobiles, with plating thickness, usually less than 0.025 mm (0.001 in.) [3].

Other more traditional methods of case hardening are *carburizing, carbonitriding, cyaniding, nitriding, flame hardening*, and *induction hardening*. In addition to the common heat sources of gas and electricity, laser beams are also used as a heat source, in surface hardening, of both metals and ceramics.

In *hard facing*, a relatively thick layer, edge, or point of wear-resistant hard metal is deposited on the surface by any of the welding techniques. Hard-facing alloys are available as electrodes, rod, wire, and powder. Typical applications of hard facing are valve seats, oil-well drilling tools, and dies for hot metalworking.

In *thermal spraying*, metal in the form of rod, wire, or powder is melted in a stream of oxyacetylene flame, electric arc, or plasma arc, and the droplets are sprayed on the preheated surface, at speeds up to 100 m/s (20,000 ft/min) with a compressed-air spray gun. All of the surfaces to be sprayed should be cleaned and roughened to improve bond strength. Typical applications of thermal spraying, also called *metallizing*, are steel structures, storage tanks, and tank cars, sprayed with zinc or aluminum up to 0.25 mm (0.010 in.) in thickness. There are several types of thermal spraying processes:

Plasma, either conventional, high-energy, or vacuum; produces temperatures in the order of 8300°C (15000° F) and very good bond strength with very low oxide content.

Detonation gun, in which a controlled explosion takes place using oxy-fuel gas mixture; with a performance similar to that of plasma.

High-velocity oxyfuel (HVOF) gas spraying, which has a similarly high performance, but it is less expensive.

Wire arc, in which an arc is formed between two consumable wire electrodes; it has good bond strength and is the least expensive process.

Flame wire spraying, in which the oxyfuel flame melts the wire and deposits it on the surface; its bond is of medium strength, and the process is relatively inexpensive.

Techniques have also been developed for spraying ceramic coatings in high-temperatures and electrical-resistance applications [3].

Vapor deposition is a process in which the substrate (workpiece surface) is subjected to chemical reactions by gases that contain chemical compounds of the materials to be deposited. The coating thickness is usually a few μm. The deposited materials may consist of metals, alloys, carbides, nitrides, borides, ceramics, or various oxides. The substrate may be metal, plastic, glass, or paper. Typical applications are coating cutting tools, drills, reamers, milling cutters, punches, dies, and wear surfaces.

There are two major deposition processes:

- *physical vapor deposition* (PVD) and
- *chemical vapor deposition*. (CVD)

These techniques allow the effective control of coating composition, thickness, and porosity. The three basic types of physical vapor deposition processes are:

- vacuum or arc evaporation (PV/ARC)
- sputtering, and
- ion plating.

These processes are carried out in high vacuum, at temperatures ranging from 200 to 5000 C° (400 to 9000 F°). In physical vapor deposition, the particles to be deposited are physically carried to the workpiece, rather than by chemical reactions as in chemical vapor deposition [3].

In *vacuum evaporation*, the metal to be deposited is evaporated at high temperatures in a vacuum and is deposited on the substrate, which is usually at room temperature or slightly higher. Uniform coatings can be obtained in complex shapes. In PV/ARC, which was developed recently, the coating material (cathode) is evaporated by a number of arc evaporators, using highly localized electric arcs. The arcs produce a highly reactive plasma, consisting of ionized vapor of the coating material. The vapor condenses on the substrate (anode) and coats it. Applications for this process may be functional (oxidation-resistant coatings for high-temperature applications, electronics, and optics) or decorative (hardware, appliances, and jewelry).

In *sputtering*, an electric field ionizes an inert gas (usually argon). The positive ions bombard the coating material (cathode) and cause sputtering (ejection) of its atoms. These atoms then condense on the workpiece, which is heated to improve bonding. In reactive sputtering, the inert gas is replaced by a reactive gas, such as oxygen, in which case, the atoms are oxidized and the oxides are deposited. Radiofrequency (RF) sputtering is used for nonconductive materials, such as electrical insulators and semi-conductor devices.

Ion plating is a generic term, describing the combined processes of sputtering and vacuum evaporation. An electric field causes a glow discharge, generating a plasma, whilst vaporized atoms in this process are only partially ionized.

Chemical vapor deposition is a thermochemical process. In a typical application, such as that of coating cutting tools with titanium nitride (TiN), the tools are placed on a graphite tray and are heated from 950 to 1050 C° (1740-1920 F°) at atmospheric pressure in an inert atmosphere. Titanium tetrachloride (a vapor), hydrogen, and nitrogen are then introduced to the chamber. The chemical reactions form titanium nitride on the tool surfaces. When coated with titanium carbide, methane is substituted for gases. Chemical vapor deposition coatings are usually thicker than those obtained from PVD [3].

In *ion implantation*, ions are introduced into the surface of the workpiece material. The ions are accelerated in a vacuum, to such an extent, that they penetrate the substrate to a depth of a few μm. Ion implantation modifies surface properties by increasing surface hardness and by improving friction, wear, and corrosion resistance. This process can be controlled accurately, and the surface can be masked so as to prevent ion implantation in unwanted places. When used in specific applications, such as semiconductors, this process is called doping.

Diffusion coating is a process during which an alloying element is diffused into the surface of the substrate, thus altering its properties. Such elements can be supplied in solid, liquid, or gaseous states. The process may take the name (e.g. carburizing, nitriding, boronizing e.t.c.), depending on the diffused element.

In *electroplating*, the workpiece (cathode) is plated with a different metal (anode) while both are suspended in a bath, containing a water-base electrolyte solution. Although the plating process involves a number of reactions, basically the metal ions from the anode are discharged, under the potential from the external source of electricity, they combine with the ions in the solution, and then are deposited on the cathode.

All metals can be electroplated, with thicknesses ranging from a few atomic layers to a maximum of about 0.05 mm (0.002 in.). Complex shapes may have varying plating thicknesses. Common plating materials are chromium, nickel, cadmium, copper, zinc, and tin. Chromium plating is carried out by first plating the metal with copper, then with nickel, and finally with chromium. Hard chromium plating is done directly on the base metal and its hardness reaches up to 70 HRC [3]. Typical electroplating applications are copper plating aluminum wire and phenolic boards for printed circuits, chrome plating hardware, tin plating copper electrical terminals for the ease of soldering, and components requiring good appearance and resistance to wear and corrosion. Noble metals (such as gold, silver, and platinum) are important electroplating materials for the electronics and jewelry industries, because they do not develop oxide films. The parts to be coated may be simple or complex, and the size is not a limitation.

The *electroless plating* process is carried out by chemical reactions, without the use of the external source of electricity. The most common application utilizes nickel, whilst copper can also be used. In electroless nickel plating, nickel chloride (a metallic salt) is reduced-with sodium hypophosphite as the reducing agent - to nickel metal, which is then deposited on the workpiece. The hardness of nickel plating ranges between 425 HV and 575 HV, and can be heat-treated to 1000 HV. The coating has excellent wear and corrosion resistance [3].

Anodizing is an oxidation process (anodic oxidation) in which the workpiece surfaces are converted into a hard and porous oxide layer that provides both corrosion resistance and a decorative finish. The workpiece is the anode in an electrolytic cell immersed in an acid bath, resulting in chemical absorption of oxygen from the bath. Typical applications of anodizing, are aluminum furniture and utensils, architectural shapes, automobile trim and sporting goods.

In *conversion coating*, also called *chemical reaction priming*, a coating is formed on metal surfaces, as a result of chemical or electrochemical reactions. Various metals, particularly steel, aluminum, and zinc, can be processed using conversion coating [3].

Stress relieving is used to reduce and compensate for internal stresses in workpieces, posed by forming conditions or by uneven heat distribution. These internal or residual stresses usually cause elastic deformation of the material, and if the component experiences loads in the same directions as the residual stresses, the elastic limit may be surpassed and the part may fail. To perform stress-relieving of the workpiece, an annealing operation can be performed at high temperatures, which causes a reduction in the yield strength of the material, without changing its lattice structure or phase. With steel, annealing usually occurs at a temperature range of 5000C to 650°C, although the duration of the heat treatment operation depends on the temperature used. Cooling at temperatures of 250 to 300°C, is best performed in a furnace, after which the component is air-cooled. In *recrystallization annealing*, a new grain structure is created in order to counteract the work hardening of the material, which can result from some forming processes. The structure to be created should be fine-grained so as to improve its mechanical properties and the conditions for subsequent forming processes. In recrystallization annealing, the temperature depends on the material properties and is usually in the range of 600°C to 700°C.

Heat treatment processes, such as *recovery annealing, stress-relieving annealing, recrystallization annealing* and *soft annealing* can also be applied to aluminum and aluminum alloys. Temperatures here are lower than those used in the heat-treatment of steels, although the results are similar.

2.6.1 Characteristics of Modifying Processes

Cost

The capital and operating costs of heat and surface treatment equipment can be high, with the cost of modern furnaces in the order of several hundred thousand dollars. Computer control on modern furnaces reduces the labor cost by minimizing the need for skilled human control of process parameters. The fuel or electrical consumption on furnaces is relatively high compared with that of other types of machine tools.

For the coating processes, there is additional cost of the special equipment required (e.g. lasers, vacuum furnaces, spray-guns e.t.c.) and for the coating material.

Production Rate

Since heat treatment processes require a rise in the temperature of the workpiece, a significant amount of processing time may be required to heat this workpiece, maintain an elevated temperature, and cool it at room temperature. The startup time for conventional furnaces may also take several hours. As a result, heat treatment processes can constitute a significant portion of the total production time of a part. However, the improvement on mechanical properties of heat-treated parts may eliminate the need for using a more expensive material.

Moreover, the production rate can be significantly increased by applying to specific areas of the part, one of the coating and/or hardening techniques (e.g. case hardening), only where needed. Thus, the processing time is significantly decreased.

Part Quality

Surface and heat treatment processes are typically used to improve the quality of the part, through material hardening (up to 65 Rockwell Hardness C), relief of residual stresses and coating of the workpiece with a material, which has favorable properties. The treated region usually extends to 0.05-1mm below the workpiece surface. Incorrect heat treatment of any metal alloy can lead to defects, such as the formation of a coarse grain, surface burning, or surface decarborization. Coarse grains may result from overheating or from keeping the part, at annealing temperature, for a long period of time. Surface or skin decarborization occurs in steels as a result of the diffusion of carbon into the atmosphere, due to excessively long annealing times. The burn structure is often a defect, which cannot be overcome. Heat treatment processes carried out at too high temperatures may lead to irreversible oxidation of the grain boundaries. Oxidation weakens their material cohesion, decreases strength and renders the material unusable.

Flexibility

Surface and heat treatment processes can accept a wide range of part geometries and sizes for processing, since there is no contact between a tool and the workpiece. However, in most conventional processes, the size of the part acceptable for processing is limited to the size of the oven or furnace used for treatment. Some state-of-the-art processes, such as that of the laser surface treatment, eliminate this restriction by directing a heat source, such as a laser beam, to the workpiece and by scanning the heat source along the workpiece surface.

	Cost	Production Rate	Quality	Flexibility
Modifying Processes	Medium to High Capital/Low Labor Cost	Low	High	Medium to High

Table 2.7. Characteristics of Modifying Processes

2.7 Nanomanufacturing

Nanomanufacturing encompasses processes aimed at building nanoscale structures, devices and systems in one, two or three dimensions. Bottom-up and Top-down manufacturing techniques are considered. Nanotechnology [21] has been defined as the technology needed to get the extra-high accuracy and ultra-fine dimensions, required in items, such as integrated circuits, opto-electronic devices, mechanical parts for pumps, bearings and computer memory devices as well as in aspheric lenses, all of which can obtain accuracies in the order of 1nm and below,. A general definition that perhaps captures the current and future thrust in the field of Nanotechnology, is the direct control of materials and devices on a molecular and atomic scale [22].

Nanomanufacturing Approaches

The field of nanomanufacturing has been approached by researchers and engineers from four different aspects, related and strongly linked to each other (Fig.2.43).

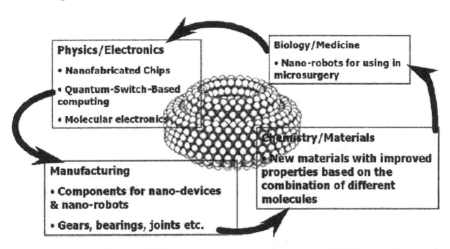

Fig. 2.43. Nanomanufacturing Approaches [26]

In physics/electronics, the field is moving towards smaller feature sizes and is already at submicron line widths. Processors in computing systems will need nanometer line widths in the future, as miniaturization proceeds. In chemistry/materials, improved knowledge on complex systems, has led to new catalyst, membrane, sensor and coating technologies which rely on the ability to tailor structures at atomic and molecular levels. In biology, living systems have sub-units, with sizes between micron and nanometer scales and these can be combined with non-living nanostructured materials to create new devices. In manufacturing, the field of nanostructures is emerging with the introduction of components for nanodevices, nanorobots and nanoscale gears, bearing, joints, pumps etc. Additionally, common manufacturing processes conventional and non, such as cutting, drilling, welding, laser material processing etc. can be studied from a nanoscale point of view

The convergence of nanomanufacturing approaches is presented in Fig. 2.44 [23].

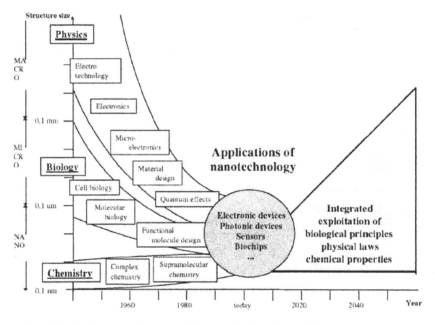

Fig. 2.44. Physics, Biology and Chemistry Meet in Nanotechnology [23]

Nanomanufacturing Techniques

Nanomanufacturing encompasses all processes aimed at building nanoscale (in 1-D, 2-D, or 3-D) structures, features, devices and systems, suitable for operation and/or integration across higher dimensional scales (micro-,meso-and macroscale) in order to provide fully functional products and useful services. The Taniguchi approach [21] has come to be known as transformative or top-down i.e. an approach from larger to smaller. An alternative concept was introduced by Drexler in [24, 25] by building larger objects from their atomic or molecular components i.e. a synthetic or bottom-up approach. This approach is more limited when compared with that of the top-down. Typical examples of the bottom-up processes include, contact printing, imprinting, spinodal wetting/dewetting, laser trapping/tweezer, assembly and joining (self- and directed- assembly), template growth, electrostatic (coatings, fibers), colloidal aggregation and 2-photon confocal processing. Typical examples of the top-down processes, include lithography (electron-beam, ion-beam, scanning probe, optical near field), thin film deposition and growth, laser beam processing, mechanical (machining, grinding, lapping, polishing) and electrochemical material removal processes (electroforming and hot embossing lithography).

In self-assembly, the atoms or molecules, required to construct the desired product, are brought together in a suitable environment in order to arrange themselves and form a product. Examples of self-assembly of large molecules, which can then assemble themselves into ordered arrays are widely found in chemistry and biology and in the field of supramolecular chemistry.

The critical question that arises, however, is how to scale up these processes so as to achieve high volume, production rate, quality and low cost. Nanomanufacturing is expected to provide the traditional industries with new capabilities and product or service lines..

Nanomanufacturing Drivers

Similar to large-scale fabrication, nanomanufacturing issues involve precursor materials; fabrication processes and characterization techniques; instrumentation and equipment; theoretical modeling and control; and design and integration of structures into devices and systems. Conventional large-scale structures can be made of poly-atomic or molecular assemblies of matter, where the multiplicity of such unit blocks, whether ordered or statistically distributed, yields the familiar averaged properties of bulk materials. Due to these macroscale properties effectively extending down to the microscale, traditional manufacturing techniques have been miniaturized for the fabrication of microstructures as in microelectronics, in a top-down approach. On the other hand, mono-atomic or molecular units, with their well-known subatomic structure in isolation, offer the ultimate building blocks for a bottom-up, atom-by-atom manufacturing synthesis. In nanomanufacturing, a decision regarding the design and operation of a system, requires technical understanding and expertise as well as the capability of satisfying certain business objectives.

The general four classes of manufacturing attributes, *cost, time, quality* and *flexibility* (Fig. 2.45.), which are considered when making manufacturing decisions, will be applied to nanomanufacturing aspects and will be contributing to the optimization of every nano oriented industrial level process, so as to receive the expected results. Furthermore, such attributes are going to drive nanomanufacturing, as they will be the key factors, from the transition of the nano era of laboratories to the production floor.

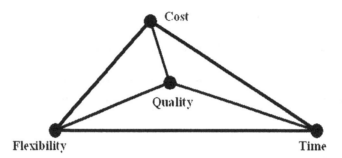

Fig. 2.45. The Manufacturing Tetrahedron

Precision is an additional key element for understanding the need of developing nanotechnology / nanomanufacturing. Precision means that there is a place for every atom and every atom is in its place. Schematics will be detailed and no unnecessary parts will exist in designing a product that is going to be fully or partially nanomanufactured. The usage of precision machines will generate products of equal precision and the cost of manufacturing will be reduced. Precision is a "key driver" of nanomanufacturing (Fig. 2.45, Fig.2.46).

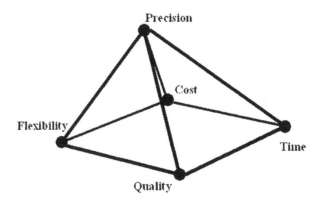

Fig. 2.46. Precision in Nanomanufacturing as Attribute [26]

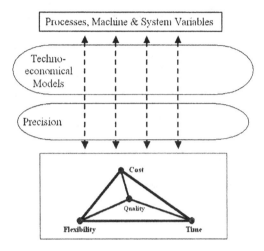

Fig. 2.47. Precision as an External Factor in Nanomanufacturing [26]

2.8 Process Attribute Comparison

The applicability of each class of manufacturing processes to the criteria, which form the manufacturing tetrahedron discussed in Chapter 1, are shown in Table 2.8. From the comparison, a trade-off exists among the attributes for any manufacturing process. Selection of an appropriate process often depends on the balance of all four criteria, based on the requirements for making a desirable product, in an efficient and profitable manner. For example, if the primary concern is production rate, then primary forming and deforming processes should be considered. However, if many different part types with small batch sizes are to be produced in the same facilities, then removing and joining processes are more appropriate. Finally, if the quality of the part is of prime concern, material removal and material property modification processes should be incorporated as secondary operations.

	FORMING	DEFORMING	REMOVING	JOINING	MODIFYING
COST	High Tooling/Low Labor Cost	High Tooling/Low Labor Cost	Medium Tooling/High Labor Cost	Low Capital/High Labor Cost	Medium to High Capital/Low Labor Cost
PRODUCTION RATE	High	High	Medium (Milling) to Low Grinding	Medium (Welding) to Low (Adhesives)	Low
QUALITY	Medium to Low	Medium to Low	Medium to High	Medium to Low	High
FLEXIBILITY	Low	Low	High	High	Medium to High

Table 2.8. Characteristics of the Classes of Manufacturing Processes

Further Reading

There have been many text and reference books written which deal specifically with the physics of manufacturing processes. Among the most widely used general texts in manufacturing processes are *Manufacturing Processes for Engineering Materials* [3] and *Manufacturing Engineering and Technology* [1] written by Serope Kalpakjian as well as *Introduction to Manufacturing Processes* [4] by John Schey. Another book, Modern Manufacturing Process Engineering [11] by Niebel, Draper and Wysk, gives a more operational perspective on manufacturing, with emphasis on the relationships among different manufacturing processes and transformations in the properties of the engineering material.

For references which are more focused on metalworking, *Processes and Materials of Manufacture* [10] by Lindberg and *Fundamentals of Manufacturing Processes and Materials* [9] by Edgar provide a comprehensive examination of removing, casting, forming and joining processes. The Lindberg book also introduces some methods of dimensional measurements for quality control applications, while the Edgar book examines the types of metals and alloys, which are producible through the various manufacturing processes.

Of the manufacturing processes discussed, the processes most frequently analyzed are those of material removal. One of the most classic books on the topic, both from a historical and analytical point of view, is *Metal Cutting Principles* [12] by Milton Shaw. Although, it was written over 30 years ago, much of the theoretical modeling of the cutting process is still relevant and widely-used. Two reference books for removal processes are the *Machining Data Handbook* [7] published by the Machinability Data Center and *Machining Fundamentals* [13] by Walker. The *Machining Data Handbook* is especially useful for providing estimates on cutting depth and material removal rate based on the process, material, feeds and speeds. Additionally, it is also a good reference for estimating the operating costs associated with particular machining operations. For references regarding Nanomanufacturing the book by Eric Drexler [25] *Nanosystems: Molecular Machinery, Manufacturing and Computation* provides a complete introduction in the field.

References

1. Kalpakjian, S., *Manufacturing Engineering and Technology*, Addison-Wesley, Reading, MA, 1989.

2. Weck, M., *Handbook of Machine Tools*, John Wiley & Sons, New York, 1984.

3. Kalpakjian, S., *Manufacturing Processes for Engineering Materials*, Addison-Wesley, Reading, MA, 1997.

4. Schey, J., *Introduction to Manufacturing Processes*, McGraw-Hill, New York, 1987.

5. Lange, K., *Handbook of Metal Forming*, Mc Graw Hill, New York, 1985.

6. Chryssolouris, G., *Laser Machining: Theory and Practice*, Springer-Verlag, New York, 1991.

7. *Machining Data Handbook*, 3rd Edition, Vol 2, Metcut Research Associates Inc., 1980.

8. Schwartz, M.M., *Metal Joining Manual*, McGraw-Hill, New York, 1979.

9. Edgar, C., *Fundamentals of Manufacturing Processes and Materials*, Addison-Wesley, Reading, MA, 1965.

10. Lindberg, R., *Processes and Materials of Manufacture*, Allyn and Bacon Inc., Boston, MA, 1977.

11. Niebel, B., A. Draper and R. Wysk, *Modern Manufacturing Process Engineering*, McGraw-Hill, New York, 1989.

12. Shaw, M., *Metal Cutting Principles*, The Technology Press, Cambridge, MA, 1954.

13. Walker, J., *Machining Fundamentals*, Goodheart-Willcox Co., South Holland, IL, 1977.

14. *Plastics Product Design Handbook*, ed. E. Miller, Marcel Dekker Inc., New York, 1983.

15. *Stamping Design Thru Maintenance*, ed. Kerl Keyes, Society of Manufacturing Engineers, Dearborn, MI, 1983.

16. Paul, F., Jacobs, *Stereolithography and other RP&M techniques*, ASME Press, 1992, New York, ISBN 0-87263-467-1.

17. Kruth, J.P., "Material Incress Manufacturing by Rapid Prototyping Techniques", *Annals of the CIRP*, (Vol.40 No. 2, 1991), pp. 603-614.

18. *Tool and Manufacturing Engineers Handbook*, 4th edition, Vol. 2, SME, 1985.

19. Dewes, R.C., D.K., Aspinwall, "A review of ultra high speed milling of hardened steels", *Journal of Materials Processing Technology*, (Vol.69, 1997), pp.1-17.

20. Wohlers, T.T., *2001 Wohlers Report*, Wohlers Associates, 2001

21. Taniguchi, N., "On the Basic Concept of Nanotechnology", *Proceedings of the ICPE International Conference on Production Eng.*, Tokyo, (1974), pp 18-23.

22. Siegel, R.W., E. Hu, and M.C. Roco, *Nanostructure Science and Technology: A Worldwide Study*, Kluwer Academic Publishers, New York, 1999.

23. Budworth, D.W., *Overview of Activities on Nanotechnology and Related Technologies*, Institute for Prospective Technological Studies, Seville, 1996.

24. Drexler, K. E., *Engines of Creation: The Coming Era of Nanotechnology*, Anchor Press / Doubleday: New York, 1986.

25.Drexler, K. E., *Nanosystems: Molecular Machinery, Manufacturing and Computation*, John Wiley, New York, 1992.

26.Chryssolouris, G., P. Stavropoulos, G. Tsoukantas, K. Salonitis and A. Stournaras, "Nanomanufacruting Processes: A Critical Review", *International Journal of Materials and Product Technology*, (Vol.21, No.4, 2004), pp. 331-348.

Review Questions

1. What is a manufacturing process and how can manufacturing processes be classified?

2. What are the two most important factors which affect cost, production rate, quality, and flexibility?

3. Explain why small lot sizes require flexible processes. Give an example of such a process.

4. What are primary forming processes? What materials can be used in them?

5. What are the primary forming processes for each class of materials?

6. Would you consider forming processes to be flexible? Give reasons.

7. What lot sizes justify the use of forming processes? Why?

8. What are the two primary forming processes for plastics? Discuss their similarities and differences.

9. What characteristics of composite materials make them desirable engineering materials?

10. What are some advantages of "cold forming" in deforming processes?

11. a) What quality problems are associated with metal casting?

    b) Which method produces better results?

    c) What process parameters can in general influence the quality of a cast part?

12. What criteria are used by the metal working industry to classify deformation processes?

13. What lot sizes justify the use of deforming processes? Give reasons.

14. What are the categories of material removal processes? Which is most widely used?

15. What order of magnitude for material removal rate (MRR) can be achieved with machining and grinding operations?

16. For what lot sizes are machining operations best suited? Why?

17. What factors can affect the surface quality of a machined part?

18. Why are joining processes important in manufacturing? List the major joining processes.

19. What joining processes are thermal processes?

20. How does welding differ from brazing and soldering?

21. a) What are the major concerns and problems associated with welding processes?

    b) What factors can affect the quality of a weld joint? How can some of these problems be reduced?

22. What is the purpose of heat and surface treatment processes?

23. What potential problems are associated with heat treatment? What is their effect on surface quality?

24. In your opinion which manufacturing process possesses the highest degree of flexibility? What criteria led you to your choice?

25. If you were asked to manufacture a high precision automobile shaft, where quality and production rate are of primary concern, which process or processes would you choose? Discuss your answer.

3. Machine Tools and Manufacturing Equipment

3.1 Introduction

This chapter deals with manufacturing equipment, particularly machine tools. It follows the chapter on manufacturing processes because machines are basically the "embodiment" of processes, and it precedes the chapters on manufacturing systems because machine tools are the "building blocks" of manufacturing systems.

The term "manufacturing equipment" covers a large variety of devices and machines, ranging from hand tools to complex automated machining centers, encountered in modern production facilities. This chapter emphasizes machine tools since they are the most important elements of manufacturing systems, particularly in discrete manufacturing parts. The chapter begins with a brief historical perspective on the machine tool industry and continues with a section describing different types and classifications of machine tools. The next section is devoted to machine tool design and analysis, in which the primary components of machine tools – frames, guideways/bearings, and drives – are discussed. The final section of the chapter concentrates on machine tool control and automation, and discusses numerical control, parts programming, process control and sensing devices.

This chapter does not attempt to be comprehensive by discussing all possible equipment and machine issues, but provides an overview of major issues and indicates fundamental concepts, useful in the manufacturing environment.

Manufacturing equipment, and particularly machine tools, have been used in industrial production for over 100 years; the evolution during this period has been quite remarkable. Indeed, considering the example of the lathe (Fig. 3.1), one can notice that in the first decade of this century, a typical lathe was powered by a motor, positioned outside of the machine itself. This configuration forced the machine to remain fixed at the site of installation, and prevented easy relocation. In the '20s, the motor became an integral part of the machine, which on the one hand allowed an easy relocation of the machine, in case of factory reorganization, but also provided a stiffer machine that could achieve a higher production rate. In the 1960s, machine tools were produced with completely enclosed frame to obtain greater stiffness, and often included template and stylus for the purpose of copying a complex shape onto the workpiece. The lathes of the '70s and '80s became numerically controlled and have the ability to be programmed with the help of computers. There is a synergy between manufacturing processes and the equipment that performs them. Developments in manufacturing processes and materials have substantially influenced the evolution of machines. The introduction, for example, of new cutting tool materials (Fig. 3.2) continuously increased the cutting speeds attainable in machining operations. Thus, machines had to be properly designed in order to accommodate the need for higher cutting speeds and production rates.

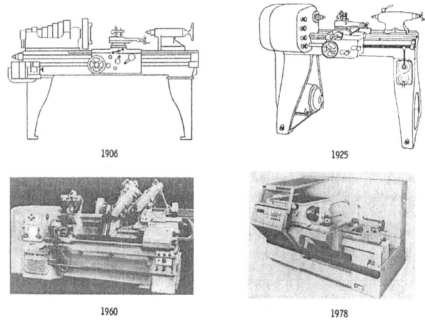

1906   1925

1960   1978

Fig. 3.1. The Evolution of the Lathe [1]

Manufacturing equipment and machine tools are particularly the critical link between intermediate products, formed from raw materials and finished discrete parts and components. These components are then assembled either into end products or yet into other machines that will in turn, make end products. Thus, virtually every manufactured product is built either directly or indirectly by machine tools [2]. It is this position at the heart of the manufacturing infrastructure that makes manufacturing equipment and the machine tool industry so critical to a nation's manufacturing competitiveness and productivity.

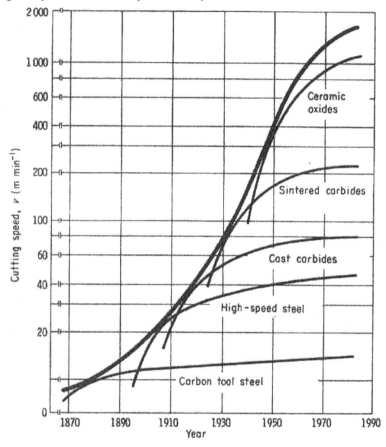

Fig. 3.2. Historical Development of Tool Materials and Cutting Speeds [1]

3.2 Machine Tool Types

In general, a machine tool can be defined as a non-portable machine with an integral power source, which causes the relative motion of a tool and a workpiece in order to produce a predetermined geometric form or shape. This "narrow" definition primarily reflects the two major classes of machines, deforming (or forming) and removing (or cutting) machines.

However, machines used for primary forming processes, such as casting, and joining processes, such as welding or assembly, can also be regarded as machine tools.

In order for machine tools to be successfully integrated in the production process, a number of criteria must be met. In modern manufacturing systems, quality is of major concern, and may be the primary driving force in machine tool design, development, acquisition, and operation. Quality in machine tools falls into two categories: *accuracy* in geometric and kinematics terms, under static, dynamic or thermal loading; and *reliability*, or the behavior of the machine over an extended period of operation time. Related to these characteristics are maintainability, work space accessibility and safety. Modern machine tools are expected to operate with accuracy, in the order of microns, in a wide range of loading conditions, some of which may exceed several tons. This accuracy, combined with the high velocity at which the machine tool elements must move in order to achieve required production rates, presents a formidable design and manufacturing challenge.

Cost is also an important factor, since machines typically represent a high capital investment, which can reach into the order of billions of dollars for mass production systems typically in the automotive industry.

Often, both cost and quality of the machine are determined by its degree of automation. Over-automated machines may be unreliable, while machines that are under-automated may have low production rates. Furthermore, the degree of automation often determines the degree of flexibility. Highly automated machines/systems, such as transfer lines, provide a very high production rate, but limited product and operational flexibility. Thus, automation of machine tools should strike a balance between cost, quality (particularly reliability) and flexibility. In general, product flexibility is high among metal removing machines, and relates to the "working envelope" of the machine, namely the maximum and minimum dimensions of the parts that the machine can handle, and the accuracy range of the different motion axes of the machine.

Nowadays, it is essential for machines to satisfy safety and environmental requirements. For example, machine tools, which operate at high noise levels, may be required to be encapsulated as a safety and environmental means of safety. Safety related issues have the highest priority and can influence the design of a machine tool considerably. However, safety considerations often have a detrimental effect on the efficiency or productivity of a machine tool.

3.2.1 Machines for Deforming

Most deforming machines are used in the metal working industry since deforming is a class of processes particularly applied to metals. Deformation occurs through the use of tools having two or more parts. A machine brings the tools together to deform the workpiece, provides the necessary force, energy and torque for the process, and ensures proper guidance of the tools. Based on the relative movement of the tool parts, deforming machines [3] can be classified into two groups: machines with linear (Fig. 3.3) relative movement, and machines with non-linear (Fig. 3.4) relative movement. Machines not belonging to either of these categories are usually considered to be special-purpose machines.

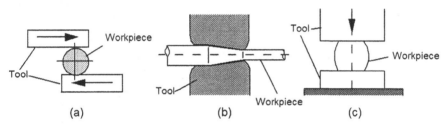

(a)   (b)   (c)

Fig. 3.3. Linear Tool Movement (a) Rolling; (b) Drawing; (c) Upsetting

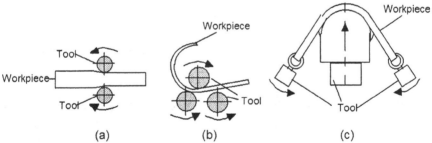

(a)   (b)   (c)

Fig. 3.4. Nonlinear Tool Movement (a) Rolling; (b) Bending; (c) Stretch-Forming

Machines for deforming can be divided into [1] energy, movement and force-constrained machines (Fig. 3.5). In energy-constrained machines, such as the belt-operated drop and the air drop hammer (Fig. 3.6), the characteristic feature is the available energy, which is converted into work on the workpiece; the deforming process is completed when this energy conversion is finished [1]. These machines are characterized by low cost, large size, and unsophisticated controls. Their production rates are low, while their forming velocity is high. The accuracy of the parts, made by drop hammers, is relatively low, and the flexibility of the machine depends on the speed at which the tools can be changed in order to produce different geometries. Hammers, particularly used in forging (Fig. 3.7), can take on a variety of designs to assist and control the forming action.

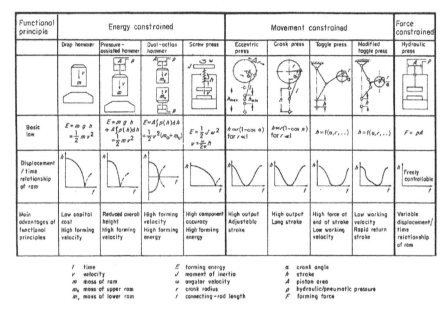

Functional principle	Energy constrained				Movement constrained				Force constrained
	Drop hammer	Pressure-assisted hammer	Dual-action hammer	Screw press	Eccentric press	Crank press	Toggle press	Modified toggle press	Hydraulic press
Basic law	$E = mgh$ $= \frac{1}{2} m v^2$	$E = mgh + A\int p(h)dh$ $= \frac{1}{2} m v^2$	$E = A\int p(h)dh$ $= \frac{1}{2} v^2(m_o + m_u)$	$E = \frac{1}{2} J \omega^2$ $v = \frac{\omega}{2\pi} h$	$h \approx r(1-\cos a)$ for $r \ll l$	$h \approx r(1-\cos a)$ for $r \ll l$	$h = f(a,r,..)$	$h = f(a,r,..)$	$F = pA$
Displacement / time relationship of ram									Freely controllable
Main advantages of functional principles	Low capital cost High forming velocity	Reduced overall height High forming velocity	High forming velocity High forming energy	High component accuracy High forming energy	High output Adjustable stroke	High output Long stroke	High force at end of stroke Low working velocity	Low working velocity Rapid return stroke	Variable displacement/ time relationship of ram

t	time	E	forming energy
v	velocity	J	moment of inertia
m	mass of ram	ω	angular velocity
m_o	mass of upper ram	r	crank radius
m_u	mass of lower ram	l	connecting-rod length

a	crank angle
h	stroke
A	piston area
p	hydraulic/pneumatic pressure
F	forming force

Fig. 3.5. Functional Principles of Forming Machines [1]

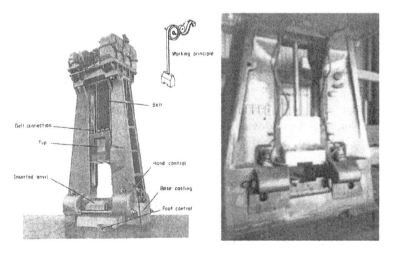

Fig. 3.6. Belt-Operated Drop [1] and Air Drop Hammer [89]

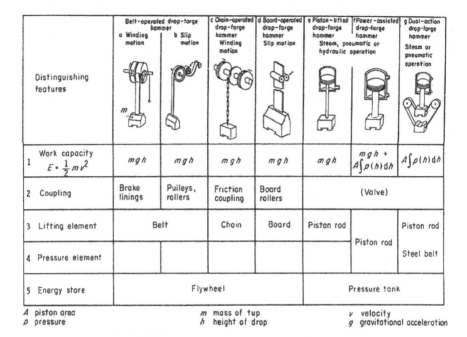

Distinguishing features	Belt-operated drop-forge hammer		c Chain-operated drop-forge hammer Winding motion	d Board-operated drop-forge hammer Slip motion	e Piston-lifted drop-forge hammer Steam, pneumatic or hydraulic operation	f Power-assisted drop-forge hammer Steam, pneumatic operation	g Dual-action drop-forge hammer Steam or pneumatic operation
	a Winding motion	b Slip motion					
1 Work capacity $E = \frac{1}{2}mv^2$	mgh	mgh	mgh	mgh	mgh	$mgh + A\int p(h)dh$	$A\int p(h)dh$
2 Coupling	Brake linings	Pulleys, rollers	Friction coupling	Board rollers	(Valve)		
3 Lifting element	Belt		Chain	Board	Piston rod	Piston rod	Piston rod
4 Pressure element							Steel belt
5 Energy store	Flywheel				Pressure tank		

A piston area
p pressure
m mass of tup
h height of drop
v velocity
g gravitational acceleration

Fig. 3.7. Different Designs of Forging Hammers [1]

Screw presses are also energy-constrained machines, which are typically driven by a flywheel (Fig. 3.8) whose rotary motion is converted into the linear movement of the ram, in the machine frame, with the aid of a lead screw. The drive is usually arranged in a triple-disk design, located at the upper part of the machine with the electric motor. On the return movement of the ram, the opposite driving disk is in contact with the flywheel; consequently, the transmission relationships are a disadvantage since the flywheel must accelerate from an almost stationary state. Designs, such as a four-disk or single-disk drive (Fig. 3.9) attempt [1] to avoid the disadvantages of the conventional triple disk (Fig. 3.8). In general, screw presses are more capital-intensive than hammers, but provide better-quality workpieces.

Eccentric presses (Fig. 3.10) belong to the category of movement-constrained forming machines, where the stroke of the machine is determined through the eccentricity of the cam. In crank presses (Fig. 3.11), which are also movement-constrained machines, the travel of the ram is determined by the geometry of the crank shaft. Crank presses are notable for their high production rate, which is typically around a thousand strokes per minute. The variability of the stroke is often reduced for simpler machine construction, greater rigidity, and higher production speed. Crank

presses are typically used for extrusion, deep drawing or bending processes, while eccentric presses are primarily used for blanking, bending and
coining processes, where only a small stroke is required.

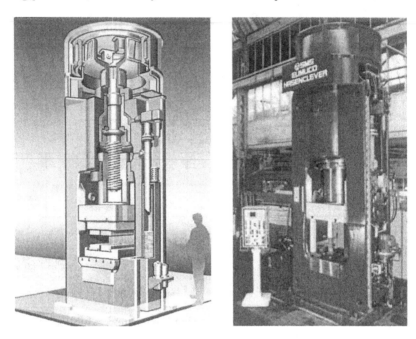

Fig. 3.8. Screw Press with Direct Motor Drive [90]

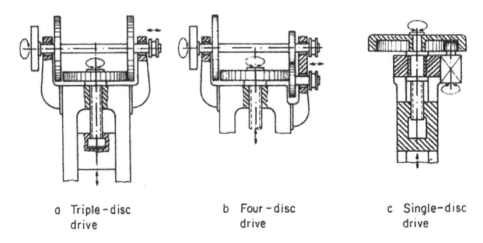

a Triple-disc   b Four-disc   c Single-disc
   drive   drive   drive

Fig. 3.9. Flywheel Drives for Screw Presses [1]

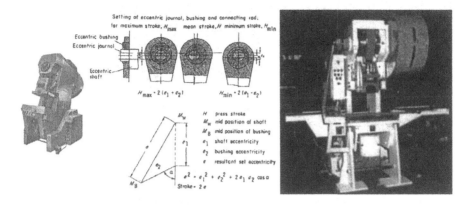

Fig. 3.10. Inclinable Eccentric Press with Variable Stroke [1, 91]

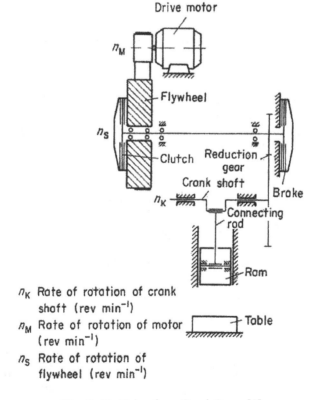

n_K Rate of rotation of crank
shaft (rev min^{-1})

n_M Rate of rotation of motor
(rev min^{-1})

n_S Rate of rotation of
flywheel (rev min^{-1})

Fig. 3. 11. Drive for a Crank Press [1]

Fig. 3.12. Forging Press with Wedge Motion [92]

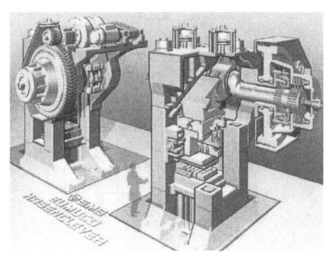

Fig. 3.13. Wedge Press [90]

Wedge presses (Figs. 3.12 and 3.13), which are mechanically or hydrau-lically operated, are able to exert very high forces. The forming movement of the machine is created by a wedge, which is placed between the upper part of the frame and the ram. The large area of contact between the wedge and the ram, and the massive structure of the frame, makes these presses particularly suitable for unevenly distributed loads. These are heavy-duty machines with a very low production rate (30 to 70 strokes per minute) but with high forming force [1].

The most widely used deforming machine is the hydraulic press (Fig. 3.14), which is a force-constrained machine and provides accurately controlled force throughout the deforming process. The frames of hydraulic presses are similar to those of mechanical presses, and typically include several hydraulic drives, which can be used for complicated deforming and cutting operations, such as fine blanking processes (Fig. 3.15), where multiple actions are performed between the tool and the workpiece in a single stroke. Hydraulic presses are expensive and robust, requiring extensive maintenance but provide highly accurate operations with complex geometries.

Fig. 3.14. Hydraulic Press [93]

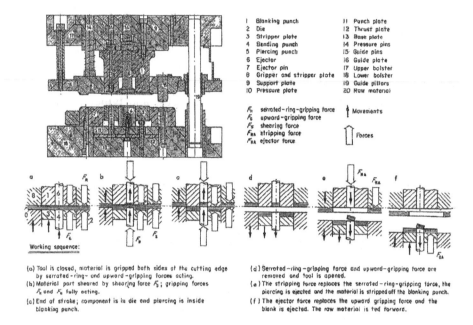

1	Blanking punch	11	Punch plate
2	Die	12	Thrust plate
3	Stripper plate	13	Base plate
4	Bending punch	14	Pressure pins
5	Piercing punch	15	Guide pins
6	Ejector	16	Guide plate
7	Ejector pin	17	Upper bolster
8	Gripper and stripper plate	18	Lower bolster
9	Support plate	19	Guide pillars
10	Pressure plate	20	Raw material

F_R serrated-ring-gripping force
F_G upward-gripping force
F_S shearing force
F_{RA} stripping force
F_{EA} ejector force

↑ Movements

⇑ Forces

Working sequence:

(a) Tool is closed, material is gripped both sides of the cutting edge by serrated-ring- and upward-gripping forces acting.

(b) Material part sheared by shearing force F_S; gripping forces F_R and F_G fully acting.

(c) End of stroke; component is in die and piercing is inside blanking punch.

(d) Serrated-ring-gripping force and upward-gripping force are removed and tool is opened.

(e) The stripping force replaces the serrated-ring-gripping force, the piercing is ejected and the material is stripped off the blanking punch.

(f) The ejector force replaces the upward gripping force and the blank is ejected. The raw material is fed forward.

Fig. 3.15. Working Sequence of a Fine-Blanking Press Tool [1]

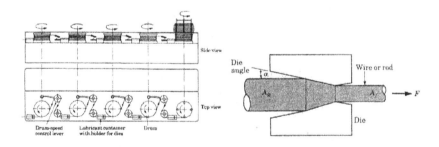

Fig. 3.16. (a) Multistage wire drawing machine and (b) wire-drawing process variables [4]

Other common material deforming equipment includes wire-drawing machines (Fig. 3.16), which are used for reducing the diameter of wires down to .01 mm. The deforming action occurs in a drawing die, usually made of carbide or diamond, depending on the application. The wire is attached to a tension drum, which pulls the wire through the die, reduces its diameter, and rolls the drawn wire onto the drum.

At the other end of the deforming machine spectrum are extrusion machines (Fig. 3.17), heavy-duty machines that force material through an extrusion die to form bars and tubes in a variety of shapes.

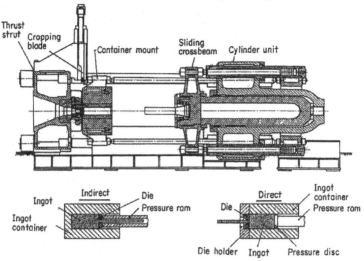

Fig. 3.17. Extrusion Machine [1]

Bending (Fig. 3.18) and rolling machines (Fig. 3.19) are used for sheet metal forming. In the former, a sheet-metal blank is clamped between the top and bottom of the clamping bar and is bent to the desired angle [1] by a bending beam (Fig. 3.18). Rotary rolling machines (Fig. 3.19) contain three rollers, two of which are driven, while the third is in a fixed position. The lower driven rollers are adjusted according to the required bending radius and metal thickness. These machines are typically used for the manufacturing of large cylindrical vessels, pressure containers, etc.

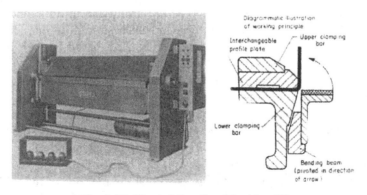

Fig. 3.18. Pivoted Bending Machine [1]

Fig. 3.19. Rotary Rolling Machine [1]

3.2.2 Machines for Material Removal

Material removal machines are the most commonly used equipment in manufacturing systems. They include machines based on *mechanical*, *thermal*, *electrochemical*, and *chemical* material removal mechanisms. Machine tools are typically understood to be mechanical removal machines, which can be divided into two categories: machines which use tools with a *single cutting edge*, such as in turning and milling, and those which use tools with *multiple cutting edges* such as in grinding. In turn, machines with single cutting edge can be classified, based on their cutting motion into *rotational* and *translational* machines.

A typical machine tool for material removal includes a *frame*, which supports the relative motions between the tool and workpiece so as for the cutting action to occur; the *main* and *secondary drives*, which provide the main cutting action and the relative motion between the tool and the workpiece; and the *auxiliary* devices, which provide coolants and other necessary functions for the machine. Finally, the machine includes *controls*, which coordinate the movement of axes so that the motions of the tool and workpiece, and the resulting cutting action, to produce accurately the required workpiece geometry.

Machines with translational main cutting action include planing, shaping and broaching equipment. Gantry or double-column planing machines (Fig. 3.20) are typical examples of machine tools with translational main cutting action, which is used for processing large, flat surfaces. The workpieces are usually long and heavy, as found in the aerospace industry and the machine/tool building industry itself. Since the cutting speed is determined by the cutting motion speed, which is provided either from the table or the cutting tool, the main drive, which may be electromechanical or hydraulic, must have a relatively wide range of speeds, adequate for different types of materials. In addition, the main drive that provides the cutting motion should be able to produce higher moving speeds, which are required for the idle movements of the table or the cutting tool in order to maximize production rate. While gantry planing machines are commonly used in small lot size or in one-of-a-kind production, broaching machines, due to their high price and the difficulty in making the broaching tool (broach) (Fig. 3.21), are most often used in mass production. Broaching machines are used both for internal and external broaching (Fig. 3.21) and are available in vertical and horizontal configurations (Fig. 3.22).

Fig. 3.20. A Double-Column or Gantry Planing Machine [94]

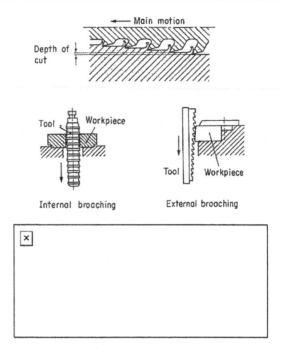

Internal broaching   External broaching

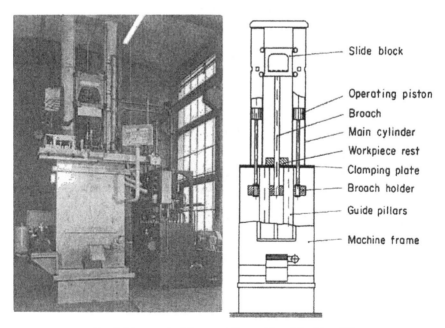

Fig. 3.21. Chip Formation in Broaching [1] and terminology of a broach [4]

Slide block

Operating piston

Broach

Main cylinder

Workpiece rest

Clamping plate

Broach holder

Guide pillars

Machine frame

Fig. 3.22. Vertical Internal-Broaching Machine [1]

The most widely used machine, whose main cutting action is rotational, is the lathe (Fig. 3.23). It can perform a variety of turning operations, such as external and internal turning, boring, threading and making cones. The lead screw of the machine has a very high precision thread and is typically used for cutting thread profiles. The feed shaft, which has a keyway along its entire length, is used to provide the normal feeding motion of the machine. Universal lathes are relatively flexible machines, which can produce a large variety of shapes determined by the relatively motion of the workpiece and the tool, but at low production rates. To improve these production rates, which in lathes are often due to tool changes, machine designs (Fig. 3.24) include turrets in which a number of tools are automatically changed. This allows the application of a variety of tools on the same workpiece, substantially reducing production time without necessarily reducing the high flexibility of the lathe. Lathes are available at a wide range of prices, starting from the low $10,000 for manual, simply constructed machines, and going up to $250,000 for highly-automated turning centers (Fig. 3.25).

Fig. 3.24. Numerically Controlled Automatic Lathe with Drum Turret [95]

Fig. 3.25. Vertical Lathe [96]

One class of machinery for producing rotational symmetric geometric features is drilling and boring equipment. Drilling machines vary from hand-operated equipment, commonly used for repair work, to radial drilling presses, which are used for larger workpieces (Fig. 3.26). During the drilling process, the area where the chips are formed is confined, making drilling a relatively slow process compared to that of turning.

Fig. 3.26. Radial Drill [97]

Chip formation also limits the generation of holes with large aspect ratios (depth to diameter). To overcome this limitation, deep hole drilling techniques provide ways of ejecting chips from the bottom of the hole by applying high-pressure coolants (Fig. 3.27), allowing high dimensional accuracy. Drilling speed can be increased with multi-spindle machines, capable of drilling many holes simultaneously (Fig. 3.28).

Fig. 3.27. Deep Hole Drilling Attachment [1]

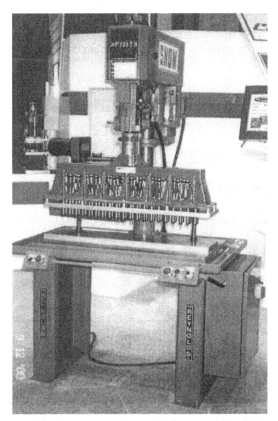

Fig. 3.28. Multi-Spindle Drilling Machine [98]

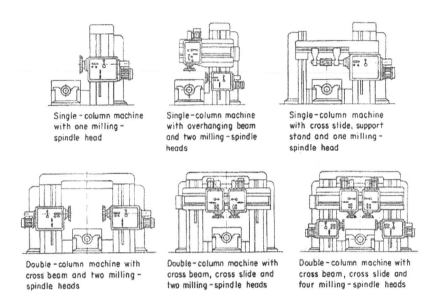

Fig. 3.29. Elements of a Single/Double-Column Milling Machine [1]

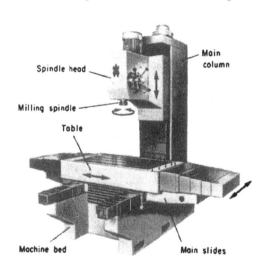

Fig. 3.30. Construction elements of a single-column bedplate milling machine [1]

In milling, the main cutting action is also rotational, although the feed motion is not in the direction of the cutting tool, but perpendicular to it. Milling machines (Fig. 3.29) are widely used in industry for both one-of-a-kind and mass production. Different designs of milling machines (Fig.

3.30) provide a variety of production speeds, flexibility and rigidity, which is required particularly for high-precision workpieces. Due to the number of axes that are involved in the relative motion between workpiece and tool (Fig. 3.31), milling machines provide a high degree of flexibility regarding workpiece geometry. This is why machining centers (Fig. 3.32) are basically milling machines with an additional tool changer, which provides a high degree of flexibility and a high production rate. Consequently, machining centers are the cornerstone of *flexible manufacturing systems*, which will be discussed later.

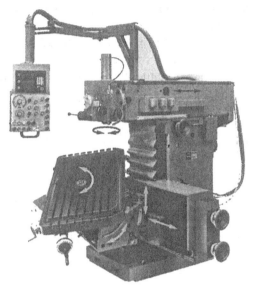

Fig. 3.31. Universal Milling Machine [1]

Fig. 3.32. Machining Center with Automatic Tool Change Unit [99]

As already discussed in Chapter 2 a number of different abrasive processes exist so as to achieve high precision, close tolerance control and good surface quality. Such processes include *grinding, honing, lapping, superfinishing, polishing, buffing* etc.

Grinding processes are categorized according to the workpiece geometry in *external, internal cylindrical* and *surface* processes. Furthermore, according to the process parameters and the special process characteristics, used the grinding processes, grinding processes can be further distinguished in *centerless grinding, creep-feed grinding, high-efficiency grinding* and *profile grinding*. In this chapter these processes are described as far as the grinding machine is concerned.

External grinding encompasses a family of processes, all of which are usually executed with an external cylindrical-grinding machine (Fig. 3.33) whereby the angle of the worktable and grinding wheel head can be adjusted (Fig. 3.34).

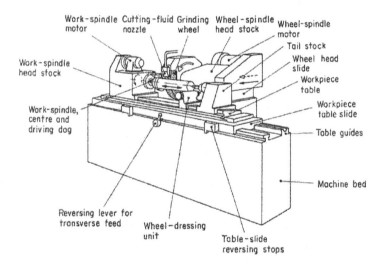

Fig. 3.33. Basic Design of an External Cylindrical-Grinding Machine [1]

When both high precision and quality are required and grinding must be used, whilst a high production rate is also required, as in the automotive industry, it is centerless grinding that provides a solution. This process is designed so as for the workpiece to rest on a support plate; and the cylindrical or hyperbolically shaped control wheel, typically made from synthetic rubber or rubberized plastic, is mounted opposite the grinding wheel. This geometric arrangement allows continuous through-feed grinding or plunge grinding of profiled workpieces (Chapter 2). When the process is used for through-feed grinding, the linear feed motion is created through an angular offset of the control wheel. Centerless grinding machines (Fig. 3.35) are considerably more expensive than external grinding machines and are a typical part of mass producing systems, particularly in the automotive industry.

Fig. 3.34. Universal External Cylindrical-Grinding Machine [100]

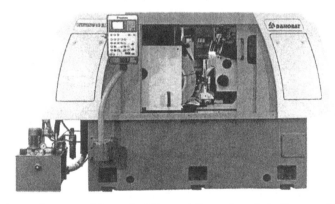

Fig. 3.35. A Computer Numerical Control Centerless Grinding Machine [101]

Internal grinding techniques (Fig. 3.36) are very important to industries, such as the bearing industry, where they are used for grinding the internal surface of hardened bearing tracks. For the grinding of internal profiles profile-dressed grinding wheels are used. The process has inherent limitations, in terms of dynamic stability, since the size of the grinding wheel and the spindle that supports it is limited by the geometry of the part to be ground. Since the size of the available grinding wheels is limited, high speed spindle is required (30.000 rpm and higher) so as to achieve the prerequisite cutting forces. The design (Fig. 3.37) of an internal cylindrical grinding machine typically provides the required transverse feed motion on the workpiece/head slide. To counter the dynamic instability on the spindle that supports the grinding wheel, adaptive control schemes (discussed later in this chapter), can be used. For the internal grinding of relatively long parts, such as needle bearings, an internal "centerless" grinding technique can be utilized (Fig. 3.38) [5].

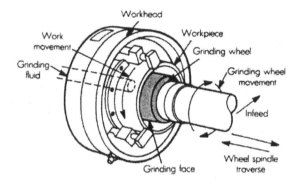

Fig. 3.36. Motion of Wheel and Work in Conventional Internal Grinding [5]

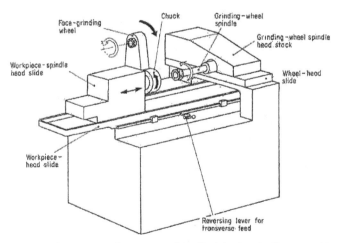

Fig. 3.37. Basic Design of an Internal Cylindrical-Grinding Machine [1]

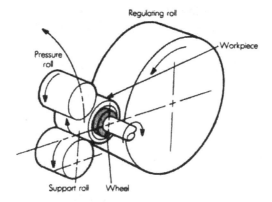

Fig. 3.38. Internal Centerless Grinding [5]

For the grinding of flat surfaces and profiled components, surface grinding techniques are used with surface grinding machines (Fig. 3.39). The design of these machines includes a table, which moves longitudinally and is mounted on a cross slide. The table drives can combine electromechanical and hydraulic units, variable DC motors or linear drives.

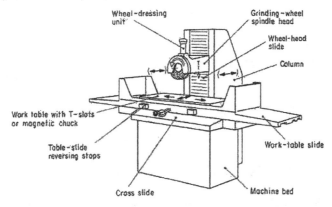

Fig. 3.39. Basic Design of a Peripheral Surface Grinding Machine [1]

A very important element in any grinding machine design is the spindle, which supports the grinding wheel. Different designs are used depending on the required grinding wheel speed and torque. Hydrodynamic bearing systems are typically used for medium speeds and large grinding wheel diameters, while rolling bearings are used for higher rotational speeds and smaller grinding wheel diameters. For very high rotational speeds the use of hydrostatic nut and spindle is very promising, due to their low backlash, little thermal losses and high stiffness. Linear direct drives are increasingly used by the machine tool builders for infeed movement at precision grinding machines. In internal grinding machines, the drives of the spindle included a belt drive in the past, however, nowadays a built-in motor or an air turbine is used, which provides the high speeds required for internal grinding, whilst they also provide reduced vibrations on the grinding wheel spindle. Centerless internal grinding machines utilize high speed spindles with magnetic bearings, capable of rotating at 110.000 rpm [6]. In cylindrical grinding machines, the workpiece is supported by a spindle and its head stock, which provides the necessary clamping and positioning of the workpiece as well as the circular feed motion. Typically, for this motion, a separate drive, such as a variable-speed DC motor is used.

The accuracy of a ground workpiece depends largely on the profile of the grinding wheel, which, however, wears continuously during the grinding process. To address this problem periodically or continuously, the grinding wheel is *dressed* (Fig. 3.40) with a variety of devices and mechanisms. Dressing is particularly important for profiled wheels, which "transmit" their profile onto the surface of the workpiece. With these, an NC wheel dressing unit, or a "crushing" unit can be used, where a diamond-impregnated roller literally crushes on the grinding wheel and imparts its profile on it. In turn, the grinding wheel imparts this profile to the workpiece (Fig. 3.41). Similar to dressing operation is the *truing* of a grinding wheel that is performed in order to create a concentric round wheel to the axis of wheel rotation.

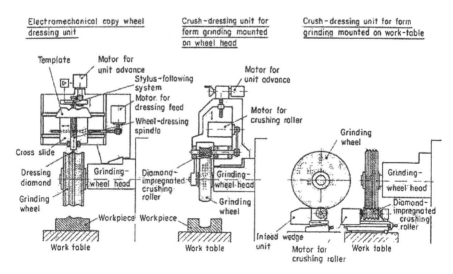

Fig. 3.40. Wheel Dressing Unit for Profile and Form Grinding [1]

Fig. 3.41. Profile gear CBN grinding [102]

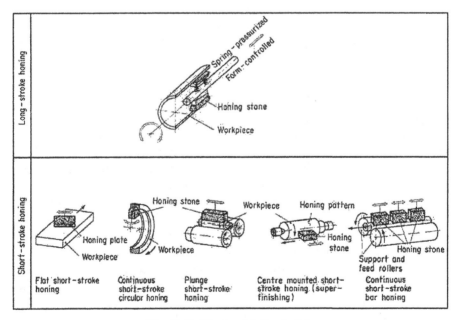

Fig. 3.42. Honing Methods [1]

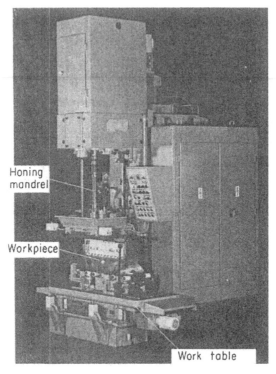

Fig. 3.43. Long-Stroke Honing Machine [1]

Honing (Fig. 3.42 and Fig. 3.43), *lapping* (Fig. 3.44 and Fig. 3.45) and *super-finishing* (Fig. 3.46) are processes which concentrate on surface finish and dimensional accuracy, but cannot address dimensional errors or eccentricity. Honing involves area contact of the abrasive surface rather than line contact [7]. Due to the rotating and oscillating motion of the honing tool, with respect to the workpiece, the resulting surface finish is cross-hatched. This surface formation is ideal for retaining a lubrication film. Lapping results in even finer surface finish than honing does. It utilizes fluid suspension of very small abrasive particles between the workpiece and the lap. The lapping process is a very slow process and should be utilized for finishing parts, requiring very smooth surface quality, such as optical components, ball bearings, gages, etc. Finally, super-finishing is similar to the honing process but it utilizes lighter pressures, more rapidly reciprocating motion and a flood of coolant [7] (fig. 3.46) resulting in optimum surface finish of the order of 0.1 to 0.01 μm. Honing, lapping and super-finishing are performed on specially designed machines.

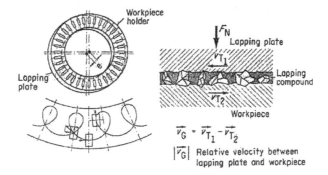

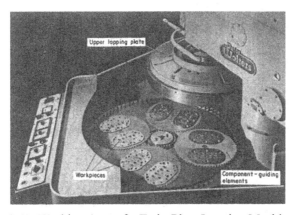

$$\vec{v_G} = \vec{v_{T_1}} - \vec{v_{T_2}}$$

$|\vec{v_G}|$ Relative velocity between lapping plate and workpiece

Fig. 3.44. Lapping process and relative motion of work between lapping plates [1]

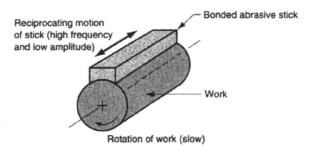

Fig. 3.45. Working Area of a Twin-Plate Lapping Machine [1]

Reciprocating motion of stick (high frequency and low amplitude)

Bonded abrasive stick

Work

Rotation of work (slow)

Fig. 3.46. Super-finishing operation

3.2.3 Laser Machines and Equipment

Present-day lasers can provide very high levels of power per unit area, while beam spot can be comparable in area with the square of the wavelength [8, 9, 10, 11 and 12]. For comparison, Figure 3.47 presents the power densities, in W/cm^2, for various energy sources.

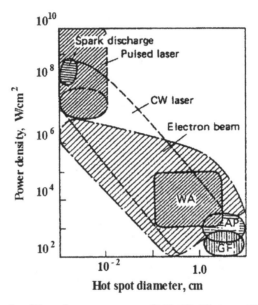

Fig. 3.47. Power densities of energy sources [13]. Welding arc (WA); arc plasma (AP); gas flame (GF)

Many applications of lasers in industry (Fig. 3.48) have proved that they are reliable, safe to operate, and simple to control. They can emit as high power in a pulse as 10^7W or more, and can process materials at an extremely high rate. A typical repetition rate varies from 0.05 to 100.00 kHz and the average power reaches 20 to 50 W with a maximum lasing efficiency close to 10% for CO_2 lasers and up to 50% for Nd: YAG lasers. These two lasers are the most popular industrial ones. There are also a few other laser sources used for industrial reasons which are used for special operations. *Excimer* lasers, for example, have extremely short pulse widths and very high peak power. Increasing lasing efficiency and power is a prerequisite for increasing quality and capacity in laser material processing.

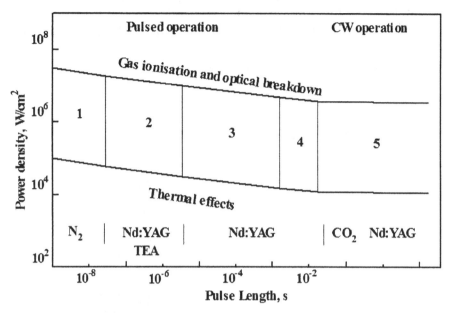

Fig. 3.48. Use of lasers for heat treatment processes: 1) thin film vaporisation; 2) scribing and film vaporisation; 3) hole drilling and punching; 4) spot welding and surface heat treatment; 5) deep melting, gas-assisted cutting, surface hardening, and splitting. TEA, transversely excited atmospheric pressure laser [13]

Laser Subsystems

An industrial laser system is composed of specific subsystems: the *laser* and its ancillary equipment, a *beam delivery* system, a *work fixture* with motion systems and associated *controller*, and a feedback system to close the process control loop. In the following paragraphs the most significant subsections will be briefly presented.

Beam delivery system for CO_2 Lasers

A *beam delivery system* leads the laser beam from the *resonator*, the CO_2 beam source, to the machine tools, by using optical systems. Typical optics for beam delivery systems include *telescopes (collimators)* for changing the beam diameter and allowing long delivery paths (more than 15m), *deviation mirrors (beam bending units)* for redirecting the incoming beam and *beam splitters* for dividing the laser beam. A beam switch leads the laser beam either to the turning machine or to the milling machine. The beam guiding system is protected by a *beam protection tubing* so as to prevent the exposure of laser radiation and further protect the laser optics and the laser beam from atmospheric contamination. A typical beam delivery system is shown in Fig. 3.49.

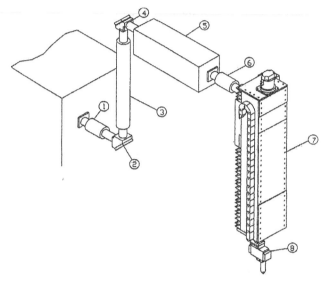

Fig. 3.49. Typical beam delivery system [12]

The beam delivery system is responsible for the exact positioning of the laser beam and thus, for the level of accuracy and precision, achieved by this machine. Therefore, it is necessary to design a system that would have high precision and would not be sensitive to vibrations. Inherent system deformations can induce positioning errors. The bending units consist of Cu mirrors that reflect approximately 99% of the irradiated laser. When the incoming laser beam exceeds 2kW, the energy absorbed by the mirrors is substantial and results in raising the temperature of the optics and thus, induces thermal deformations. Therefore, the optics has to be directly cooled, usually with water flowing through coolant channels. The pressure of the cooling water itself, however, causes deformations as well. Experimental examinations have shown that the deformations of the optical systems, during processing, are in the range of 10 μm and higher. These values are much higher than those of the manufacturing quality of the optics, around 0.01 μm [14, 15, 16 and 17]. Although a deformation of a few micrometers for one mirror seems to be very low, the deformations of all optics in a beam guiding system together, result in a misalignment of the focal point of the laser beam on the workpiece, in the order of 0.1 mm, which is obviously not acceptable for precision manufacturing. It is possible, though, to reduce the thermally induced deformations of the mirrors, by a special design of the cooling system (Fig. 3.50). The precision of the beam delivery system can be optimized by using *adaptive mirrors* that can change their surface curvature and thus, compensate for the displacement of the focal position, due to the thermal load of the optical components [12] (Fig. 3.51).

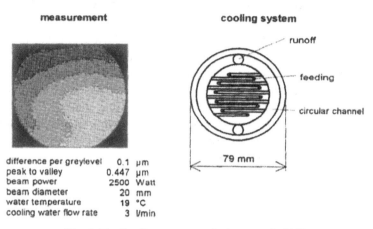

measurement   **cooling system**

difference per greylevel	0.1	µm
peak to valley	0.447	µm
beam power	2500	Watt
beam diameter	20	mm
water temperature	19	°C
cooling water flow rate	3	l/min

Fig. 3.50. Cooling system of a laser optic [18]

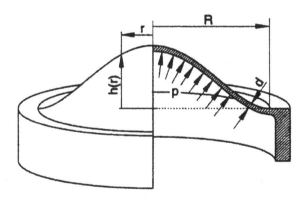

Fig. 3.51. Schematic of working principle of hydraulic adaptive mirror [12]

YAG-lasers with fiber glass delivery

Applications, such as hardening, cannot be realized with CO_2 lasers without having deposited an absorption enhancing coating on the workpiece in advance. Drilling and caving applications require pulse lasers with high peak powers. Nd: YAG-lasers with a shorter wavelength than CO_2 lasers offer these attributes as well as higher efficiency, compact size, higher reliability and higher absorption to perform a laser hardening without additional coatings. Furthermore, the laser beam of Nd: YAG-lasers can be delivered via *glass fibers* so as to allow the integration of a laser "tool" on robotic arms.

The Nd: YAG-lasers with a glass fiber delivery system available (Fig. 3.52) cover the whole range of applications: from fine cutting up to welding and surface treatment. Single-mode fibers, with diameters of 200 to 500 μm, can be used for low and medium power Nd: YAG-lasers (50-500 W) in pulsed or Q-switched mode with more than 50 kW pulse peak powers. For higher average powers, the fiber diameter has to be increased up to 600 μm. Furthermore, for delivering high power beams (today up to 4000 kW), multi-mode fibers should be used. Additionally, multi-mode fibers enable the use of simple optics for coupling the laser to fiber and the coupling efficiency can reach 80%.

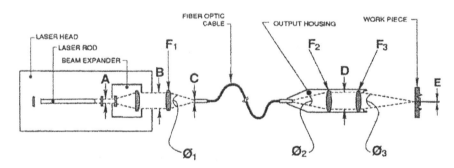

Fig. 3.52. Fiber optic delivery and focus system [12]

2D laser processing systems

The 2D laser systems (Fig. 3.53) are used mainly for cutting, welding and hardening operations. The first industrial laser systems, the so called *laser machining centers*, were occupied with a fixed beam delivery system. The workpiece was positioned relatively to the laser spot. Typical applications were turbine blade drilling, contact welding, etc. Nowadays, the laser systems available, can be used either as flexible manufacturing cells or as fully automated components of a production chain. The current 2D systems can be categorized according to the laser source used and whether or not the relative motion of the laser spot to the workpiece, is provided by the laser head, the bed or a combination of these [12]. The existing CO_2 laser systems can cut efficiently mild steel of thickness up to 22 mm and aluminum up to 12 mm [103].

Fig. 3.53. Laser cutting system [103]

3D laser processing systems

Although lasers can be easily and accurately manipulated with optical components, only in the last years have the processing and producing three-dimensional shapes become efficiently. The development of advanced optical scanning systems, in combination with laser sources of increased power; quality and overall efficiency paved the way for the so-called 3D laser materials' processing [19].

The 3D laser processing systems can be classified according to the intended process in four main categories:

- 3D laser machining with two converging laser beams,
- Laser assisted machining (LAM),
- 3D remote laser processing and
- 3D laser processing with multi-Axes laser head

The primary aspect of the *3D laser processing with two converging laser beams* (Fig. 3.4) is the laser grooving process [8, 20 and 21]. Each beam creates a groove in the workpiece through single or multiple passes. A volume of material is removed when the two produced grooves intersect. Turning, milling and gear cutting operations can be accomplished with the use of this system.

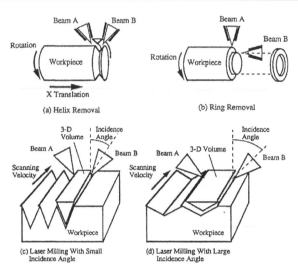

Fig. 3.54. Three dimensional laser machining using two converging laser beams
[8]

Laser Assisted Machining uses the laser beam only as a supporting or assisting tool in a conventional material removal process so as to enable the processing of materials of poor machinability [22]. The laser beam heats the material directly in front of a single point cutting tool (Fig. 3.55), reducing this way, the yield strength of the material, in the outer layer of the workpiece, to a value below the fracture strength. This allows visco-plastic flow, rather than brittle fracture during machining, without fusion or sublimation of the material [23 and 24].

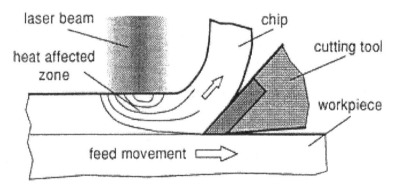

Fig. 3.55. Laser Assisted Machining [8]

The principle of 3D remote laser processing is based on the implementation of high quality laser sources and optical assemblies that combine lenses with relatively large focal lengths (up to 1.600 mm), along with one or two (Fig. 3.56) bending mirrors [19]. The most significant applications that utilise the 3D remote application include remote keyhole laser welding, laser hardening, laser alloying, laser soldering and laser sintering.

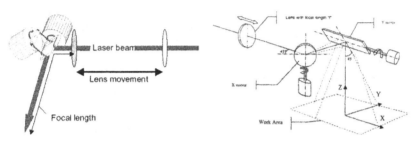

Fig. 3.56. Principle of remote laser processing [19]

The concept of *3D laser processing with multi-axes laser heads* is based on the combination of optical assemblies with special motion controlling systems that can provide up to 5 DOF in the laser head. Combined with workpiece movement, up to 10 axes of relative motion, between the workpiece and the laser head, can be provided. Lasers can be coupled on the end effector of a robotic arm. For the case of CO_2 lasers this coupling is achieved through the use of telescopic articulated arms, whereas in the case of Nd: YAG lasers the beam is delivered through fibers. Robotic arms coupled with laser beams are widely used in the automotive industry for precision welding and cutting operations.

3.2.4 Additive Processes Machines

The processes that rely on additive methods are known as Rapid Prototyping or Solid Free Form Fabrication methods (Chapter 2). As it has already been mentioned, the RP technologies are used to generate 3D solid models for prototyping and tooling reasons. A number of different processes exist, and epigrammatically are these of Stereolithography, Fused Desposition, Laminated Object Manufacturing, Selective Laser Sintering, 3D printing, LENS etc. In this section, the most widely used machines will be briefly presented.

Stereolithography systems

The Stereolithography systems [104 and 25] are the most widely used systems today, since they can introduce solid imaging for building small and detailed precision parts. Applications are limited production runs, for rapid tooling, master patterns for investment casting, form and fit as well as for function testing.

In general, Stereolithography systems use Nd: YVO$_4$ lasers, with power up to 800mW for the SLA7000 system and minimum beam diameter of 0.075 mm for Viper si2 system. The re-coating system comprises of a Zephyr blade capable of producing minimum build layer of 0.1 mm. The maximum part drawing speed, depending on the small or large laser spot size, varies between 2.54 m/sec and 9.52 m/sec. The maximum build envelope can reach 508 x 508 x 600 mm for SLA 7000 (fig. 3.57). The materials used by the Stereolithography machines are epoxy-based photocurable resins.

Fig. 3.57. SLA 7000 series

Fused deposition modeling systems

The latest fused deposition modeling (FDM) systems can deliver large, highly accurate and dimensionally stable parts in reduced time [s17]. Technological advances, such as high-speed electro-magnetic drives, have been introduced so as to move and control the extrusion heads. The electro-magnetic motion control unit eliminates bearings, cables, and gears, resulting in maximum building speeds of 7 in/sec. Various machines with different working envelopes are available. The maximum build chamber of 600x500x600 mm is attributed to the Stratasys Maxum system whereas, all available machines can produce models with an accuracy of ± 0.127 mm in their "fine" operation mode.

Fig. 3.58. FDM Titan

The FDM process uses a spool-based filament system to feed the material into the machine thus facilitating its being quickly changed. Material options currently include ABS, investment casting wax, Polycarbonate (PC) and Polyphenylsulfone.

Laminated object manufacturing systems

Laminated object manufacturing (LOM) systems provide one of the largest working envelopes in rapid prototyping industry [106]. Since Helisys folded in 2000, the LOM machines are developed and manufactured by Cubic Technologies Inc. The servo-based X-Y motion control system and new software algorithms result in reduction in the part build times. They comprise a 50W CO_2 laser system, producing a beam diameter between 0.203 and 0.254 mm. Working envelope size is 813x559x508 mm and the maximum cutting speed is 457 mm/sec. The material used by LOM systems is thin sheets of paper so as to build wood-like components. However, with slight modifications the LOM systems can also handle plastics, polymer composites prepregs, metals, ceramics and even ceramic matrix composites [26, 27]. The parts produced can be used for sand casting, silicon rubber molding, thermoforming, spray metal tooling, plaster and spin casting, investment casting, and epoxy tooling for injection molding.

Fig. 3.59. LOM-2030H

Selective laser sintering systems

The Selective Laser Sintering (SLS) process creates solid 3-D objects, from plastic, metal or ceramic powders that are "sintered" or fused using 25 or 100 Watt CO_2 laser with a maximum scanning speed of 10,0 m/sec. The working envelope size is 380x330x447 mm. A number of plastic-based powders are used to produce functional models having excellent mechanical integrity, heat resistance, and chemical resistance. They can often be used for testing, in environments, similar to those intended for the final product. The SLS process can be used for small batch prototype manufacturing [107].

Fig. 3.60. 3D Systems SLS Vanguard HS

3D Printers apparatus

The 3D Printing technology is the most promising Rapid Prototyping technology [108]. The building speed of the 3D printers may vary between 25 and 50 mm/hour. The maximum build volume for Z810 System is 500 x 600 x 400 mm. The layer thickness available, can be either 0.076 mm or 0.254 mm. The system prints colored complex parts layer by layer without any support structures. It uses low cost, environmentally friendly materials (scratch-based, plaster-based or composite based powders) that can be easily used in an office environment. The final parts can be further infiltrated with waxes or resins so as to enhance their strength, improve their temperature resistance and increase their durability.

Fig. 3.61. Z406 3D printer

3.3 Machine Tool Design and Analysis

Machine tool design, in terms of overall structure and kinematics, is dictated primarily by the process to be performed by the machine. The design is also substantially influenced by the range of workpiece sizes, the required accuracy, operation mode (manual, semi-manual, fully automated, etc.), maintenance procedures, cost, and a host of other factors, related to the requirements and attitudes of the machine end user. Since machine tools are complex electromechanical structures, which include many interconnected devices, their design involves a number of disciplines ranging from applied mechanics to ergonomics. This unique level of complexity, coupled with strict requirements, in terms of kinematic accuracy, static and dynamic behavior, etc., causes the machine tool design process to rely heavily on empirical knowledge and expertise. Engineering analysis is typically confined to the identification of trends and the creation of a framework for machine design, instead of delivering exact solutions for the structure of the machine, the size of its components, etc. In general, stand-alone machines are designed to accommodate a variety of conditions in terms of workpiece size, etc., while dedicated machines, such as transfer lines are designed for reliability rather than flexibility.

Examining the historical development of machine tool designs and their slow evolution, one must consider two factors: first, machines require a large initial capital investment, and second the production and ultimately the revenues of the end user depend on the machine's proper functionality. An experimental, unproven machine design usually involves the high risk of not achieving the required production quotas, which may explain the conservative attitude of many machine tool builders throughout the world, in terms of introducing radically new ideas to their machine tool designs. On the other hand, the lack of an innovative approach to machine tool design may lead to a loss of competitive advantage. The correct attitude in machine tool building is to strike a balance among new ideas for the future and well-proven designs of the past. Furthermore, since the machine tool design substantially influences the production capabilities of the end user, there is need for close cooperation between the end user and the machine tool builder.

The remainder of this section provides an overview of design issues and some analytical approaches, which, as mentioned earlier, are not used so as to identify solutions for machine tool design problems, but rather in order to present overall trends and provide a general framework for their solution. The discussion is focused on the three major elements of any machine tool: *frames*, which are load carrying bodies supporting the individual elements and devices of the machine; *guideways and bearings*, which support and guide the motion of the different machine elements; and *drives and transmissions*, which provide the motions of the machine and the co-ordinated movements between tool and workpiece that produce the shape, contour or form of the workpiece.

3.3.1 Frames

Machine tool frames are typically modular structures consisting of a number of elements such as *baseplates*, *beds*, *columns*, and *crossbeams*, whose size, shape and material depend primarily on:
- the position and length of the moving axes of the machine;
- the direction and magnitude of the anticipated process forces;
- accessibility and safety of the working space,
- workpiece characteristics such as the initial raw material geometry, the needed tools, the debris that will be generated, etc.,
- manufacturability and cost of the machine and last but not least,
- environmental considerations as well as energy consumption minimization.

A structural element in order to be used in a machine tool should meet a number of requirements such as:
- mechanical strength,
- static stiffness,
- dynamic stiffness,
- wear strength,
- thermal stability and
- lightness.

These requirements can be achieved through the proper selection of materials and the thorough designing of the frame shape.

Material used in the construction of the machine tool frame

Traditionally, machine tool frames were manufactured from four basic materials (soft steel, carbon steel, machinery iron and bronze), whereas nearly fifty materials are used nowadays (*steel, cast iron, aluminum, copper, brass, titanium, composites, ceramics, polymer concrete,* etc.). None of the materials available, does it present superiority to all the requirements, considered for machine tool frames. For example, steel presents the best mechanical characteristics such as strength and stiffness, whereas, polymeric concrete that presents the least mechanical characteristics, has the best internal damping behavior. In the same context, polymeric concrete is superior compared with other available materials as far as internal stresses are considered. Finally, cast iron presents superiority in relation to the wear behavior due to its microstructure and hardness.

Since no material is ideal, from all points of view, and after considering economical factors dealing both with the material and the manufacturing of the frame elements, the state of the art of the material utilisation for machine tool frames can be summarised into [7]:

- Cast iron is most frequently used in small to medium size machine tools for material removal processes. Its combined characteristics of strength, stiffness, damping, thermal stability, forming and cost, make cast iron the best solution in most of the cases.
- Welded structures are used whenever cost saving is critical, the material damping requirements are not critical and large machines are considered.
- Polymeric concretes have a limited application to structural elements of foundation (bed) in small and medium size machines.
- Machine tools, utilizing ceramic elements exclusively, have been developed so as to present superior performance; however, their cost renders them prohibitive. It is likely though, that in future, more and more machine tools will be adopting ceramic components.
- Composite machine tool structures have been presented, resulting in weight reduction and increased dumping capacity [28].

Frame shape

In deciding on the shape and design of the machine tool frame, there is typically a tradeoff among the different factors influencing the frame. If, for example, accessibility of the working space is an important factor, then a *C-shaped* frame (open section frame) is preferable to an *O-shaped* frame (closed section or bridged or portal frame) (Fig. 3.62). The lack in symmetry of the C-shaped frame constitutes it vulnerable to thermal gradients and bending moments, which result in large displacements, making it therefore, not ideal for precision machines. A O-shaped frame provides better stiffness and is used for large machines, but reduces the accessibility of the workspace. These two frame shapes are most frequently found in machine tools; however, there are some "exotic" frame formations providing unique characteristics, such as 6 DOF. Such frames are the *Tetrahedral frames* [7] and the *Hexapod frames* used for material removal processes as well as for CMMs, which nevertheless, are still in the development phase.

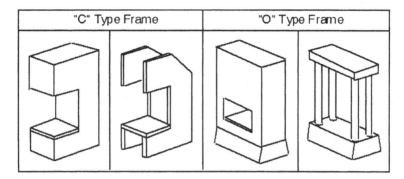

Fig. 3.62. "C" and "O" Types of Machine Frames

The frame geometry is dependent on a number of factors, such as the degree of automation adopted and the generated debris. For example, in order to remove more easily the chips during turning operations, inclined bed (Fig. 3.63) is preferred for most CNC lathes.

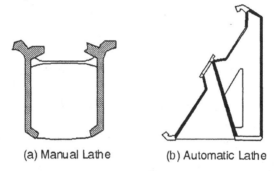

(a) Manual Lathe   (b) Automatic Lathe

Fig. 3.63. Bed Construction for (a) Manual and (b) Automatic Lathes

In selecting materials, shapes and sizes of the different frame elements, the designer must keep in mind that the goal is to reduce unwanted displacement between the tool and workpiece during the manufacturing process. During the manufacturing process, the machine frame is exposed to process forces, which vary with time, and tend to move the workpiece relatively to the tool. This motion has an unwanted effect on workpiece accuracy, and is to be restrained. In the following, the behavior of the machine frame under static, dynamic and thermal loading will be briefly discussed in order to provide a framework for understanding some of the solutions encountered in machine tool design. One must bear in mind, however, that the analytical approaches, described below, provide an order of magnitude analysis of frame design problems since, due to the joints that connect the different machine tool elements, frames are highly non-linear structures.

Static Behavior

The static behavior of a machine tool is governed primarily by the process forces and secondly by the masses of the workpiece and machine components. The static analysis of a machine tool usually begins with an analysis of the force flux through the machine (Fig. 3.64). The forces exerted by the tool on the workpiece, are successively transmitted to the machine tool table, then to the slides, and finally to the foundation of the machine. Forces acting on a machine tool [1] will cause deflection of all machine components (Fig. 3.65); all these deflections summed up, present the maximum geometric error.

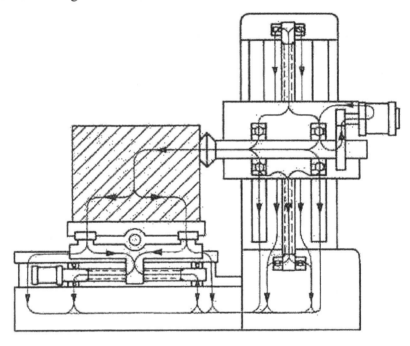

Fig. 3.64. Force Flux in a Horizontal Milling Machine [1]

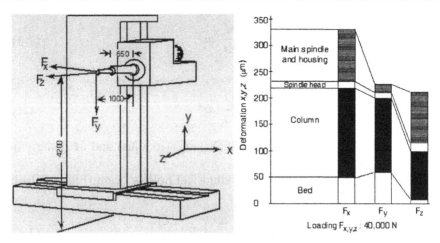

Fig. 3.65. Forces and Deflection on a Horizontal Milling Machine

Any machine tool frame will deform, to a certain degree, under static load. The stiffness of the machine is a measure of the magnitude of the deformations. The most frequent modes of machine tool loading are bending and torsion. The resistance to bending is related to the second moment of area (moment of inertia) about the neutral axis, while the resistance to torsion is related to the polar moment of the area [29]. The second moments of area for a rectangular beam model (Fig. 3.66a) are determined as:

$$J_x = \int_A y^2 dA \qquad\qquad (3\text{-}1)$$

$$J_y = \int_A x^2 dA \qquad\qquad (3\text{-}2)$$

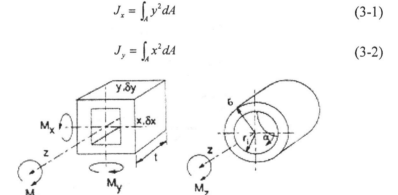

Fig. 3.66. Rectangular (a) and Cylindrical (b) Beam Models

The deflections δx and δy (at $z=t$) in the x and y directions resulting from an applied bending moment are found using the following equations [29]:

$$\delta x \big|_{z=l} = \int_0^l \int_0^l \frac{M_y}{EJ_y} dz^2 \qquad (3\text{-}3)$$

and

$$\delta y \big|_{z=l} = \int_0^l \int_0^l \frac{M_x}{EJ_x} dz^2 \qquad (3\text{-}4)$$

where E is the modulus of elasticity (Young Modulus) and M is the moment applied to the beam.

The polar moment of area for a cylindrical hollow beam (Fig. 3.66b) is determined as [29]

$$J_T = \frac{\pi r_0^4}{2}\left(1 - \frac{r_i^4}{r_0^4}\right) \qquad (3\text{-}5)$$

The angle of twist resulting from an applied torsional moment M_Z is then [29].

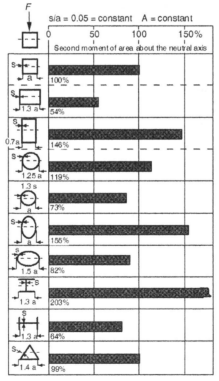

Fig. 3.67. Second moment of area for different cross sections

$$a\big|_{z=l} = \int_0^l \frac{M_z}{EJ_T}dz \qquad (3\text{-}6)$$

The second moment of area (Fig. 3.67) and polar moment of area (Fig. 3.68) depend on the beam cross-section. Since machine tool frames are exposed to both bending and torsional loads, a method of obtaining the required stiffness of the frame elements, without significantly increasing the weight of the structure, is to add ribbing (Fig. 3.69). However, ribbing increases the cost of the machine and should be applied with restraint [1].

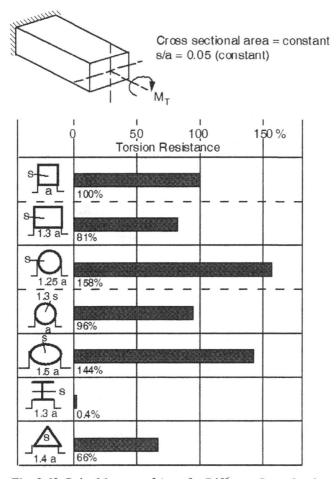

Fig. 3.68. Polar Moment of Area for Different Cross Sections

The methods employed for joining the elements of a machine tool can have a great influence on the performance of the machine. The dynamic loads developed during the operation of the machine propagate through the joints to the different elements of the machine tool. Deflections of the joints can have a significant influence on the accuracy and surface quality of the parts, produced by the machine. Since joints can account for up to 90% of the total deflection of machine tool structures, the design goal is to maximize the stiffness of the joints (Fig. 3.70). Since the force-deflection characteristics of joints are highly non-linear they are generally difficult to analyze.

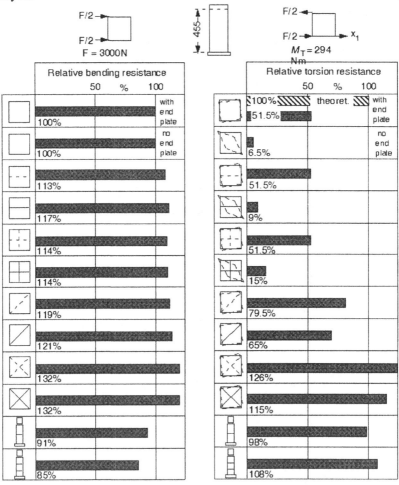

Fig. 3.69. Ribbing of Machine Tool Columns

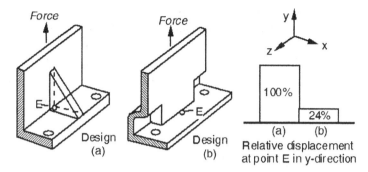

Fig. 3.70. Stiffness Comparison of Two Joint Designs

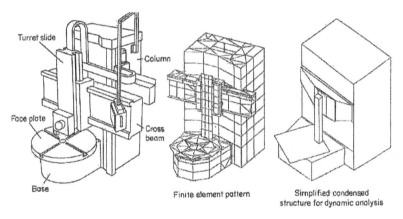

Fig. 3.71. A Finite Element Model of a milling machine [1]

Numerical techniques, such as finite element analysis (FEA), are frequently employed in analyzing machine tool frames. The first step in applying finite element analysis is to approximate the geometry of the structure using elements, which have simple boundaries and easily definable dimensions (Fig. 3.71). The expected locations of the structural deformations should be considered during this geometric approximation. There are several types of elements, which can be used in the finite element analysis of machine tool frames (Fig. 3.72). After the geometric model of the machine tool has been completed, a discrete mathematical model is developed relating the machine tool deformations with the applied external loads. These equations are based on Hooke's Law:

$$\{F\} = [K]\{U\} \tag{3-7}$$

In this relationship, $\{F\}$ is the vector containing the applied forces and moments; $[K]$ is the stiffness matrix, and $\{U\}$ is the displacement vector. In order to provide a basic understanding of how finite elements are used in context of machine tool analysis, the derivation of the stiffness matrix for a single truss element will be discussed briefly (Fig. 3.73). The deformation behavior between the two nodes of the truss can be expressed by a linear polynomial equation:

$$u_x(x) = a_0 + a_1 x \tag{3-8}$$

From the displacement of the nodes, the truss displacement can be determined at any location x, according to the following relationship.

$$u_x = \left[1 - \frac{x}{L} \quad \frac{x}{L}\right]\left\{\begin{matrix} U_1 \\ U_2 \end{matrix}\right\} \tag{3-9}$$

The strain of the element can be determined through the partial differentiation of the displacement relationship, which in the case of the depicted truss, gives the following relationship.

$$\varepsilon_x = \left[-\frac{1}{L} \quad \frac{1}{L}\right]\left\{\begin{matrix} U_1 \\ U_2 \end{matrix}\right\} = [b]^T \{U\} \tag{3-10}$$

Assuming that the displacement is elastic, the stress-strain (σ-ε) relationship is governed by Hooke's Law.

$$\{\sigma\} = [H]\{\varepsilon\} \tag{3-11}$$

where $[H]$ is the modulus of elasticity matrix. Thus, the stiffness matrix can be derived from the principle of virtual work, which states that the external work dW_a equals the internal work dW_i. The external work is the product of the external load F and the actual displacement U whereas the internal work results from the stress and the actual strain. Using Equations (3-10) and (3-11) the following integral for internal work is obtained.

$$\delta W_a = \delta\{U\}^T\{F\} = \delta W_i = \int_V \delta\{U\}^T[b]^T[H][b]\{U\}\,dV \tag{3-12}$$

From Equation (3-12), the stiffness matrix [K] for this element is derived to be

$$[K] = \frac{AE}{L}\begin{bmatrix} 1 & -1 \\ -1 & 1 \end{bmatrix} \tag{3-13}$$

where A is the cross sectional area of the truss.

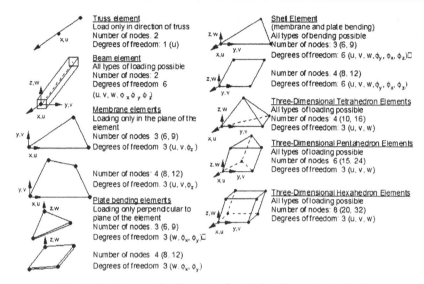

Fig. 3.72. Geometric Elements for Finite Element Analysis

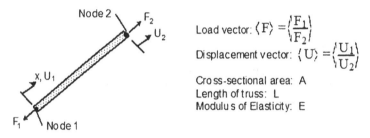

Fig. 3.73. Analysis of a Truss Using Finite Elements

FEA is a widely used powerful tool for machine design. FEA is typically available in a commercially available software package, used in personal computers [109, 110, 111 and 112]. A thorough treatment of FEA and its theoretical underpinnings are provided in [30].

Dynamic Behavior

The machine, the tool, and the workpiece form a structural system with complicated dynamic characteristics. Vibrations occur because of the interaction between the cutting process and the machine tool structure. These vibrations may be classified [31] as:

- *Free or transient vibrations* resulting from intermittent impulses transferred to the structure from rapid reversals of reciprocating masses (such as machine tables), or from the initial engagement of the tools. The structure is deflected and oscillates in its natural modes of vibration until the damping of the structure causes the vibration to diminish.

- *Forced vibrations* resulting from periodic forces within the system, such as unbalanced rotating masses, eccentricity of rotating tools (for example grinding wheel) or the intermittent engagement of multitooth cutters (milling). Forced vibrations may also result from hydraulic devices integrated into machine tools as well as from vibrations, transmitted through the foundation from nearby machinery. The machine tool oscillates at the forcing frequency, and if this frequency corresponds to one of the natural frequencies of the structure, the machine will resonate.

- *Self-excited vibrations* usually resulting from a dynamic instability of the process. This phenomenon is commonly referred to as machine tool *chatter* and the structure will oscillate in one of its natural modes of vibration.

It is important that vibrations of the machine tool structure be limited, as their presence results in poor surface finish, tool damage, undesirable noise, and machine component fatigue. While the causes and methods of controlling free or forced vibrations are generally well understood and these types of vibration can generally be avoided, chatter is less easily controlled.

The dynamics of a single mass vibrator, including mass, spring and damper system, are discussed briefly in order to illustrate the effects of the design parameters, such as mass, stiffness, and damping on the dynamic response of the system (Fig. 3.74).

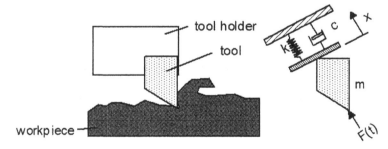

Fig. 3.74. Spring-Damper System Used for Chatter Modeling

The response of a dynamic system is the sum of its transient response, $x(t)_{trans}$, (free response) plus the response due to an external excitation, $x(t)_{forced}$, (forced response).

$$x(t) = x(t)_{trans} + x(t)_{forced} \tag{3-14}$$

For a single degree of freedom system (Fig. 3.74) the equation of motion is [32]:

$$m\ddot{x}(t) + c\dot{x}(t) + kx(t) = F(t) \tag{3-15}$$

where m is the mass, c is the viscous damping coefficient, k is the spring stiffness and $F(t)$ is the time varying excitation force.

Free Response

The free response is determined from the homogenous equation:

$$\ddot{x}(t) + \frac{c}{m}\dot{x}(t) + \frac{k}{m}x(t) = 0 \tag{3-16}$$

Introducing the non-dimensional *viscous damping ratio*,

$$\zeta = \frac{c}{2\sqrt{km}} \tag{3-17}$$

and the natural frequency of the system given by:

$$\omega_n^2 = \frac{k}{m} \tag{3-18}$$

Equation (3-16) can be rewritten as,

$$\ddot{x}(t) + 2\zeta\omega_n\dot{x}(t) + \omega_n^2 x(t) = 0 \tag{3-19}$$

The standard method for solving Equation (3-19) is by Laplace transformation, where a solution of the following form is assumed:

$$x(t) = Ae^{st} \tag{3-20}$$

where s is a complex variable which defines the Laplace domain. Inserting Equations (3-20) in (3-19) yields:

$$Ae^{st}\left[s^2 + 2\zeta\omega_n s + \omega_n^2\right] = 0 \tag{3-21}$$

From Equation (3-21) s can be determined as

$$s_{1,2} = \left(-\zeta \pm \sqrt{\zeta^2 - 1}\right)\omega_n \tag{3-22}$$

Three distinct possibilities exist for the values of ζ [32, 33]:

1. $\zeta > 1$ yields real roots and an over-damped system (Fig. 3.75a):

$$x_{trans}(t) = \left(c_1 e^{\omega_d t} + c_2 e^{-\omega_d t}\right) e^{-\zeta \omega_n t} \qquad (3-23)$$

where c_1 and c_2 are unknown coefficients to be determined from the boundary conditions and where ω_d is the frequency of the free damped vibration given by

$$\omega_d = \sqrt{1 - \zeta^2}\, \omega_n \qquad (3-24)$$

2. $\zeta = 1$ yields two identical roots and a critically damped system (Fig. 3.75b):

$$x_{trans}(t) = \left(c_1 + c_2 t\right) e^{-\omega_n t} \qquad (3-25)$$

3. $\zeta < 1$ yields imaginary roots and an under-damped system (Fig.3.75c):

$$x_{trans}(t) = A e^{-\zeta \omega_n t} \cos\left(\omega_d t - \psi\right) \qquad (3-26)$$

where ψ is the phase angle.

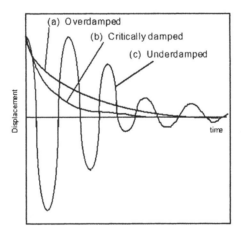

Fig. 3.75. Transient Response of (a) Overdamped, (b) Critically Damped and (c) Underdamped Single Mass Vibrator.

Forced Response

The forced response of a system describes the motion of the single mass vibrator under a specific type of loading. The most common types of loads encountered in machine tools are *impulses*, *step* function loads, and *periodic* or *harmonic* loads (Fig. 3.76). The first two types of loads are usually due to tool engagement, where a sudden increase in force is imposed on the machine tool structure and the structure oscillates in its natural modes until the motion decays from inherent damping.

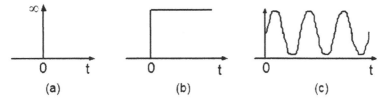

(a)   (b)   (c)

Fig. 3.76. (a) Unit Impulse, (b) Step, and (c) Periodic loads

The unit impulse is a pulse with an infinitesimal time duration that has a net area of unity. In practice, an input with a very short duration can be considered to be an impulse [32]. The unit impulse response of a single mass vibrator is given below (Fig. 3.77) [32].

For $\zeta<1$,

$$x(t)=\frac{\omega_n}{\sqrt{1-\zeta^2}}e^{-\zeta\omega_n t}\sin\left(\omega_n t\sqrt{1-\zeta^2}\right)\quad(t\geq0)\tag{3-27}$$

For $\zeta=1$,

$$x(t)=\omega_n^2 t e^{-\omega_n t}\tag{3-28}$$

For $\zeta>1$,

$$x(t)=\frac{\omega_n}{2\sqrt{\zeta^2-1}}\left[e^{-\left(\zeta-\sqrt{\zeta^2-1}\right)\omega_n t}-e^{-\left(\zeta+\sqrt{\zeta^2-1}\right)\omega_n t}\right]\quad(t\geq0)\tag{3-29}$$

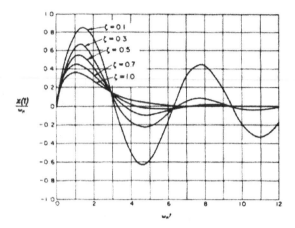

Fig. 3.77. Unit Impulse Response Curves [32]

The unit step response (Fig. 3.78) of the single mass vibrator is:

For $\zeta < 1$,

$$x(t) = 1 - \frac{e^{-\zeta\omega_n t}}{\sqrt{1-\zeta^2}} \sin\left[\omega_d t + \tan^{-1}\left(\frac{\sqrt{1-\zeta^2}}{\zeta}\right)\right] \quad (t \geq 0) \tag{3-30}$$

For $\zeta = 1$,

$$x(t) = 1 - e^{-\omega_n t}(1 + \omega_n t) \quad (t \geq 0) \tag{3-31}$$

For $\zeta > 1$,

$$x(t) = 1 + \frac{\omega_n}{2\sqrt{\zeta^2-1}}\left(\frac{e^{s_1 t}}{s_1} - \frac{e^{s_2 t}}{s_2}\right) \quad (t \geq 0) \tag{3-32}$$

where $s_1 = \left(\zeta + \sqrt{\zeta^2-1}\right)\omega_n$ and $s_2 = \left(\zeta - \sqrt{\zeta^2-1}\right)\omega_n$.

In designing machine tools, tool vibration should be minimized as soon as a force impulse or step has been encountered. However, the critical damping ($\zeta = 1$) conditions should not be exceeded since critical damping will minimize the detrimental effects of vibration and chatter, while returning the system to equilibrium conditions without overshoot or oscillation.

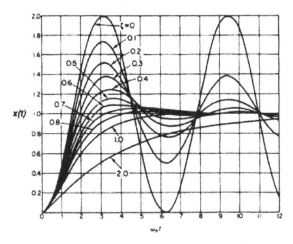

Fig. 3.78. Unit Step Response Curves [32]

The tool rotation and the reciprocation of other masses during the machining process may result in dynamic instabilities due to the periodic nature of the excitation. This type of load will cause unacceptable structural vibration when the forcing frequency coincides with one of the natural frequencies of the structure. This phenomenon is known as *resonance*. During the designing of a machine tool, it is critical to consider the dynamic response of the machine due to harmonic loading.

In order to better introduce the resonance notion, the response of a physical system owing to periodic excitations, is analyzed mathematically by considering the response of a single mass vibrator. The equation of motion for such a system is given by:

$$\ddot{x}(t) + \omega_n^2 x(t) = 0 \tag{3-33}$$

Introducing the general solution for the response Equation (3-20) into Equation (3-33), the following characteristic equation is obtained,

$$s^2 + \omega_n^2 = 0 \tag{3-34}$$

which yields two complex roots $\lambda_{1,2} = \pm i\omega_n$. This gives the following solution

$$x(t) = A_1 e^{i\omega_n t} + A_2 e^{-i\omega_n t} \tag{3-35}$$

where A_1 and A_2 are the constants of integration. For $x(t)$ to be real, A_1 must be the complex conjugate of A_2. This consideration allows the solution in Equation (3-35) to be expressed alternatively as a sinusoidal function.

Knowing that

$$e^{i\omega t} = \cos(\omega t) + i\sin(\omega t) \tag{3-36}$$

Equation (3-35) can also be written as,

$$x(t) = B_1 \sin(\omega_n t) + B_2 \cos(\omega_n t) \tag{3-37}$$

where B_1 and B_2 are constants of integration.

The above equations demonstrate that a loading situation of periodic nature can be expressed both as a sinusoidal function in time-displacement axis, and as a vector quantity in the complex plane. If a sinusoidal excitation force is assumed, the expression for the load can be written as:

$$F(t) = F_0 \cos(\omega_f t) \tag{3-38}$$

then the response of the system is given by:

$$x(t)_{force} = \frac{F_0 M}{K} \cos(\omega_f t + \phi_f) \tag{3-39}$$

where ω_f is the forcing frequency and K is the effective spring constant of the machine structure. ϕ_f is the phase angle which indicates how much the output lags the input, and is given by:

$$\phi_f = \tan^{-1}\left(\frac{-2\zeta\omega_f\omega_n}{\omega_n^2 - \omega_f^2}\right) \tag{3-40}$$

The amplitude or magnification ratio M is the ratio of the output to the input amplitude of oscillation (Fig. 3.79)

$$M = |G(i\omega)| = \left\{\left[1 - \left(\frac{\omega_f}{\omega_n}\right)^2\right]^2 + \left(\frac{2\zeta\omega_f}{\omega_n}\right)^2\right\}^{-\frac{1}{2}} \tag{3-41}$$

At the resonant frequency

$$\omega_r = \omega_n\sqrt{1 - 2\zeta^2} \tag{3-42}$$

the peak value for the magnification ratio is given by:

$$M_r = \frac{1}{2\zeta\sqrt{1-\zeta^2}} \quad \text{for } \zeta < \frac{1}{\sqrt{2}} \tag{3-43}$$

and

$$M_r = 1 \text{ for } \zeta \ge \frac{1}{\sqrt{2}}$$
(3-44)

The rather elementary equations presented above for harmonic excitation provide the basics for understanding the important dynamic characteristics of machine tool structures, arising from periodic loadings. Excessive vibrations can lead to poor part quality, machine component fatigue, and reduced tool life. Thus, harmonic excitation, as well as impulse and step responses need to be analyzed during the design phase in order to construct a machine tool that will produce quality parts reliably.

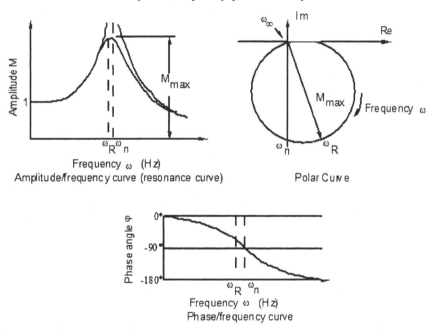

Fig. 3.79. Amplitude and Frequency Response for a Single Mass Vibrator

A *Polar Frequency Response Locus* (Fig. 3.79) can be used to graphically present the above information in a single diagram, which represents the system response in vector form. The equation of motion for a second order system, subjected to harmonic loading, can also be written as

$$\ddot{x}(t) + 2\zeta\omega_n \dot{x}(t) + \omega_n^2 x(t) = A\omega_n^2 e^{i\omega_n t}$$
(3-45)

using Equation (3-36), which yields a solution for the response in vector form:

$$x(t)_{force} = AMe^{i(\omega t + \phi)} \tag{3-46}$$

where A is a constant having units of displacement, M is the magnification, and ϕ is the phase angle.

Chatter

Chatter is caused by the dynamic interaction of the cutting process and the machine tool structure and can result in reduced machining accuracy and productivity. The magnitude of the cutting force, generated by the tool on the workpiece, depends largely on the feed and depth of cut. A disturbance in the cutting process (because of a hard spot in the work material, for example) will cause a deflection of the structure, which may alter the undeformed chip thickness and in turn alter the cutting force. The initial vibration may be self-sustaining and cause the machine to oscillate in one of its natural modes of vibration. Two effects can cause this instability: the *regenerative effect*, which is the dominant phenomenon, and the *mode-coupling effect* [31].

Regenerative Instability

Regenerative instability occurs when a disturbance in the cutting process generates a wavy surface on the workpiece. During successive passes of the tool, the undeformed chip thickness will depend on the current relative vibration of the tool (inner modulation) and the wave cut on the surface during the previous pass (outer modulation). Depending on the phase among these waves, the force variation may increase after successive passes of the tool and the vibration will be building up until it is interrupted by some nonlinearity, such as the tool leaving the workpiece during a part of the vibration cycle.

A lot of research is being conducted on chatter vibrations suppressing methods [34, 35]. These methods can be classified in the ones that prevent the onset of the chatter and the methods that monitor the process, the machine behavior, the time of the chatter's initiation and alter the process characteristics so as to suppress it. Examples of the former methods are presented hereafter. Milling cutters, with a variable pitch between successive teeth, can lead to improvements on stability. Such cutters result in different phasing among the waves, cut by successive teeth and can reduce the regenerative effect. A second way to achieve irregular pitch for helical slab milling cutters is to use different helix angles on successive teeth. Another method of altering the phase between successive waves, cut on the work surface, is by superimposing a variation in speed on the steady spindle rotation. Although it is difficult to be achieved in practice, this approach can be effective for processes with a slow buildup of vibrations, such as some grinding operations.

The chatter initiation can be identified through the monitoring of process quantities, cutting tool or workpiece [34]. Once the chatter has been initiated, a number of in-process methods can be adapted for suppressing it. For example in grinding processes, the grinding wheel unbalance can be detected with a vibration sensor and be balanced automatically. In general, the self initiated vibrations can be suppressed by modifying the process conditions, by increasing the dynamic stiffness of the mechanical system as well as by disturbing the regenerative effects [34].

Mode-Coupling Instability

Mode-coupling instability can occur even when successive passes of the tool do not overlap, such as in screw cutting, and results in an undesirable motion of the tool relative to the workpiece. This type of instability may happen when the structure has closely coupled modes of vibration. For example, the characteristics of the structure can be such that a free vibration of the structure could cause the tool to follow an elliptical path relative to the workpiece when the structure is disturbed [31].

Machine Tools as Multi-Mass Vibrators

The accuracy of the work produced on a machine tool, the wear behavior of the tools, and the ability of the machine tool to function well, are related to the behavior of the machine tool frame under dynamic loads. Since machine tools are assemblies of individual units, they can be regarded as multi-mass vibrators with respect to their dynamic behavior. By neglecting the effect of joints, a machine tool structure could be regarded as a system of unconnected single mass vibrators. It is possible, therefore, to simulate the behavior of these structures to a certain extent, by using different masses, springs, and damping elements that are connected together; although in practice, choosing the appropriate values to represent each element, is difficult.

As the frequency of the exciting force is varied, the vibrations of a machine tool will have the following characteristics [31]:

• A complex machine structure could exhibit several resonances or natural frequencies and the frequency response curve would have several peaks corresponding to each resonance (Fig. 3.80).

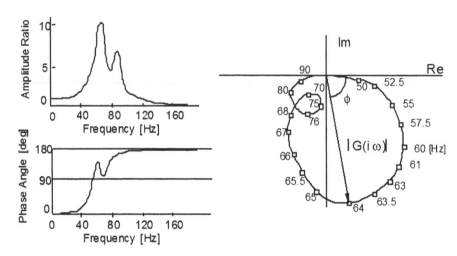

Fig. 3.80. Frequency Response of a Typical Machine Tool Structure

- At each resonant frequency the contribution of the various elements of the structure to the overall response will vary. Some parts will move with large amplitudes at some frequencies, but other parts will hardly move. Each resonance will have a corresponding mode of vibration or mode shape, which the structure will adopt. The resonant frequencies and corresponding modes can be estimated analytically by constructing and solving the eigenvalue problem [31]. Each resonant frequency is an eigenvalue and the associated eigenvector describes the mode or shape of the vibration. As an example, in fig. 3.81 the modes of vibration of a surface grinder are represented graphically. The machine consists of the column, the wheel head carriage and the wheel head. The carriage and the head are considered as one body, so no elasticity elements (springs) are added between them. Spring elements were added from the column's base to the ground and between the carriage and the column.
- The points of application and the direction of the exciting forces and measured vibrations must be carefully considered for a three-dimensional structure. For a single-degree-of-freedom system, the vibrations are measured at the point where the exciting force is applied. This resulting frequency response is referred to as a direct response. However, in a complex three-dimensional structure, it is obviously possible to apply exciting forces at particular points and in particular directions and then to measure the displacement at other points, in different directions. The resulting frequency response is referred to as a cross-frequency response.

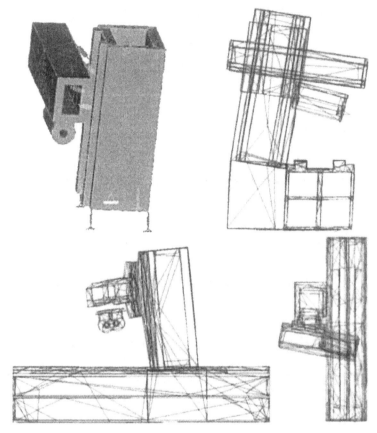

Fig. 3.81. Mode Shapes for a Horizontal Milling Machine (a) Grid for Describing Modal Shapes, (b) Mode of Vibration at 21.1 Hz, (c) Mode of Vibration at 24.7 Hz, and (d) Mode of Vibration at 28.2 Hz

The dynamic characteristics of a machine tool are usually determined experimentally (Fig. 3.82) since the structures are highly non-linear and cannot be easily studied analytically. They can also be determined using a hammer, equipped with piezoelectric elements (Fig. 3.83).

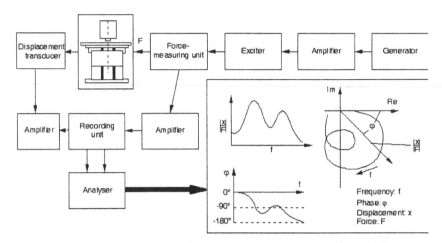

Fig. 3.82. Measuring Scheme for the Dynamic Characteristics of a Frame

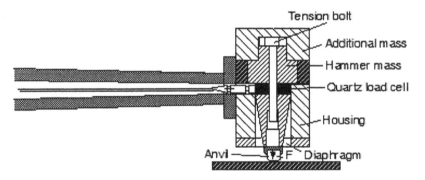

Fig. 3.83. A Piezoelectric Hammer

Improving the Machine Tool Dynamic Behavior

During the operation of a machine tool, particularly with numerical control, it is not always possible to avoid the conditions during which chatter occurs. Improving the stability of the machine may therefore be necessary. The modes of vibration which cause significant relative motion between the tool and workpiece need to be considered. These modes may be suppressed by adding more stiffness and damping to the structure. However, determining where to apply additional damping most effectively may require careful analysis of the structure. It might be possible, by redesigning portions of the structure, to alter the principal directions of the major modes of vibration so as to prevent them from coinciding with the cutting force and/or the normal-to-the-cut surface directions.

Another way of enhancing the dynamic stability of a machine tool is by using a vibration absorber, a mass-spring system added to the structure and tuned to reduce the effect of the major modes of vibration. Vibration absorbers can be considered for machine tools, but may be difficult to use if several major modes of vibration exist. For machines with a single dominant mode of vibration these devices may be effective [31].

Thermal Behavior

The thermal loads affect the function and accuracy of the machine tool and are one of the major reasons of dimensional and geometric errors in the workpieces produced. There are a number of factors in the working environment of a machine tool which can contribute to thermal load (fig. 3.84). One of the main sources of thermal loads is the heat sources, which can be classified into two categories - external and internal. External heat sources refer to the environmental changes, including the temperature of the surroundings, heating units close to the machine, other machines which generate heat, the sun, and heated objects such as lubricators, cooling fluids, etc. Internal heat sources account for the heat generated by spindles, drives, gear boxes, bearings and guideways, and for the process itself, including the heat generated upon the workpiece and conducted to the tool, toolholder and clamping device. The latter is easily controlled through the use of adequate coolant fluid [36].

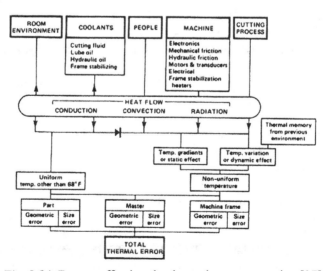

Fig. 3.84. Factors affecting the thermal error generation [37]

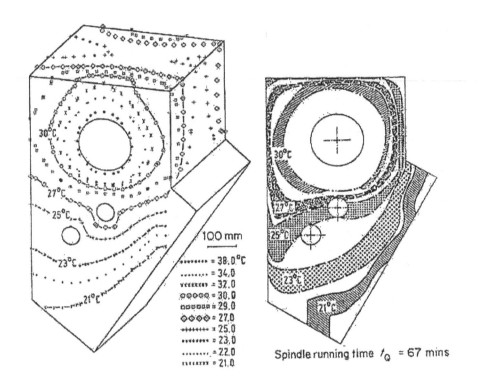

100 mm

•••••••• = 38.0 °C
•••••••• = 34.0
ʏᴄᴄᴇxxᴀʏ = 32.0
ǫǫǫǫǫ = 30.0
ɒɒɒɒɒɒ = 29.0
◇◇◇◇ = 27.0
++++++++ = 25.0
•••••••• = 23.0
••••••••= 22.0
ʀʟᴇᴀʏᴛʏ = 21.0

Spindle running time t_Q = 67 mins

Fig. 3.85. Isothermal Lines in the Head Stock of a Lathe [1]

The heat sources, whether external or internal, may create a temperature field around particular components (Fig. 3.85). The thermo-elastic deformations of a component depend upon the temperature field, the geometry of the component, the way the component is joined to the machine, and the material of the component itself. Displacement between the tool and workpiece is the summation of the deformation of all the components subjected to the temperature field. The influence that each heat source has on the displacement, can be determined experimentally through thermal drift tests. During the design phase of a machine tool, the displacement can be predicted analytically using Finite Element Analysis (thermal modal analysis). Once determined, the effect of deformation created can be reduced by locating all internal heat sources in a way that will have minimum effect on the machine components, or by locating them in a way that their thermal influence will be compensated by the influence of another heat source. Often, main drives are externally mounted and the machine is designed with heat isolators. Providing sufficient cooling lines for the bearings and drives is another measure that allows the frictional heat to be adequately dissipated. Expansion joints, which absorb the thermal expansion of components while attempting to maintain acceptable static and dynamic behavior of the machine, can also be used. Alternatively, for the manufacturing of machine tool elements, materials showing favorable thermal properties (low coefficients of expansion and low thermal conductivities), such as ceramics and fiber reinforced plastics, can be used [38]. Finally, for the case of high precision machine tools the effect of thermoelastic deformation is countered with the use of temperature closed loops in which feedback control is applied in tandem with compensating methods. The compensating methods are classified into direct and indirect ones [36]. The direct compensating methods rely on the measurement of the relative displacements between the tool and the workpiece so as to either precisely set the reference points or to compensate them by means of the CNC. The indirect methods are based upon mathematical models (e.g. neural networks) for determining the relative displacement. Thermal displacements can be reduced up to 90% with the aid of these methods [36].

3.3.2 Guideways and Bearings

Guideways and bearings are used for the movement of slides and work tables as well as for supporting the spindles of machine tools. Guideways are characterized by their unrestrained degrees of freedom, type of movement (linear, rotary, or a combination thereof) and functions during the manufacturing process. For example, a slide moving in a straight line has one degree of freedom, while a spindle has two degrees of freedom since the spindle rotates and moves along one of the linear axes (Fig. 3.86).

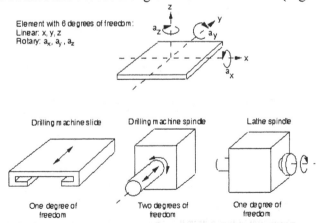

Fig. 3.86. Degrees of Freedom in Guideways and Bearings

Guideway functions include *working* and *setting*. Working guideways are active during the manufacturing process, whereas setting guideways are only used before and after the manufacturing process for tasks, such as tool positioning. Working guideways usually sustain high stresses, so lubrication and wear resistance issues must be considered. There are several types of guideways and bearings (3.87) each with its own characteristics (Fig. 3.88). The selection of the proper guideway or bearing for a specific use, depends upon a great number of design considerations (fig. 3.89).

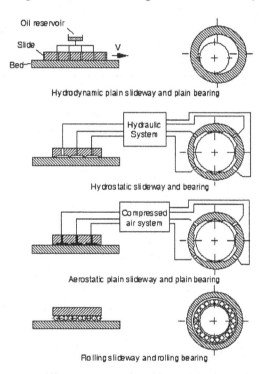

Fig. 3.87. Different Types of Guideways and Bearings

Characteristic	Hydrodynamic Bearing	Rolling Bearing	Hydrostatic Bearing
Damping	●	○	●
Running Accuracy	●	◑	●
Speed Range	○	◑	●
Wear Resistance	◑	◑	●
Power Loss	●	○	◑
Installation Costs	◑	○	●
Cooling Capacity	◑	◑	●
Reliability	●	●	◑

Evaluation of characteristics
● high   ◑ medium   ○ low

Fig. 3.88. Characteristics of Bearings for Machine Tools

• Speed limits	• Environmental sensitivity
• Acceleration limits	• Sealability
• Range of Motion	• Size and configuration
• Applied loads	• Weight
• Accuracy	• Support equipment
• Repeatability	• Maintenance
• Resolution	• Material compatibility
• Preload	• Mounting requirements
• Stiffness	• Required Life
• Vibration absorbance	• Availability
• Shock absorbance	• Designability
• Damping capability	• Manufacturability
• Friction	• Cost
• Thermal performance	

Fig. 3.89. Bearing design considerations [7]

Hydrodynamic Guideways

Hydrodynamic guideways are widely used in machine tools and do not require external source of pressure for the lubrication oil. Instead the pressure is generated by the sliding surface, moving over a lubrication film on small wedge-shaped pockets. These pockets can be modeled as wedge-shaped films (Fig. 3.90). A variety of gap shapes for hydrodynamic pressure formation are used in machine tool construction (Fig. 3.91). A brief derivation for the pressure distribution acting along the sliding surface is presented below. This derivation considers a slide, which is moving with velocity v over a film of fluid (Fig. 3.92).

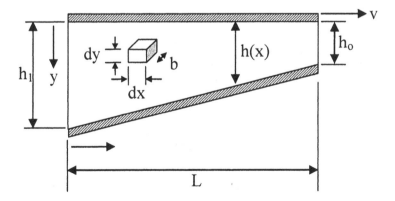

Fig. 3.90. Fluid Wedge

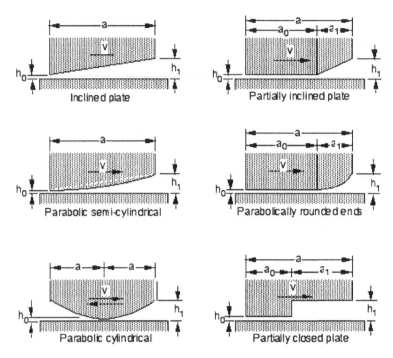

Fig. 3.91. Various Hydrodynamic Wedges

Assuming that a control volume with dimensions dx, dy, and depth b (fig. 3.92) is in equilibrium, the momentum equation for a uniaxial flow reduces to

$$\frac{dp}{dx} = \frac{d\tau}{dy} \tag{3-74}$$

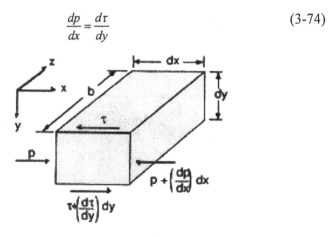

Fig. 3.92. Control Volume

For Newtonian fluids the shear stress is proportional to the velocity gradient,

$$\tau = n\frac{dv_s}{dy} \tag{3-48}$$

where τ is shear stress, n is viscosity, and v_s is the flow velocity. By differentiating Equation (3-48) we obtain

$$\frac{d\tau}{dy} = n\frac{d^2v_s}{dy^2} \tag{3-49}$$

From Equations (3-47) and (3-49):

$$\frac{d^2v_s}{dy^2} = \frac{1}{n}\frac{dp}{dx} \tag{3-50}$$

Since the pressure gradient (dp/dx) does not change with respect to y, the second order differential Equation (3-50) has the solution

$$v_s(y) = \frac{1}{2n}\left(\frac{dp}{dx}\right)y^2 + C_1 y + C_2 \tag{3-51}$$

with the following boundary conditions and values for the constants

$$\text{at} \quad y = 0 \text{ and } v_s(y) = v \Rightarrow C_2 = v$$

$$\text{at} \quad y = h(x) \text{ and } vs(y) = 0 \Rightarrow$$

$$C_1 = -\frac{v}{h(x)} - \frac{1}{2n}\frac{dp}{dx}h(x) \tag{3-52}$$

Thus,

$$v_s(y) = \frac{1}{2n}\frac{dp}{dx}\left\{y^2 - h(x)y\right\} + v\left\{1 - \frac{y}{h(x)}\right\} \tag{3-53}$$

The maximum pressure P will occur at the point where $(dp/dx)=0$.

Therefore at this point,

$$v_s(y) = v\left(1 - \frac{y}{h^*}\right) \tag{3-54}$$

where h^* is the wedge gap at maximum pressure (Fig. 3.93).

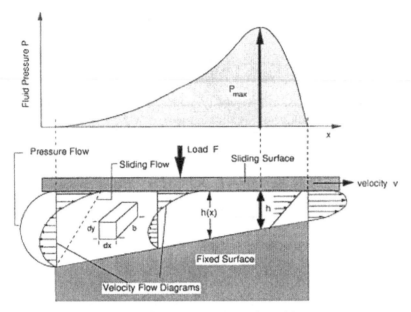

Fig. 3.93. Principle of Hydrodynamic Guideways

The flow Q through any cross-section of the wedge is

$$Q = \int_{y=0}^{y=x(x)} v_s(y)bdy \tag{3-55}$$

and the flow at the maximum pressure point is given by:

$$Q^* = \frac{1}{2}vbh^* \tag{3-56}$$

Since the flow is steady we have

$$Q = Q^* \tag{3-57}$$

Hence, for the wedge shown in Fig. 3.90

$$h(x) = h_0 + \frac{L-x}{L}(h_1 - h_0) \tag{3-58}$$

and

$$\frac{dp}{dx} = \frac{6nv}{h_0^2}\left\{\left(1 + m\frac{L-x}{L}\right)^{-2} - k^*\left(1 + m\frac{L-x}{L}\right)^{-3}\right\} \tag{3-59}$$

where $m = \dfrac{h_1}{h_0} - 1$ and $h^* = k^* h_0$.

After integrating equation (3-59),

$$p(x) = \frac{6nv}{h_0^2}\frac{L}{m}\left\{\frac{1}{1+m\dfrac{L-x}{L}} - \frac{k^{*}}{2\left(1+m\dfrac{L-x}{L}\right)^2} + C\right\} \tag{3-60}$$

The boundary conditions and the constants are:
$$x = 0 \Rightarrow p = 0$$
$$x = L \Rightarrow p = 0$$

$$C = \frac{-1}{2+m}, \quad k^{*} = \frac{2m+2}{2+m}$$

Therefore,

$$p(x) = \frac{6nvL}{h_0^2} K_p(x,m) \tag{3-61}$$

in which

$$K_p(m,x) = \frac{1}{1+m\dfrac{L-x}{L}} - \frac{1}{2+m} - \frac{2m+2}{2(2+m)\left(1+m\dfrac{L-x}{L}\right)^2} \tag{3-62}$$

The load that can be supported by the wedge is
$$F = p_m Lb \tag{3-62}$$

where

$$p_m = \frac{1}{L}\int_{x=0}^{x=L} p(x)\,dx \tag{3-63}$$

This elementary analysis provides an order of magnitude estimation of the load that a hydrodynamic guideway can carry [1]. In a practical setting, such estimation provides an overall orientation for the design of slideways but for detailed designs, numerical techniques or experience are typically used.

The frictional characteristics of hydrodynamic slides are critical for machine tool accuracy. Irregular sliding motion may occur at low sliding velocities. This irregular motion is due to the *stick-slip* effect (Fig. 3.94) namely the variation of the coefficient of friction, μ, with respect to the sliding velocity, which is described by the "*Stribeck* curve" (Fig. 3.95).

The stick-slip effect can be minimized by properly selecting slide materials and with lubrication.

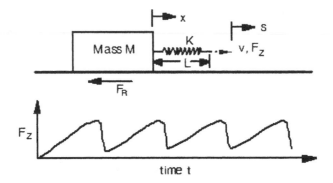

Fig. 3.94. Analogy of the Stick-Slip Effect

The stick-slip effect can be better comprehended by analyzing a mass that is pulled across a surface (Fig. 3.96). Equating the forces acting on the mass along the surface

$$F_z - F_R = m\ddot{x} \qquad (3\text{-}64)$$

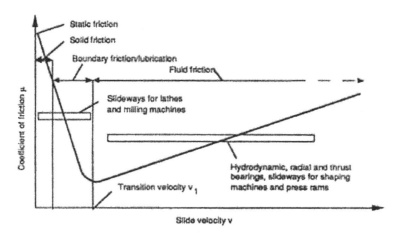

Fig. 3.95. Stribeck Curve

F_Z the frictional force, and $F_{R,}$ the spring force is defined below:

$$F_R = c\dot{x} + F_{static} \qquad (3\text{-}65)$$

$$F_Z = K(s - x) \qquad (3\text{-}66)$$

where

K is the effective stiffness coefficient,
c is proportional to the slope from the Stribeck curve, and
F_{static} is the force to overrun the static friction.

The distance that the end of the spring travels is given by

$$s = v.t \qquad (3\text{-}67)$$

where v is a given velocity. Substituting Eqs. (3-65), (3-66), and (3-67) into (3-64), the dynamics of the mass is represented by the following differential equation:

$$m\ddot{x} = K(s-x) - c\dot{x} - F_{static} \qquad (3\text{-}68)$$

Lubricants, utilized for hydrodynamic bearings and guideways, vary from light oil to solid lubricants, such as graphite or PTFE polymer. The material combinations used for the sliding contact surfaces affect greatly the behavior, the wear and the load carrying capacity of the guideways. Cast iron on cast iron was exclusively used in the past but nowadays, the trend is to have a surface harder than the other so as to decrease the wear [7]. Cast iron on steel and brass on steel material combinations are widely used. Polymer coatings applied upon the metallic contact surfaces can almost minimize the stick-slip effect and are thus, widely used for machine tool applications. Ceramic materials, such as aluminum oxide and silicon carbide, although they are more expensive to be used in a bearing, they present excellent hardness, minimum wear danger and very low friction, consequently are ideal for precision machine tool applications [7].

Just like hydrodynamic guideways, hydrodynamic bearings, also known as journal bearings, are less appropriate for precision applications than hydrostatic or magnetic bearings are.

Hydrostatic Guideways and Bearings

In hydrostatic guideways and bearings (Fig. 3.96) the lubricating gap is created by a thin film of oil, pressurized by an external pump. Since the pressure is maintained constant (in most cases at about 3.5 MPa with a maximum of 21 MPa), the thickness of the lubricating film is independent of the sliding speed. The nearly constant lubrication film thickness, provided by hydrostatic guideways and bearings, minimizes wear and friction, whilst they also provide very high damping along the constrained axes. Furthermore, since there is very little variation in the coefficient of friction, there is no significant stick-slip effect. By minimizing static friction and the stick-slip effect, these guideways and bearings provide high accuracy. Hydrostatic bearings offer the advantages of having extremely low friction and the ability of achieving very high stiffness.

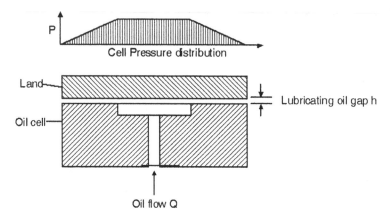

Fig. 3.96. The Principle of a Hydrostatic Guideway or Bearing

A hydrostatic guide consists of a number of recesses known as oil cells (Fig. 3.97). The space between the land and the sliding surface is filled with the lubricating oil film with a thickness ranging from 1 to 100 μm. The gap forms a hydraulic resistance, causing a pressure build-up to carry the external load. The analysis of the flow in such hydrostatic guides follows the Hagen-Poiseuille law [1]. Different combinations of pumps per cell and different arrangements of the cells can be utilized to meet the flow, required by a specified external load. Five basic types of hydrostatic bearing arrangements exist: single pad, opposed pad, journal, rotary thrust and conical journal/thrust bearings (Fig. 3.98) [7]. Cells for guideways are usually situated parallel to the direction of motion. A large number of cells can be used to compensate for vibration or deviation from a linear path during motion. The cell depth should be 10 to 20 times greater than that of the oil film gap for the even distribution of the lubricant to the lands.

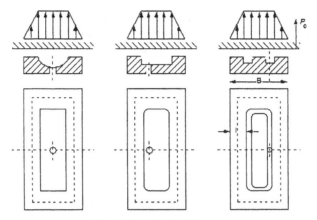

Fig. 3.97. Oil Cell Shapes

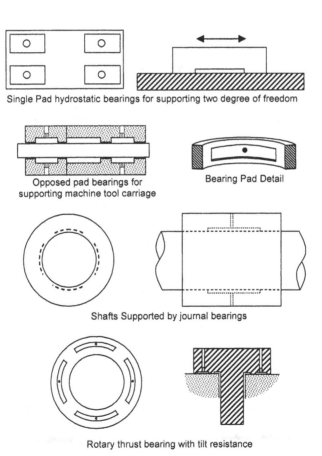

Single Pad hydrostatic bearings for supporting two degree of freedom

Opposed pad bearings for supporting machine tool carriage

Bearing Pad Detail

Shafts Supported by journal bearings

Rotary thrust bearing with tilt resistance

Fig. 3.98. Various hydrostatic bearing configurations

Hydrostatic guideways can support extreme loads due to the fact that they distribute it over a long area. Off shore oil platform decks, weighing up to 20,000 tons, are transferred upon hydrostatic bearings [7].

Hydrostatic guideways and bearings are both costly and complex since each pressure pad needs an elaborate hydraulic system (pumps, oil lines, etc.) to maintain the necessary fluid flow and pressure. The use of hydrostatic guideways and bearings generally requires a high level of maintenance. Furthermore, running hydrostatic guideways and bearings, at high speeds, may increase considerably the fluid temperature generated by fluid shear friction, changing thereby the physical characteristics and possibly reducing the effectiveness of the guideway or bearing. This may induce the need for closed loop temperature control of the oil, increasing the cost of the bearing. Since hydrostatic guideways require a high-pressure hydraulic system, the power consumption of the pump may be substantial. Flow control valves can be used to give the bearings the desired stiffness while minimizing the power consumption.

Aerostatic Guideways and Bearings

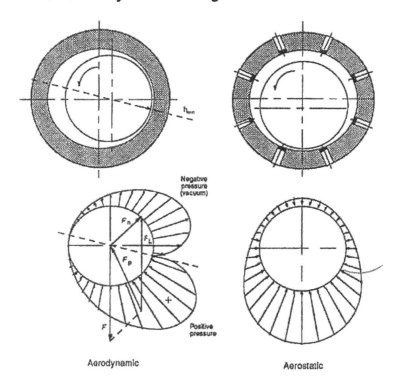

Aerodynamic   Aerostatic

Fig. 3.99. Principles of aerostatic bearings

Aerostatic guideways and bearings use high-pressure air (typically 690 kPa) as lubricating medium between the sliding surfaces. These guideways and bearings are suitable for high operating speeds because they provide low friction and minimal heat generation. They are relatively simple in terms of construction because there is less possibility for contamination of the lubricating medium as compared with that of lubricating fluids. Furthermore, they are not significantly affected by drastic changes in the operating temperature. Aerostatic guideways and bearings, like their hydrostatic counterparts, provide minimal static friction and very low wear, and the load carrying capacity is independent of the sliding velocity. The function of aerostatic bearings and guideways is based on the drag effect in the small gap (on the order of 1 to 10 μm) between two moving parts, and consequently, the load carrying capacity of these bearings and guideways, is moderate (Fig. 3.99). The calculations for the air gap dimensions, air consumption and load carrying capacity of an aerostatic bearing or guideway are far more complicated than those for a hydrostatic one because of additional factors, such as the possibility of air turbulence and the shape of the air nozzles. Due to the compressibility of air, self-excited vibration (pneumatic hammer instability) may also occur in aerostatic bearings or guideways. To avoid this effect, inherently compensated orifices (Fig. 3.100) or a porous material, such as a sintered sleeve or graphite, can be used between the air inlet and the bearing or guideway clearance gap. The sliding surfaces of aerostatic guideways and bearings may corrode owing to moisture in the air. This problem may be resolved by using corrosion-resistant materials or air dryers but these solutions may be expensive to implement.

Hydrostatic and aerostatic bearings and guideways have many similarities in their operation principles. The selection among these two types is based upon the particular application intended for. Aerostatic bearings and guideways are ideal for moderate loads and moderate stiffness, at high speeds, while hydrostatic bearings are ideal for high load and high stiffness at moderate speeds [7].

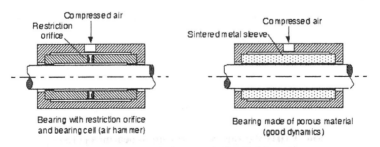

Fig. 3.100. 3.102 Aerostatic Radial Bearing

Rolling Contact Guideways and Bearings

Rolling contact guideways and bearings are widely used in machine tools because they provide low starting and running friction, minimal stick-slip, high stiffness and relatively trouble-free operation. Their sizes and specifications are standardized and they are available by all suppliers in different specified forms (fig. 3.101). Their basic scope of application is the transmission of the loads between surfaces, moving in opposite direction through *rolling elements*. Besides the rolling elements, the rolling contact bearings consist of an *outer* and an *inner ring,* formed with raceways, where the rolling elements roll, and a *cage* that protects and restrains the rolling elements (fig. 3.102). According to the rolling element used, the bearing can be classified into ball, cylindrical, needle, tapered, symmetrical barrel and asymmetrical barrel roller. In general it can be stated that ball bearings present low to moderate load carrying capacity but can operate at very high speeds. In contrast, the roller bearings can withstand higher loads whilst they operate at lower nominal speeds.

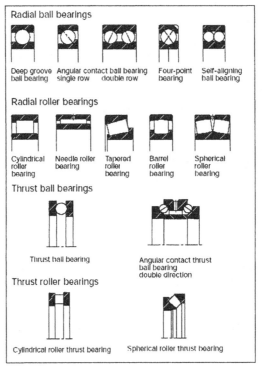

Fig. 3.101. Various types of rolling bearings [113]

Fig. 3.102. Elements consisting a rolling element bearing (1. outer ring, 2. inner ring, 3. rolling elements & 4. cage) [113]

Rolling guideways (known also as linear motion rolling element bearings) usually employ a roller chain with recirculating rolling elements (Fig. 3.103). The accurate operation of a guideway not only does it depend on the quality of the guiding surfaces but also on the geometric accuracy and precision of the rolling elements and their position in the cage. A variety of basic designs for rolling guideways are available, including flat and V guides (Fig. 3.104).

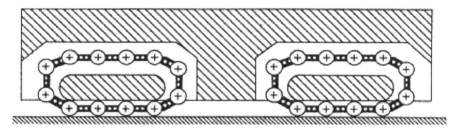

Fig. 3.103. Linear Rolling Guideway

In machine tool construction, rolling contact bearings are frequently used in the spindle drive and as shaft supports for the feed drives. Both ball and roller bearing designs are available. Spindle bearings are critical elements in machine tools because they must provide high accuracy under a wide variety of loads. Spindle bearings are often preloaded; some designs use a hydraulic system to maintain a constant preload. There are three basic spindle bearing designs which are widely used (Fig. 3.105). *Tapered roller* bearings are adequate for low rotational speeds, provide good stiffness, and are usually applied to spindles in turning and milling machines. *Double row parallel roller* bearings are adequate for medium rotational speeds and provide good stiffness; this type of bearing works well for lathes, milling, drilling and grinding machines. *Inclined angular contact ball* bearings are used for very high rotational speeds. However, they provide relatively poor stiffness and are used in internal grinding machines or in machines used to turn or mill non-ferrous metals, which do not require high cutting forces.

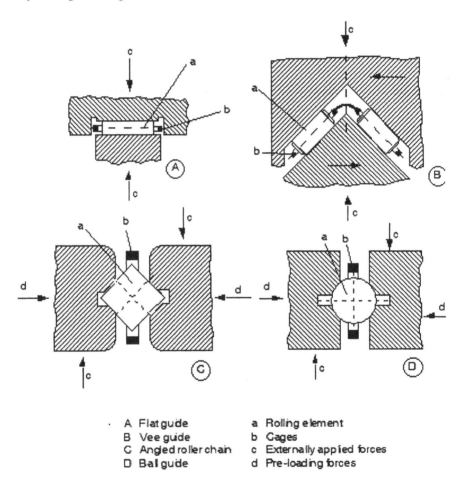

```
·  A  Flat guide           a  Rolling element
   B  Vee guide            b  Cages
   C  Angled roller chain  c  Externally applied forces
   D  Ball guide           d  Pre-loading forces
```

Fig. 3.104. Rolling Guideway Design

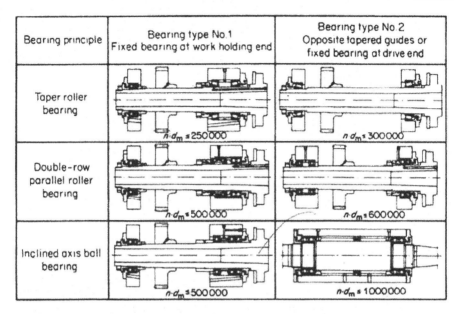

Bearing principle	Bearing type No.1 Fixed bearing at work holding end	Bearing type No.2 Opposite tapered guides or fixed bearing at drive end
Taper roller bearing	$n \cdot d_m \leq 250\,000$	$n \cdot d_m \leq 300\,000$
Double-row parallel roller bearing	$n \cdot d_m \leq 500\,000$	$n \cdot d_m \leq 600\,000$
Inclined axis ball bearing	$n \cdot d_m \leq 500\,000$	$n \cdot d_m \leq 1\,000\,000$

Fig. 3.105. Spindle Bearing Designs [1]

Magnetic Suspension Guideways and Bearings

Magnetic suspension offers the advantages of insignificant static and viscous friction as well as excellent thermal isolation properties. The principle of operation is based upon that an electromagnet will attract ferromagnetic material. A ferromagnetic rotor can thus be supported in a magnetic field generated in the bearing electromagnet stator (fig. 3.106). Magnetically suspended spindles have been constructed, rated at a maximum speed of 100,000 rpm. Virtually, any radial and axial load can be supported. Typically, the commercial magnetic bearings are rated for radial loads up to 75 kN and axial loads up to 50 kN. Their main advantages are the contactless operation, the auto balancing capabilities [39], the monitoring ability of the operational state and their almost infinite life time. On the other hand, their high cost is the only significant drawback. Magnetically suspended slides are used for precision applications where extremely high velocities and high accelerations are required [40]. The position of a slide is constrained to one axis by magnet pairs, which are horizontally and vertically opposed to each other. The relative attractive strength, between two opposing magnets, controls the slide's position in the axis. Capacitance gauges measure the position between slide and magnet surfaces for controlling the slide position.

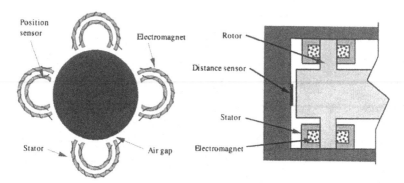

Fig. 3.106. Magnetic bearing principle [7]

Flexural Bearings

The motion in mechanical bearings is obtained utilizing sliding or rolling solid bodies. An alternative to these are the flexural bearings (also known as flexure pivots) which rely on the inert properties of the material. The motion is achieved through the stretching of atomic bonds during elastic motion [7]. The range of motion is very small, whereas their size is significant; therefore, their application is limited to precision machine tools in alignment mechanisms. The maximum speed of operation is determined by the system's natural frequency and the stress levels in the bearing. The advantages of flexural bearings include high precision, no friction, no wear, no need for lubrication and no jamming risks. There are two kinds of flexural bearings according to the manufacturing approach, monolithic (fig. 3.107) and clamped-flat-spring.

Fig. 3.107. Various types of monolithic flexural bearings [7]

3.3.3 Drives and Transmissions

Drives

Machine tools use hydraulic actuators, DC motors, AC motors, or stepper motors for drives. The type of drive used is determined by the power requirements of the machine tool, the power sources available, and the desired dynamic characteristics.

Stepper motors are limited both in power and in available torque so they are suitable only for small machine tools. DC motors provide excellent speed regulation, high torque and high efficiency, and therefore, are ideally suited for control applications. DC motors can be designed to meet a wide range of power requirements and are utilized in most small to medium-sized machines. Hydraulic systems may range in size up to hundreds of horsepower. They are well suited for machine tools where power requirements are high. The cost of a hydraulic drive is not proportional to the power required, and, thus, it is expensive for small to medium-sized machines [32].

Hydraulic Systems

Hydraulic systems are used extensively for driving high-power machine tools since a relatively small hydraulic system can deliver a high level of power. They can develop much higher maximum angular acceleration than that of DC motors of the same peak power. They have small time constants and therefore, can provide smooth operation of the machine tool slides. Hydraulic actuators have been also introduced to precision machine tools, since they can be designed with zero friction and can have very low backlash transmission effect [7].

Hydraulic systems, however, present some problems in terms of maintenance and oil leakage from the transmission lines and the system components. The oil must be kept clean and protected against contamination. Other undesirable features of hydraulic systems are the dynamic lags, caused by the transmission lines and variations in the oil viscosity as a function of the oil temperature.

Hydraulic motors are manufactured similarly to pumps. They can be classified into linear and rotary actuators. Units of piston and cylinder are commonly used in linear as well as in precision applications with limited range of motion; metal bellows actuators are also available and there is a great variety of rotary hydraulic actuators. Gear motors are widely used in high forces applications, vane type actuators whilst on the other hand, provide high torques, requiring very little space and rack. Pinion rotary actuators are used when torque requirements vary from very small (in the range of N.m) to very high ones (in the range of MN.m) [7].

Hydraulic systems (Fig. 3.108) are generally comprised of the following components [41]:

1. A hydraulic power supply
2. A servo-valve for each axis of motion
3. A sump
4. A hydraulic motor for each axis of motion
5. Support equipment including fluid lines, filters, flow control devices, seals and pumps.

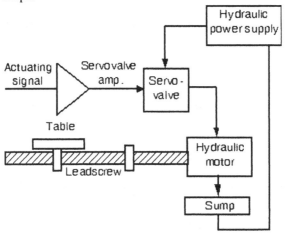

Fig. 3.108. Hydraulic System

DC Motors

DC motors allow precise control of the speed over a wide operating range by manipulation of the voltage applied to the motor. These motors are ideally suited for driving the axes of small to medium-size machines. DC motors are also used to drive the spindle in lathes and milling machines when continuous control of the spindle speed is desired. The DC motor can function either as a motor or as a generator.

The DC motors can be classified into brushed and brushless servomotors. The principle of operation of a DC brushed motor is based on the rotation of an armature winding within a magnetic field (Fig. 3.109). The armature is the rotating member, or rotor, and the field winding is the stationary member, or stator. The armature winding is connected to a commutator, which is a cylinder of insulated copper segments, mounted on the rotor shaft. Stationary carbon brushes, which are connected to the machine terminals, are held against the commutator surface and enable the transfer of a DC current to the rotating winding. In a motor, electrical energy is supplied to the armature from an external DC source, and the motor converts it into mechanical energy [1].

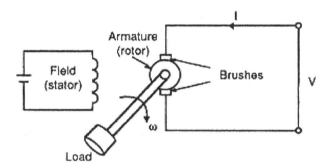

Fig. 3.109. Schematic Diagram of a Separately Excited DC Motor

Two equations are required to define the behavior of a DC machine: the torque and the voltage equations. The torque equation relates the torque to the armature current:

$$T = K_f \phi I \tag{3-69}$$

and the voltage equation relates the induced voltage in the armature winding to the rotational speed:

$$E = K_f \phi \omega \tag{3-70}$$

where

T	=	magnetic torque, N·m
ϕ	=	flux per pole, Wb
I	=	current in armature circuit, A
E	=	induced voltage (emf), V
ω	=	angular velocity, rad/s
K_f	=	constant determined by design of winding

For a motor, an input voltage, V, is supplied to the armature, and the corresponding voltage equation becomes

$$V - IR = K_f \phi \omega \tag{3-71}$$

where R is the resistance of the armature circuit and IR is the voltage drop across this resistance. The armature inductance is negligible in Equation (3-71).

Multiplying Equations (3-69) and (3-71) yields the power equation

$$P = \omega T = VI - I^2 R \qquad (3-72)$$

where P is the mechanical output power, VI is the electrical input power, and $I^2 R$ is the electrical power loss.

DC motors are classified as separately excited, shunt-, series-, and compound-connected, according to the method of field connection. The separately excited DC motor with constant field excitation is well suited for control applications, since it provides smooth control of speed over a wide range. The motor field is excited by a separate constant DC voltage supply, and consequently the flux f becomes constant as well; thus, Equations (3-69) and (3-71) are written as

$$T = K_t I \qquad (3-73)$$

$$V - IR = K_v \omega \qquad (3-74)$$

The parameters K_t and K_v are referred to as the torque and voltage constants.

DC brushed motors are characterized by their simplicity of operation and their easy braking without any additional power input. Among their disadvantages are included the non uniform wear of the brushes, the excess heat generation that can lead to thermal drift and the generation of sparks that can be dangerous in explosive environments [7].

Modern DC motors use a permanent-magnet (PM) field (*brushless servomotors*), rather than an externally excited field. Both types are referred to as DC servomotors and are characterized by equations (3-73) and (3-74). The PM does not require a field voltage source and results in higher efficiency and fewer thermal problems. Brushless servomotors' primary disadvantages include the possibility of the rotor becoming demagnetized in the case of current overload, high temperatures and strong magnetic fields. Furthermore, when used as high speed spindles they require a lot of power to stop [7].

If a DC servomotor drives a mechanical load, consisting of dynamic and static components, equating the torques acting on the load yields:

$$T = J \frac{d\omega}{dt} + T_s \qquad (3-75)$$

where J is the combined moment of inertia of the motor and load, and T_s is the static load due to friction and cutting forces in machining operations.

Elimination of J and T from Eqs. (3-73) through (3-75), and rearrangement of the terms so as to separate the independent variables, gives the speed equation:

$$\tau_m \frac{d\omega}{dx} + \omega = \frac{1}{K_v} V - \frac{R}{K_t K_v} T_s \qquad (3\text{-}76)$$

where τ_m is the mechanical time constant of the loaded motor and is defined by

$$\tau_m = \frac{JR}{K_t K_v} \qquad (3\text{-}77)$$

The Laplace transformation of Equation (3-76) is

$$\omega(s) = \frac{K_m V(s) - R \dfrac{K_m}{K_t} T_s(s)}{1 + s\tau_m} \qquad (3\text{-}78)$$

where K_m is the gain of the motor and is defined by $K_m = 1/K_v$ The solution of Equation (3-78) in the time domain depends on the applied voltage and load torque. For example, assuming that the motor is initially at rest, $T_s = 0$, and a step voltage of V volts is applied at the armature terminals, the solution is

$$\omega(t) = K_m V \left(1 - e^{t/\tau_m}\right) \qquad (3\text{-}79)$$

Thus, the motor response is described by a steady-state speed $K_m V$ and a decaying exponential with a time constant τ_m given by Equation (3-77).

In many machines a leadscrew is driven through a gear box. The gear ratio K_g is defined as the ratio between the speed of the leadscrew ω_L and the speed of the motor ω:

$$K_g = \frac{\omega_L}{\omega} \qquad (3\text{-}80)$$

In order to calculate the time constant by Equation (3-77), the inertia of the leadscrew should be referred to the motor shaft. Consequently, the inertia J in Equation (3-73) is

$$J = J_r + K_g T_l \qquad (3\text{-}81)$$

where J_r is the inertia of the rotor and J_l is the inertia of the leadscrew and load. Note that load torques should also be referred to the motor shaft:

$$T_s = K_g T_l \tag{3-82}$$

where T_l is the load torque at the leadscrew [41].

Stepper Motors

The stepper motor (SM) is an incremental digital drive. It translates an input pulse sequence into a proportional angular movement and rotates one angular increment, or a *step*, for each input pulse. The shaft position is thus, determined by the number of pulses without the use of a feedback sensor, unless the maximum torque is exceeded. The step angle is at the order of few milliradians. The shaft speed is proportional to the pulse frequency. The shaft speed in steps per second is equal to the input frequency in pulses per second (pps).

To obtain optimal stepping motor performance, an electronic switch, or translator, is required as part of the drive unit. The drive unit contains a steering circuit and a power amplifier. The steering unit translates the incoming pulses into the correct switching sequence required to step the motor. The steered pulses are converted to power pulses, with appropriate rise time, duration, and amplitude for driving the motor windings. A typical stepper motor with its components is depicted in fig. 3.110.

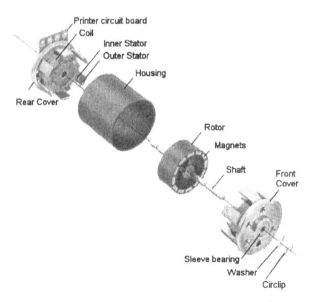

Fig. 3.110. Schematic diagram of stepper motors [114]

AC Motors

Some CNC manufacturers use alternate-current (AC) synchromotors as drives for machine tools [41]. As per DC brushless motors, AC motors operate without brushes, thus eliminating one of the main maintenance problems associated with DC motors. The velocity of the AC synchromotor is controlled by manipulation of the voltage *frequency* supplied to the motor, rather than the voltage *magnitude*, as in DC servomotors. The frequency manipulation requires the use of an electrical *inverter*. The inverter contains a DC power supply and a circuit that inverts the resultant DC voltage into AC voltage with a continuously controllable frequency [41]. Among their disadvantages are the significant cogging torque, the heat generation in the rotor conductors and the need of additional power for braking. On the other hand, simplicity of operation as well as low cost, are their advantages.

Transmissions

Transmissions are mechanical systems that provide motion to other mechanical components using power from drives. Transmissions include *gearboxes*, *rack* and *pinions*, and *leadscrew and nuts* [42].

Leadscrew and Nut Drives

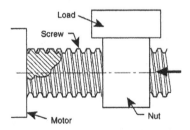

Fig. 3.111. Leadscrew and nut drive

Leadscrew and nut drives (fig. 3.111) are standard rotary-to-linear transmission components that can be purchased off-the-shelf. They have constant, low transmission ratios to allow for slow, smooth and accurate motions of feed drives. Furthermore, the self-braking ability of these drives makes them applicable for horizontal and vertical positioning. The low efficiency associated with sliding friction of the lead screws and nuts makes them impractical for use as main drives but for feed and auxiliary drives this drawback is minimal. Furthermore, their load capacity is approximately ten times lower than that for ballscrews.

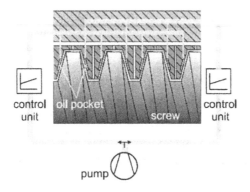

Fig. 3.112. Hydrostatic leadscrew and nut drive

Leadscrews that are lubricated hydrostatically (fig 3.112) present improved behavior in terms of starting and running operation. Since a lubricant film of constant thickness interferes between the leadscrew and the nut contact areas, the stick-slip effect is eliminated and thus, the wear and friction is minimized. Furthermore, high damping is provided between the threads. These transmission elements can be implemented in high speed precision machine tools since they work with no backlash, generate only little thermal losses due to the lubricant interface and can be built with high stiffness.

Leadscrew and nut drives use ball bearings, so the design issues are similar to those of ball bearings: ball wear, precision, heat generation, preloading, etc. [42]. They can be selected by the grade of precision, which is usually directly proportional to price. Stiffness of a leadscrew transmission is controlled by the preload between the nut and the screw. There is a tradeoff when considering stiffness since by increasing the preload, the static and viscous friction is also increased.

Ball Screw and Nut Drives

Ball screw and nut drives have low friction, high efficiency and can be preloaded to eliminate backlash for precise alternating motions [42]. In this drive, balls run along the thread between the screw and the nut and recirculate through a return passage that can be either an external integrated tube or an internal channel (Fig. 3.113) [41]. The achieved repeatability of the ball screws is in the order of 1 μm and after special post processing they can attain submicron resolution [7]. The advantages associated with ball screws have made them popular for many machine tool drives.

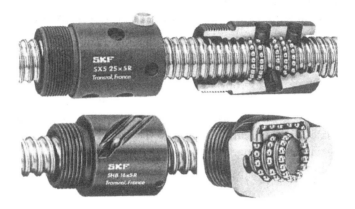

Fig. 3.113. Ball Screw and Nut Drive [115] with an external integrated tube (a) and an internal channel (b)

Roller Screws and Nut Drives

Roller screws utilize multiple threaded rollers assembled in a planetary arrangement for the transmission of the load from the nut to the shaft. Due to this engagement, the area of the contact surface is substantially larger compared to that of the ball screws and thus, very high loads (a maximum dynamic load of 753 kN has been reported in [7]) can be transmitted at high rotational speeds (up to 6,000 Rpm), with almost no axial backlash or wear. Other advantages of these transmission elements are their ability to absorb shock loads, their high reliability and their low maintenance cost. The roller screws can be classified in *planetary* and *recirculating* roller screws according to whether the roller axis is fixed or free (fig. 3.114).

Fig. 3.114. Roller screws and nut drives [116]

Rack and Pinion Drives

Rack and pinion drives (fig 3.115) are typically used for quick, power-ful, but relatively imprecise motions [42]. They are inexpensive and easy to manufacture but they often have high transmission ratios that are not uniform due to errors in the gearing. The high efficiency of these drives enables them to transmit a considerable amount of power in large dis-tances. Thus, they are appropriate for main drives in machine tools. They are mainly used in machining centers with long travel ranges (greater than 5 m) [7]. The lack of self-braking ability of these drives prevents them from being used for vertical positioning. However, a lead screw and nut drive can be installed in parallel with a rack and pinion drive in order to obtain a self-braking feature.

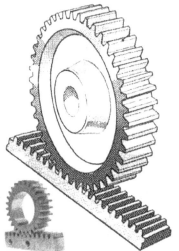

Fig. 3.115. Rack-and-pinion drive

Worm and Rack Drives

Worm and rack drives (fig. 3.116) have much lower transmission ratios than rack and pinion drives have and provide smoother motion [42]. How-ever, they tend to be more difficult to manufacture than rack and pinion drives. Worm and rack drives are primarily used as auxiliary drives.

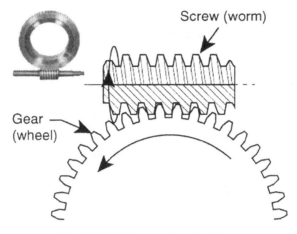

Fig. 3.116. Worm and rack drives

Small Displacement Drives

Small displacement drives produce minute displacements and are typically used on grinding machines [42]. Most of these drives are used for extremely precise applications that neither do they require much rigidity nor displacement. Some types of small displacement drives are:

- *Thermal-Expansion Drives*: Heat causes a thermal element to deform to provide small displacements
- *Magnetostriction Drives*: The displacement of a ferro-magnetic rod is varied by varying the strength of the magnetic field surrounding the rod.
- *Elastic-Link Drives*: An elastic link, such as a leaf spring, is initially bent using hydraulic power. The hydraulic oil is allowed to drain freely through an aperture to gradually straighten out the elastic link and cause small motions.

Linear motor drives and guides

Linear drives are mainly used for positioning systems and as feed drives in high performance CNC machine tools, when conventional rolled or ground ball screw drives have reached their operation limits. Linear motor drives provide high speed, highly dynamic and precise positioning. The accuracy achieved is superior compared to that of conventional ball screw driven machines, since linear motors eliminate inherent inaccuracies from wind-up, heat, lost motion and backlash. They consist of a short number of components: the primary part which contains the current carrying coil, the secondary part, which is a permanent magnet, the guideway, a linear measuring system and a power train (fig. 3.117). Early linear motor drives presented overheating problems, which nowadays, have been minimized. However, their high cost prohibits their wide adoption and limits their use to some special applications, such as high speed machining [43].

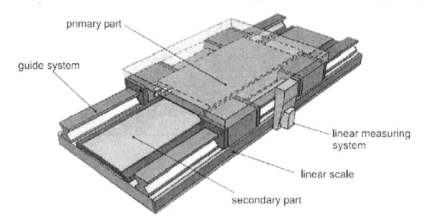

Fig. 3.117. Linear motor drive

3.4 Machine Tool Control and Automation

The most important reasons for introducing automation in a manufacturing system are:

- *Higher productivity as a result of higher production rate.* In mass production systems, such as those used in the automotive and consumer goods industries, complex parts, which would take a few hours to manufacture in manually or semi-automated systems, can be made in seconds by automated systems, such as transfer lines. Such "hard wired" automation is designed to reduce material handling time, or the time the part spends between operations. In the metalworking industry, parts often spend only 5-15% of their time in added-value operations, while the rest is spent in handling and waiting. Automation may considerably increase the production rate of the system, but it also increases the required capital investment, whilst substantially reduces the product, operation, and volume flexibility of the system.
- *Consistent quality.* This objective involves elimination of human errors and inconsistencies, which result in quality and dimensional deviations. Manually performed manufacturing processes are influenced by the

skills and abilities of the human operator; consequently the quality and the dimensional accuracy of the finished parts also depend on the skill and ability level of the operator. Implementing automation for the purpose of consistent quality may reduce the flexibility of the system, though, since the human factor is one of the most flexible, in the production process.

- *Better work content.* Repetitive or hazardous tasks can be automated to relieve humans. This "humanistic" automation also streamlines the production process, making it more efficient.

Since the introduction to automation provides higher productivity while, in general, reduces flexibility, considerable research and development has been devoted to the area of "flexible automation," which provides high production rates and consistent quality, while being flexible enough to accommodate a variety of tasks. A result of these efforts is the establishment of Flexible Manufacturing Systems (FMS), and the introduction of robots, which can be programmed to perform different tasks. Despite this technological progress, measures of automation are often difficult to be economically justified, particularly when the removal of direct labor is considered the primary reason. Indeed, in most industries, the manufacturing cost is influenced very little by direct labor. For example, in the metal working industry, direct labor accounts, on average, for 10% of the entire manufacturing cost. Thus, introducing automation, just to reduce or eliminate direct labor, is not prudent from an economical point of view. Instead, the benefits of automation should take into account the reasons for automation discussed above, rather than solely direct labor savings.

Machine tools, which perform a variety of manufacturing processes, are a cornerstone in the introduction of automation, so numerical control and part programming in machine tools will be discussed in the remainder of this chapter, together with process control and computer integrated manufacturing.

3.4.1 Numerically Controlled Machine Tools

One of the most important developments in manufacturing automation is *numerical control* (NC). NC has been defined by the Electronics Industries Association (EIA) as "a system in which actions are controlled by direct insertion of numerical data at some point. The system must automatically interpret at least some portion of this data." The data required to produce a part is called a part program. The part program is a group of instructions, read by the control system, and is converted into signals that move the drives. The focus of NC machines has traditionally been on complex parts manufactured in large volumes. However, because of the development of more efficient programming languages, NC is now employed for smaller volume sizes. A numerical-control machine tool system includes the machine-control unit (MCU). The MCU is further divided into two elements: the data-processing unit (DPU) and the control-loops unit (CLU) [41]. The DPU processes the coded data read from the tape or other data-store media and passes information on the position of each axis, its direction of motion, feed, and auxiliary function controls signals to the CLU. The CLU operates the drive mechanisms of the machine, receives feedback signals about the actual position and velocity of each of the axes, and indicates an operation's completion time. The DPU sequentially reads the data when each line has completed execution as noted by the CLU.

The motion control of NC machine tools is completed by translating NC codes into machine commands. The NC codes can be broadly classified into two groups: (1) commands for controlling individual machine components, such as motor on/off control, selection of spindle speed, tool change, and coolant on/off control (these tasks are accomplished by sending electric pulses to the relay system or logic control network) and (2) commands for controlling the relative movement of the workpiece and the tools. These commands consist of information, such as axis and distance to be moved at each specific time unit. They are translated into machine-executable motion-control commands that are then carried out by the electro-mechanical control system.

Currently, NC controllers are all built with the use of computer technology. Such controllers are called CNC (computer numerical control). CNC systems are more flexible than their NC counterparts because they allow programs to be edited, stored in memory and recalled instantly. CNC machines can generally machine more complex shapes than NC machines can and also provide circular interpolation and canned programming cycles. CNC controllers have been applied to nearly every kind of machine tool: lathes, milling machines, drill presses, grinders, etc. Features available to modern machines include tool, pallet, and workpiece changers. Controller features include interpolators, graphics interfaces, interactive operator programming, and data communication.

In numerical control, data concerning all aspects of the machining operation, such as locations, speeds, feeds, and cutting fluids, are stored in compact discs, floppy or computer's hard disks,. The concept of NC is that specific information can be relayed from these storage devices to the machine tool's control unit. On the basis of input information, relays and other devices (called hard-wired controls) can be actuated to obtain a desired machine setup. Complex operations, such as turning a part having various contours or die sinking in a milling machine, can be carried out. If a single computer is used to download part programs to a variety of NC machines, this system is referred to as a direct numerical-control (DNC) system. This computer may also be used to download instructions to the material-handling system equipment.

A number of specific functions are executed by the components of an NC-controlled servo drive system (Fig. 3.118). An operating panel/keyboard allows the alpha-numeric instructions to be entered in the machine. A decoder/encoder receives the data, entered from the computer and divides them into two sections; one for the part geometry data and the other for the process data which includes information about feed rates, spindle speeds, and other machining parameters. The geometric data also contains information about tool motions. The same set of data is also used to determine the tool-length, tool-radius, tool-compensation, etc, required in the process. Process data consists of switching functions for adjusting feed rates, spindle speeds, tool changes, cutting fluid application, etc. The switching functions are initiated by switching commands to an interface unit, where they are compared with feedback signals from the machine tool and then are translated into appropriate control signals for the particular device to be operated. In addition, a linkage is provided with safety switches as a hazard precaution, in order to stop the machine in case of conflicting instructions, which may damage the machine or cause injury. The geometric data can be used only after adjustments are made in order to fit the particular tool-workpiece relationship. This "fitting" allows the programmer to edit a program irrespective of the actual position of the workpiece, etc. Corrective calculations, such as the length of the drill, the size of the turning tool or the diameter of the milling cutters, etc. could be made regarding the dimensions of the tool.

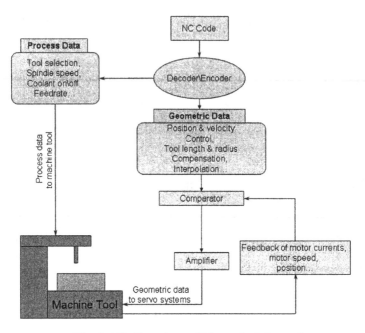

Fig. 3.118. Functions of Numerical Control

The complexity of the control system is directly related to the type of motion that can be performed. The main classifications (Fig. 3.119) of control strategies and interpolators, are point-to-point (positional control), line motion control (linear interpolation), and continuous path or contour control. Point-to-point control can be used when the requirements are simple positioning of a tool. The principal function of the point-to-point positioning control is to position the tool from one point to another within a coordinate system. The positioning may be linear in the X-Y plane, or linear and rotary if the machine has a rotary table. Each tool axis is controlled independently; therefore, the programmed motion may be simultaneous or sequential, but always in rapid traverse [44]. When positional control is used, the tool is not active while moving from one position to the other and the path is a random one [45]. The programmer needs to know that he will not encounter any obstacles while using this control. The most common applications of the point-to-point control are drilling, boring, tapping, riveting, pipe bending, sheet metal punching and spot welding. When linear control is applied, the final position of the movement is reached in a straight line and the tool may be in operation during the positioning motion. Tool and machine table movements, in any desired circular or curved path, can be performed with continuous path control where the machine slides and the tool move simultaneously in a coordinated fashion to achieve the desired path. Most continuous path controls are based on straight-line movements and curves, which are produced by the combination of a series of cords. An interpolator performs the necessary calculations to convert given displacement information into a series of coordinate movements along the axis with appropriate direction, velocity, and feed rate for the feed drive units. The information from the interpolator is fed to the comparator, where it is compared with that from the positional measurements. The deviation is amplified by an amplifier and is fed back to different motors of the machine for obtaining the desired motions.

Type of Control	Problem	Tool Action	Application
Point-to-point or positional control	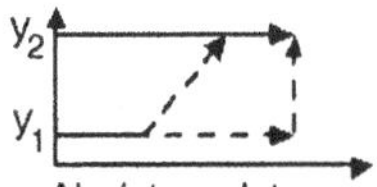 y_2 y_1 No interpolator	Not cutting during table movement	Drilling, spot welding
Line-motion control (simple)	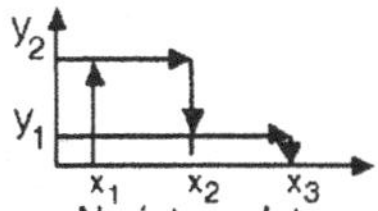 y_2 y_1 x_1 x_2 x_3 No interpolator	Cutting during table movement	Parallel turning, milling
Line-motion control with linear interpolation	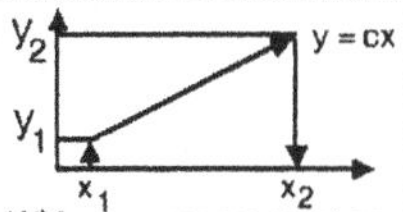 y_2 y_1 $y = cx$ x_1 x_2 With gear engagement or linear interpolator	Cutting during table movement	Turning Milling
Continuous path or contour control	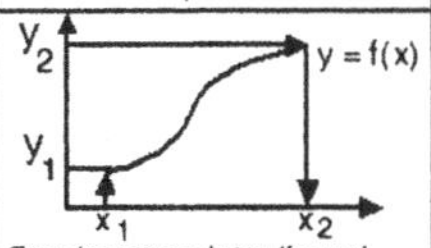 y_2 y_1 $y = f(x)$ x_1 x_2 Circular interpolator (based on equation of order 2 or higher)	Cutting during table movement	Turning, Milling, Flame cutting (any contour)

Fig. 3.119. Classifications of Control Strategies

Interpolation usually makes up a significant portion of the effort needed to produce complex contours. In interpolation, the geometric information related to a linear or continuous path of control is translated into axis-oriented coordinate commands of movement so that the velocity vectors of the control axes correspond to the required contour. This function is incorporated into a geometric processor, which has the major task of calculating a large number of positional coordinate values. One way to go about creating this necessary interpolation is shown in Fig. 3.120. The contour pattern is defined by input values P_1, P_2, and P_3. At the first stage, the interpolator calculates the reference points P_{11}, P_{12}, P_{13}, and so on, while a second, finer interpolation scheme determines the intermediate coordinate values x and y between the reference points. The two main interpolation schemes, used in machine tools, are linear and circular interpolations.

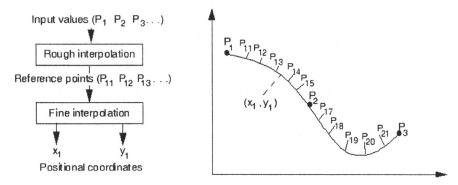

Fig. 3.120. Interpolation in Two Stages

Many interpolation techniques have been developed for determining the positional values on each individual axis. One such technique is presented in Fig. 3.121, where the straight line shown is to be interpolated in a positive x direction between the points P_S and P_E. Starting from P_S, it is incremented on the x-axis to a new position P'. A negative deviation (F<0) indicates a correction in the y direction, while a positive deviation (F>0) indicates a correction in the x direction. If F is zero, the next increment can be either in the y or x direction. The advantage of this method is that the error is related only to the individual incremental length, and not propagated throughout the interpolation process. This method can also be applied to circular interpolation (Fig. 3.122).

Positioning accuracy in NC machines is defined by how accurately the machine can be positioned to a certain coordinate system. An NC machine usually has a positioning accuracy of at least $\pm$ 3μm (0.0001 in.). Repeatability, defined as the closeness of agreement of repeated position movements under the same operating conditions of the machine, is usually around $\pm$ 8μm (0.0003 in.). Resolution, defined as the smallest increment of motion of the machine components, is usually about 2.5μm (0.0001 in.).

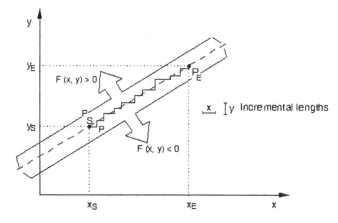

Fig. 3.121. Interpolation Method for Linear Interpolation

The stiffness of the machine tool and backlash in its gear drives and leadscrews are important for accuracy. Backlash can be eliminated with special backlash take-up circuits, whereby the tool always approaches a particular position on the workpiece from the same direction, as it should be done in traditional machining operations. Rapid response to command signals requires that friction and inertia be minimized–for example, by reducing the mass of the machine's moving components.

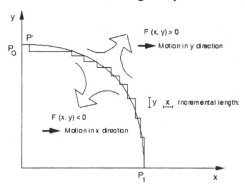

Fig. 3.122. Interpolation Method for Circular Interpolation

Controllers can be implemented using a programmed processor, a dedicated motion controller, or in simple cases, an analog circuit. Digital controllers have three main attributes:

- Speed: A/D and D/A conversions and control law calculations require time and restrict the speed at which a controller can update a control signal. The benchmark for controllers is the servo loop time (the period between updates of the control signal).

- Resolution: All digital signals are quantized. Quantization is limited by the complexity of the D/A and A/D conversions. The quantization in the processor is often no longer an issue in most modern controllers.
- Cost: This is usually directly proportional to controller speed and resolution. Another factor affecting cost is the flexibility of the controller.

The dedicated motion control chip is the most commonly used point-to-point controller. The controller receives its commands from the process controller, which can make real-time adjustments of the control law parameters and the command inputs. Some controllers will also perform functions, such as interpolation and motion profile generation.

After the input data has been transformed into specific codes and signals, it is used to drive the motors to position the machine slides to the programmed position [44]. There are two basic types of motion control systems: open loop and closed loop. *Open loop control* relies on a predetermined input with no feedback of the actual motion (output) of the system. A typical application of the open loop control system is the NC drilling machine [44]. In the open loop control systems, the motor continues turning until the absence of power indicates that the programmed location has been attained and the driving mechanism is disengaged. In *closed loop control*, the motion of the tool is monitored by measuring devices and is compared with the desired motion by the presence of feedback. The controller acts to minimize the error between the actual and desired motion. Most actuators require closed loop control for satisfactory performance. The goal of the closed loop system is to complement the system's dynamics and to force the error between desired and actual motion to zero. More advanced closed loop control systems will also compensate for unpredictable disturbances and nonlinearities in the plant.

CNC systems use two different feedback principles. The indirect feedback, which monitors the output of the servomotor as shown in fig. 3.123a. The direct feedback monitors the load condition in the feedback loop (Fig. 3.123b) and that is the reason that is more accurate than the indirect feedback [44].

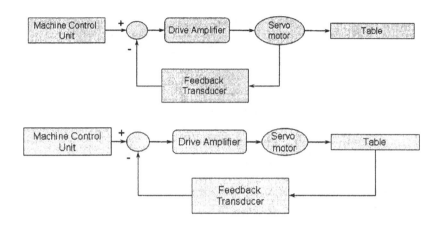

Fig. 3.123. a) Indirect and b) direct feedback principles used in closed loop systems

For precision applications, high end processors, such as digital signal processing (DSP) chips, replace the motion processor. DSP chips are generally fully programmable and can be implemented practically with any control algorithm. These chips perform arithmetic functions in the hardware and are consequently extremely fast. A low end DSP processor can perform a PID calculation in less than 3 µs, an order of magnitude faster than that of the fastest dedicated programmed controller. Machine tool manufacturers that provide 32 bit control often use DSP technology. Use of these advanced processors allows the control system to fully take advantage of very precise actuators and high resolution sensors, such as laser interferometers.

The point-to-point control of a machine tool slide (Fig. 3.124) requires a control law which will compensate for disturbances, such as the tool dynamics. The goal of any algorithm is to achieve the fastest and most precise point-to-point movement. From a control perspective a machine tool is essentially a "black box" or "plant", which contains the actuator and all the associated mechanical parts. The input is the actuator control signal, and the output is the information on the motion of the plant (position, velocity, end-point interrupts). Most control methods require a nearly exact input/output model or transfer function of the plant in order for the plant to perform effectively.

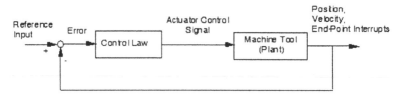

Fig. 3.124. Point to Point Control for Machine Tool

The most common controllers today are *fixed gain linear controllers*. Given a good model of the plant, a controller can be designed to attain optimum response to a commanded input to the system. Standard fixed gain linear controllers are:

- PID
- Lead-lag or other pole zero filter
- State feedback
- State feedback with full state or reduced order observers

Adjustable gain linear controllers are more complex controllers, which fall under the category of adaptive control. If the dynamics of the plant change, the controller gains are adjusted in real time, based on a recursive estimation of the plant parameters, in order to compensate for these disturbances. However, these controllers only perform adequately for slowly varying parameters.

The dynamic feedback control can be contrasted with adaptive control as follows: (1) Dynamic feedback control has a fixed controller mechanism that adapts or adjusts controller signals in response to measured changes in system behavior. A constant-gain control is a special case of dynamic feedback control, gain being defined as the rate of output to input in an amplifier. (2) Adaptive control adjusts not only the controller signals, but also the controller mechanism.

Gain scheduling is perhaps the simplest form of what is now called adaptive control. In gain scheduling, a different gain for the feedback is selected depending on the measured operating conditions. A different gain is assigned to each region of the system's operating space. With advanced adaptive controllers, the gain may vary continuously with changes in operating conditions.

The development of control motor technology is critical for machine tools to execute high precision manufacturing operations. Servo systems, based on digital technology, have increasingly been used instead of analog servo systems. Digital servo systems lack the servo delay of analog systems and enhance the operating precision and high speed response. An additional advantage of digital servo control is that it allows the use of nonlinear control to minimize the effects of static friction. Recent advances in digital servo technology include improvements in acceleration/deceleration control, at high speeds, in order to minimize shock to the machine.

3.4.2 Parts Programming

A part program can be defined as a set of alpha-numeric instructions, which indicate the type and the order of the individual operations for the production of a workpiece on a NC machine tool. The existence of the part program ensures that the same part will be produced the same way every time. The cost of the part programming process can be divided into direct and indirect costs. The direct cost is proportional to the length of the editing time. The associated indirect costs include the cost of verifying the program, the quality of the finished parts, the accuracy of the machined part compared to its design, and the cost of programming errors. For parts that require tens or hundreds of hours of machining time, the indirect costs can be substantial.

Before writing a part program, a complete engineering design (e.g. part geometry and tolerances), and a process plan (cutting speeds, feeds, sequence of operations, etc.) must be constructed. The procedure for producing a part program consists of a number of steps (Fig. 3.125).

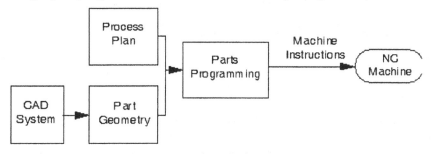

Fig. 3.125. Process of Producing a Part Program

Part programming methods can be classified according to the degree of automation involved:

- Manual part programming
- Computer-assisted part programming
- NC programming using CAD/CAM
- Computer-automated part programming

Manual Part Programming

In manual part programming, the NC program is entered directly into the NC controller by the machine operator or part programmer. Programming a contour is done by stating its beginning and ending coordinate positions and how these points are connected.

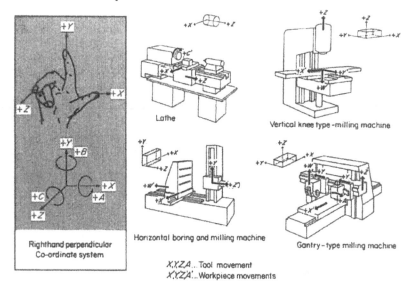

Fig. 3.126. Coordinate System for Machine Tools [1]

For describing the motions during machining, it is necessary to define a coordinate system and a set of datum points within the working space (Fig. 3.126). Usually the X, Y, and Z axes describe a Cartesian coordinate system with the Z-axis being parallel to the machine's spindle and the X-axis being the main axis of the positioning plane which generally is parallel to the clamping surface of the workpiece and runs horizontally whenever possible [1]. If there are axes parallel to the X, Y, and Z, they can be designated U, V, and W while the letters A, B, and C are used for identifying rotary axes where the positive direction of a rotation is taken to be that of a right-handed screw. In order to produce a relative change in the position between the tool and the workpiece, either the tool or the workpiece must move. If the tool is moved, then a positive direction of the movement coincides with the positive direction of the corresponding axis and these are identified with +X, +Y, etc. If, on the other hand, the workpiece moves, a positive movement will be opposite to the axis convention and the movement will be designated + X', +Y', etc. The first case is the workpiece co-

ordinates and the second one is the machine coordinates. There is a similar convention for rotary axes. However, the program assumes that the
tool will move in relation to the coordinate system of the workpiece. In
addition to the coordinate systems [41], a number of different datum (Fig.
3.127) and reference points are defined for every numerically controlled
machine.

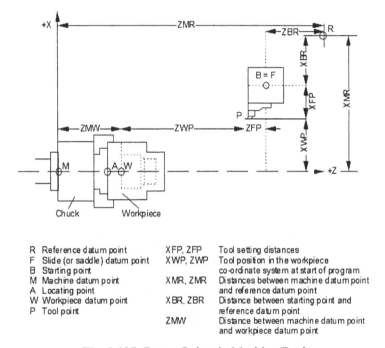

R	Reference datum point	XFP, ZFP	Tool setting distances
F	Slide (or saddle) datum point	XWP, ZWP	Tool position in the workpiece
B	Starting point		co-ordinate system at start of program
M	Machine datum point	XMR, ZMR	Distances between machine datum point
A	Locating point		and reference datum point
W	Workpiece datum point	XBR, ZBR	Distance between starting point and
P	Tool point		reference datum point
		ZMW	Distance between machine datum point
			and workpiece datum point

Fig. 3.127. Datum Points in Machine Tools

While structuring a part program for an NC machine, one must consider
issues related to the workpiece and the machine. The workpiece information to be considered includes both the geometric (e.g. the unmachined and
finished dimensions of the part, the dimensional tolerances, the type and
order of the required operations, etc.) and the technological information
(e.g. the workpiece material, the cutting tool material, the stiffness of the
workpiece and machinability data, related to the workpiece and cutting
tool material etc.). As far as the machine is concerned, it is important to
consider the working space, the type of clamping and fixturing devices, the
speed and feed range, the shape and form of the different tools, the precision of the machine, and any special auxiliary equipment specific to the
machine. Having this information, one can start structuring the part program by building the NC blocks, which collectively make up the entire

program (Fig. 3.128). There is a set of standard commands whose meaning follows the international standards (Fig. 3.129).

```
:02
N05 G21
N10 G91 G28 X0 Y0 Z0
N15 G40 G49 G80
N20 M06 T01
N25 S4550 M03
N30 G90 G00 G43 X-10 Y-10 Z35 H01
N35 G01 Z0 F682 M08
N40 G42 X0 Y0 D01
N45 X56
N50 Y55
N55 X12
N60 X0 Y43
N65 Y0
N70 G40 X-10 Y-10
N75 Z35 M09
N80 G28 G91 X0 Y0 Z0
N85 M02
```

Fig. 3.128. Part Program

Code	Function
G00	Rapid traverse
G01	Linear interpolation(feed)
G02	Circular interpolation CW
G03	Circular interpolation CCW
G04	Dwell
G07	Imaginary axis designation
G09	Exact stop check
G10	Offset value setting
G17	XY plane selection
G18	ZX plane selection
G19	YZ plane selection
G20	Input in inch
G21	Input in mm
G22	Stored stroke limit ON
G23	Stored stroke limit OFF
G27	Reference point return check
G28	Return to reference point
G29	Return from reference point
G30	Return to 2nd, 3rd and 4th reference point
G31	Skip cutting
G33	Thread cutting
G40	Cutter compensation cancel
G41	Cutter compensation left
G42	Cutter compensation right
G43	Tool length compensation+direction
G44	Tool length compensation-direction
G45	Tool offset increase
G46	Tool offset decrease
G47	Tool offset double increase
G48	Tool offset double decrease
G49	Tool length compensation cancel

Code	Function
M00	Program stop
M01	Optional stop
M02	End of program, no rewind
M03	Spindle CW
M04	Spindle CCW
M05	Spindle stop
M06	Tool change
M07	Mist coolant ON
M08	Flood coolant ON
M09	Coolant OFF
M19	Spindle orientation ON
M20	Spindle orientation OFF
M21	Tool magazine right
M22	Tool magazine left
M23	Tool magazine up
M24	Tool magazine down
M25	Tool clamp
M26	Tool unclamp
M27	Clutch neutral ON
M28	Clutch neutral OFF
M30	End program, rewind stop
M98	Call sub-program
M99	End sub-program

Fig. 3.129. Example of Program Commands (G-Code and M-Code)

The major effort in producing an NC program is spent on determining tool movements. The coordinate values of the end of a movement within the workpiece coordinate system are programmed, using the workpiece datum point, W, as a reference (Fig. 3.127). The position of the workpiece, and the position of each datum point in the working space of the machine is determined by the dimensions of the workpiece and the way it is fixtured. Therefore, the location of the workpiece datum point is one of the first instructions to the controller. Referencing is achieved by programming a datum shift (zero offset) on machines, which have absolute measuring systems (dimension ZMW in Fig. 3.127). Furthermore, tool setting dimensions such as the XFP and ZFP must also be programmed. If the programmer knows the clamp position of the workpiece, the position of the tool, and the tool length, the distances of the tool point from the workpiece datum point can be calculated. Once these dimensions – XWP and ZWP – have been fed into the controller, as positional values before the first axial movement, then the workpiece datum is fixed. Thus, with every tool change, a new tool position needs to be reprogrammed. In general, all the dimensions of the part to be machined are given from a datum face, edge, or point. In addition to the geometric statements, an NC program contains the necessary technological statements, which state the process parameters, such as cutting speeds and feeds. These parameters usually come from process planning.

Computer-Assisted Part Programming

In order to reduce the programming effort, computer-aided programming techniques have been developed that describe the work to be done on the different parts/components, in a problem-oriented language, using mnemonic technical expressions. The most universal programming language is APT (Automatically Programmed Tools), which is suitable for both simple and complex production processes and can be used for machines with up to five control axes. With APT, machine control instructions can be produced for machines operating in either a continuous path or in point-to-point manner [44]. For a continuous path operation, the geometry of the part must be described in terms of curves whereas for point-to-point operation, the geometry of the part must be described in terms of points. In addition to APT programming language, other languages, operating in the same way in general, are ADAPT (ADaption of APT), AUTOSPOT (AUTOmaticSyste, for POsitiong Tools), EXAPT (Extended subset of APT), COMPACT and SPLIT (Sundstrand Processing Language Internally Translated). The part program is developed independently of the machine tool and, subsequently, is translated into a machine code suitable for the particular machine and controller combination with the help of a program called *post processor*. The majority of the part programming software products available today uses wireframe geometry and requires interaction with the programmer to generate the NC code. For the majority of software producers, the main emphasis within the last several years, has been given on making the software more user-friendly.

NC Part Verification

In using an NC milling machine, the consequences of an error can be very costly. One incorrect character in a 10,000 character NC program (typical size of an average part program) can break a tool, ruin a work-piece, cripple the machine, or possibly cause injury. Some of the very large dies used to stamp exterior automobile body panels can require up to three weeks of continuous machining. Any errors require expensive re-working and can result in significant delays in production schedules. Efforts to detect and prevent these costly errors, during the machining process, fall under the heading of NC Verification. The cost of NC verification has traditionally been a large portion of the cost of NC machining.

While it is true that new techniques in automatic generation of complex NC tool paths simplify the NC verification process, current generation techniques still cannot guarantee that the tool path generated will properly produce the desired surface during the first iteration. Improper part programming or flaws in the path generation software can result in work pieces being gouged (cut too deeply) or undercut. Also, there is the possibility that the tool may interfere with the holding fixtures or other potential obstacles. Errors of any type can add significantly to material cost, setup time and machining time and therefore, have to be eliminated.

A program verification procedure is the simulation of the process by the visual display of the tool paths on the computer's graphic display. During this procedure, moves in the axes do not take place, allowing the programmer to diagnose any alarm and thus, fix the problem quite fast.

Currently, the common practice for machine shops is to manually check NC programs before use. This checking may take the form of "cutting air," cutting a soft material or cutting at a reduced speed with an operator closely monitoring the situation.

3.4.3 CAD/CAM

Computer-Aided Design (CAD) is the technology concerned with the use of computer systems to assist in the creation, modification, analysis and optimization of a design [46]. The basic concepts of computer-aided design (CAD) have begun with the development of interactive computer graphics. The Sage Project at the Massachusetts Institute of Technology

(MIT) aimed at developing CRT displays and operating systems. A system called Sketchpad [47] was developed under the Sage project. A CRT display and light pen input were used to interact with the system. These developments occurred at about the same time that NC and APT first appeared. Later, X-Y plotters were used as the standard hard-copy output device for computer graphics. An interesting note is that an X-Y plotter has the same basic structure as an NC drilling machine except for a pen substituted for the tool on the NC spindle.

In the beginning, CAD systems were nothing more than graphics editors with some built-in design symbols. The geometry available to the user was limited to lines, circular arcs, and to the combination of the two. The development of freeform curves and surfaces, such as Coon's patch, bicubic patch, Furguson's patch, Bezier's surface and B-Spline surface, enabled CAD systems to be used for sophisticated curves and surfaces design. With the development of three-dimensional CAD systems CAD models contained enough information for NC cutter-path programming. The link between CAD and NC began with the so-called turnkey CAD/CAM systems, which were developed in the 1970s and 1980s.

In the 1970s three-dimensional solid modelers were developed. Prior to the development of solid modelers, three-dimensional wire-frame models represented an object by bounding only its edges. Wire frame models are ambiguous in the sense that several interpretations might be possible for a single model. In addition, there is no way of determining volumetric information from a wire-frame model because these kinds of models do not contain any information about the surfaces themselves and in addition, do not differentiate between the inside and the outside of designed objects. Three-dimensional wire frame modeling represents part shapes with interconnected line elements, giving the simplest geometric representations of objects. Although wire-frame models provide accurate information about the location of a surface discontinuity on a part, they usually do not provide a complete description of the part. After wireframe modelers, surface modeling was developed. Surface modeling is the creation of a surface, or profile, through a series of points, curves and/or lines. In surface modeling, in addition to the information about the characteristic lines and their points, the mathematical description contains information, regarding which surfaces are connected and the way they are joined. Surface based CAD systems can represent many types of surfaces such as flat, analytical, swept and free form. Bezier curves, B-splines and Non-Uniform Rational B-Splines (NURBS) are the most common techniques used for surface modeling. Solid models contain complete information; engineering analysis can be performed on the same models that are used and produce engineering drawings. It was not until the mid-1980s that solid modelers were used in the design environment. It was during the 1990's that the maturity of the computer hardware allowed for evolutionary applications of the CAD systems. Solid modeling became the standard way of modeling even in small platforms [48]. The solid modeling system provides a complete unambiguous and realistic description of solid objects. Object models can be rotated, shaded, or even cut by a cross section to show their interior details. Solid objects can also be combined with other parts stored in the database to form a complex assembly of the part, whose design has to be carried out. Furthermore the solid model, not only does it depict the interior properties such as size and shape, but also depicts interior properties, such as mass or even finite element analysis data.

The computer's visualization power allowed for the photo-realistic inspection of very big and very complex designs. Design functionalities have been greatly evolved with the use of parametric design and feature based modeling. The advanced CAD systems of the 1990's provide a fully integrated environment with a plethora of CAM/CAE tools, thus allowing the engineer to perform a variety of simulations upon his/her designs. The basis of any manufacturing effort is the design drawing. Painstakingly prepared drafts and blueprints are used in every step of a product's inception, design, and construction. But like the old-fashioned office, drafting is paperbound and is very labor intensive. Unlike the electronic data in a computer, paper drawings cannot be manipulated, processed, and quickly analyzed.

CAD can be most simply described as "using a computer in the design process." In the design process, a computer can be used in both the representation and the analysis steps. The application of CAD for representation is not limited to drafting.

Most of today's CAD/CAM tools use solid modeling. Solid modeling is implemented mainly by two methods: Boundary representation (B-Rep) and constructive solid geometry (CSG). In B-Rep systems, the objects are represented by their bounding faces while in CSG systems, the objects are modeled using primitive shapes as building blocks.

Parametric modeling allows the user to create product models with variational dimensions. Dimensions can be linked via (conditional) expressions. Bi-directional associativity, between the model and the dimensioning scheme, allows the automatic regeneration of product models after changes in dimensions and in the automatic updating of the related dimensions have taken place. In this way, a "flexible" product model can be created. Many CAD systems offer limited 2D parametrics. Recent full parametric 3D systems have been introduced. Both are based on B-rep and enable feature based modeling with a set of primitive features, which can be combined to more complex "user features".

Feature modeling is based on the idea of designing with "building blocks". Instead of using analytical shapes such as boxes, cylinders, spheres and cones as primitives, the user creates the product model, using higher level primitives, which are more relevant to the specific application. This approach should make solid modeling systems easier to use. Generic feature-shapes can be formalized in a canonical form. A combination of generic parameters and lists of geometrical and topological entities and relations can describe Generic feature-shapes. The engineering significance of a feature may involve the function, which the feature serves, the way it can be produced and the actions its presence must initiate, etc. Features can be thought of as "engineering primitives" relevant to some engineering task.

To aid in engineering analysis, there are packages that perform kinematic simulation and finite-element modeling for different engineering analyses. CAD systems frequently consist of a collection of many application modules under a common database and graphics editor.

A CAD system is a tool used to automate design work. Instead of using a pencil and paper, the designer uses a sophisticated computer graphics display. Solid shapes, called from a computer library of geometric primitives, can be assembled to form complex shapes. A CAD system can eliminate the tasks of searching through blueprint libraries, transcribing what has been drawn before, or redrawing existing figures or part assemblies. Using simple commands, a designer can display a cutaway of a part, a closeup view, or an exploded view of all the parts in an assembly. Using a CAD system, the complete geometry of a part can be visualized without the necessity of viewing different layers of a drawing. Any product or part that was drafted on paper can be drafted using a CAD system [49]. Some CAD systems are more complex than others, and their design capabilities are limited by the machine itself. However, even the most limited CAD system typically offers significant productivity gains over traditional manual methods of hand drawing.

A great number of CAD tools exist in the market today. Rather than single s/w tools, most of them are comprised of a family of design and simulation tools. Those tools can work independently on the same designs or integrated forming a modular extensive application. Some of the most important ones are: CATIA by Dassault Systems, a tool with many facilities in the design of difficult three-dimensional representations by using surface or solid modeling. It provides solutions in a great number of areas, with solutions in the industrial design through a set of application specific modules for CAM/CAE.

Pro/Engineer by Parametric Technology Corp., is a tool based upon parametrical design, offering a big set of design functionalities and engineering analysis tools. I/DEAS by SDRC, is a tool with many features for surface design, having lots of applications in mechanical and architectural engineering. Mechanical Desktop by AutoDesk, is a tool for PC platforms, having reached to a level that is close to its competitors on the big Unix workstations. It offers a variety of functionalities for parametric and feature based designing and of course, many other equally important ones (more than 100 tools are currently available on the market).

Many manufacturers of discrete products have employed NC machine tools for decades. For machine tools, such as a milling machine or lathe, the part program describes the path that the cutter will follow, as well as the direction of rotation, rate of travel, and various auxiliary functions such as coolant flow. Whether it is a simple two-axis drill or a complex five-axis machining center, all NC machines require part programming. It often takes several hours or days to program a part on a machine tool [50].

Traditionally, programs for NC machine tools have been created using manual part programming or computer-assisted part programming. Simple programs are often created manually, perhaps with the aid of a calculator, and more complex programs are usually created with the use of a computer and a part-programming language, such as APT. Since part programmers have difficulty in keeping up with the demand for new part programs, there is a strong incentive to develop efficient procedures capable of replicating the human part programming process and of installing these procedures on a CAD/CAM system.

Creation of NC programs from CAD files allows a part programmer to access the computer's computational capabilities via an interactive graphics display console. The part-programming operation typically starts with a design in the form of a CAD drawing or model. After a review by a production planner, the tool design/selection process is completed, often with the assistance of the CAD system. Part programming through CAD allows the part geometry to be described in the form of points, lines, arcs, etc., just as it is on an engineering drawing, rather than be translated into a text-oriented notation. CAD/NC systems allow the user to rapidly define the geometry as well as to use powerful graphics display capabilities to quickly define, verify, and edit the actual cutter motion. The programmer generates a cutter path by selecting geometric elements with a digitizer (such as a light pen, mouse, etc.) attached to the display terminal. Software subroutines are often used to execute common machining operations (e.g., milling a pocket in a part) thus simplifying tool path generation. Various auxiliary and postprocessor commands can also be entered at the terminal. A CAD system can display the cutter-path geometry, allowing early verification of a program without actually using valuable machine tool time. The computer can assist a part programmer by simulating the entire tool path on the display terminal by having graphic animations, showing the location of the cutter visually, and by displaying the X, Y and Z coordinates (Fig. 3.130). Editing may be done interactively during the replay to correct errors or make changes to the program.

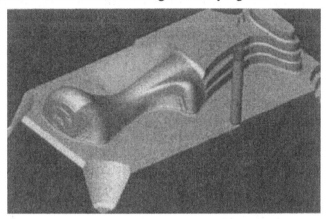

Fig. 3.130. Example output of NC tool path simulation [46].

After verification, the various cutter passes are combined to form a program, which is then stored for processing by different postprocessors. These post processors produce programs that can be interpreted by a specific machine tool (Fig. 3.131). Finally, the NC program is produced.

Several CAD/CAM systems, such as CADAM, Computervision, CATIA, etc., have the capability of generating NC machining instructions, based on the geometric definition of a workpiece. Facilities currently exist where parts are designed on a CAD system, production plans are created from the CAD data, using an automated process-planning system, numerical-controlled part programs are created on a CAD system, using the tool-path requirements, and parts are manufactured under the control of a computer.

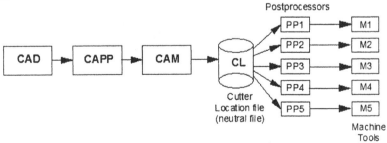

Fig. 3.131. Creation of Part Program from CAD File

Virtual Reality and Manufacturing

In the early 90s, the concept of Virtual Manufacturing (VM) was introduced. Defined as the concept of executing manufacturing processes in the computer as well as in the real world, VM has provided the theoretical infrastructure upon which new approaches to manufacturing simulation have been presented. VM is a concept, which summarizes computerized manufacturing activities dealing with models and simulations instead of objects and their operations in the real world. Within this concept, Virtual Reality has been recognized as the appropriate technology for the simulation of the physical aspects, behaviors and interactions of a Virtual Manufacturing System [51].

Virtual Reality (VR) is a means by which humans visualize, manipulate, and interact with computers and extremely complex data. Methods of human-computer interaction are called computer interfaces, and VR can be considered as the newest, in a long line of interfaces [52]. Virtual environment systems differ from traditional simulator systems in that they rely much less on physical mock-ups for simulating objects within the reach of the operator, being in this way more flexible and reconfigurable. Virtual environment systems differ from other previously developed computer-centered systems in the extent to which real-time interaction is facilitated, the perceived visual space is three-dimensional rather than two-dimensional, the human-machine interface is multimodal, and the operator is immersed in the computer-generated environment [53].

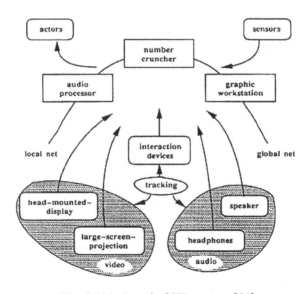

Fig. 3.132. A typical VR system [54]

Virtual Reality applications need all the system power they can get. Therefore VR systems have to deal with a number of different and physically displaced hardware devices in order to exploit existing resources. Fig. 3.132 shows a typical set of devices and computing nodes that are handled by such systems. The typical hardware infrastructure consists of a heterogeneous set of processing nodes with a fast local connecting net, such as ethernet or FDDI, and access to the world through slow global nets, namely the internet. Processing nodes range from simple personal computers, multimedia workstations, high performance graphic workstations to special-purpose hardware devices [54]. Such devises support the immersive and interactive interfacing of the user with the virtual environment and include Head Mounted Displays, Data Gloves, Position/Orientation Trackers, 3D Mice, Stereoscopic Glasses, Haptic / Force Feedback devises etc.

While industrial applications of VR are approaching a first level of maturity, a number of manufacturing related demonstrators have been recently presented in literature. A Virtual Factory has been developed to provide capability of defining, modeling, and carrying out manufacturing processes in a virtual environment [55]. The environment enables engineers to produce the designs of the components manufactured, part programs and process plans for their manufacture. A VR based system, integrating virtual product development tools, called VEDAM (Fig. 3.133), has been also presented [54]. One of the major components of the system, the Virtual Manufacturing Environment, is an immersive virtual environment that allows the user to analyze and develop process plans using a virtual factory that replicates and functions as the real factory. The Virtual Machine Shop (Fig. 3.134) is another virtual environment developed to provide the capability of virtually performing machining processes [56]. The features of the environment enable the user to set-up a process, operate a machine tool and execute NC programs in an immersive and interactive way. Additional functions are provided to support verification in terms of geometrical, technical and economic characteristics.

Fig. 3.133. A table-top milling machine of the VEDAM environment

Fig. 3.134. NC Part Programming in the Virtual Machine Shop

3.4.4 Process Control

Basic Concepts

Machine control includes measurement and control of the position and velocity of the spindle and feed drives. Automatic process control implies the use of sensors to monitor the process behavior and adjusts parameters, such as the speed and feed, in order to obtain adequate operation of the process.

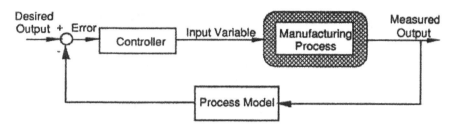

Fig. 3.135. Adaptive Control Scheme

Parameters of numerically controlled machine tools, such as the speed and feed, are usually selected based on previous experience and handbook information. However, a number of disturbances can occur during a process including chatter, vibrations of the machine tool frame, or variations in the workpiece material properties. *Adaptive control* attempts to adjust the process parameters in order to adapt to such disturbances (Fig. 3.135). There are two kinds of adaptive control approaches. The first is *adaptive control constraint* (ACC), which attempts to maximize the use of the machine tool and/or the tool capabilities by constraining the process within certain boundaries. An example of such a constraint is *maximum load*, to prevent machine overloading. The other type of control system is *adaptive control optimization* (ACO), which attempts to optimize the production process with respect to a set of optimization criteria. Commercially available systems implement adaptive control constraint and not adaptive control optimization because the latter requires more elaborate mathematical process models and more sophisticated approaches to process monitoring.

In adaptive control [57], the four critical elements involved are the *sensors, actuators, controller*, and *process models*. Sensors on machine tools usually measure process variables such as power, force, acoustic emissions, surface temperature, vibrations, etc. Sensors are one of the most critical elements necessary for the successful application of adaptive process control. However, the development of robust sensors, which can accurately and reliably monitor the process has proven to be a difficult task. Actuators are hardware devices that accept the in-process control system commands as their input and convert these commands into adjustments of the process parameters. The controller is normally an electronic hardware device that receives information from the system's sensors and utilizes a process model to compute responses to the system's actuators. Process models are the mathematical descriptions which correlate the process variables measured by the sensors (e.g. cutting force) with the process parameters (e.g. feed and/or speed).

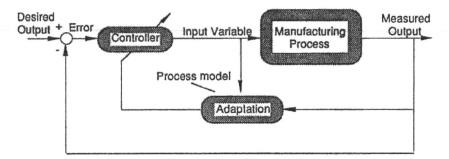

Fig. 3.136. Adaptive Control Scheme with Model Parameter Adaptation

An adequate mathematical description of the process may be very complex because a number of physical phenomena interact during the actual process. This problem has been addressed by means of adaptation (Fig. 3.136) whereby efforts are made to adapt the model parameters in such a manner that the model behavior closely reflects the behavior of the actual process. Typically an adaptive control approach consists of two control loops. The first loop is the ordinary feedback loop consisting of a controller and the process. The parameters of the controller are adjusted to the second loop so that the error between the outputs of the process and a reference model is minimized.

Sensing Techniques and Devices

The overall goal of automation for manufacturing processes necessitates the establishment of control loops by providing some feedback information about the state of the manufacturing process. This function is carried out by human supervision in conventional process control. With increasing automation, effective sensor systems are needed in order to achieve the automatic control, supervision and error recovery tasks, required to realize the full potential of intelligent manufacturing.

In the conventional case, most of the sensing functions are fulfilled by human perception and evaluation. The push toward unattended machining and "lights-out" factories requires high levels of automatic machine monitoring. A sensor system must reach at least the level of effectiveness of the human sensing capabilities in order to replace a human operator (Table 3.1). Sensors having sensing and processing capabilities equal to or exceeding those of humans can be called "intelligent" sensors.

Process sensors can be divided into three categories according to their function: tool monitoring, workpiece monitoring, and machine performance monitoring.

Human Sense	Human Monitoring Activity	Possible Machine Sensors
Vision	Establish datum points between incoming work material stock, fixtures, and cutting tools.	Touch-trigger probe
	Watch for sudden breakage of cutting tools.	Machine vision and touch-trigger probes
	Carry out between-pass visual checks of trends in flank and crater wear.	Machine vision
	Detect tool temperature by studying chip color.	Thermocouples
	Watch for changes in part surface finish.	Machine vision
Hearing	Listen for excessive vibrations between tool and part.	Accelerometers and dynamometers
	Listen for sounds of tools near or at failure.	Accelerometers, dynamometers, and acoustic emission devices
Touch	Feel excessive vibrations through floor or by touching fixtures.	Accelerometers
	Approximately sense excessive cutting forces by touching fixtures or tool-holders.	Dynamometers and strain gages
	Touch surface finish and approximately gage quality.	Stylus measurements of surface finish
Smell	Excessive tool temperatures sometimes change the smell of cutting fluids.	Chemical sensors

Table 3.1: Machine Sensors Related to Human Senses

Tool Monitoring

Tool sensors typically perform five tasks: (1) tool identification, (2) tool grip confirmation, (3) automatic tool change (ATC) and tool magazine operation confirmation, (4) tool tip position confirmation, and (5) tool condition monitoring [58].

1. Bar code sensors are most widely used for tool identification.

2. Proximity switches and limit switches are used for tool grip confirmation
3. ATC operation confirmation is performed using proximity switches and limit switches.
4. Mechanical contact sensors are used for tool tip position detection.
5. Various sensors are used for detecting tool wear and tool damage:
   - Load cells, proximity switches, and AE (acoustic emission) sensors are used for lathes.
   - AE and touch sensors are used in machining centers.
   - Some manufacturers have developed sensors specifically for monitoring grinding wheel wear.

An area currently receiving significant attention is tool condition monitoring. The objective is to diagnose and recover from problems without human intervention. Current research is focused on sensors and methods for monitoring tool wear.

Tool wear and failure detection methods can be divided into two approaches, direct and indirect. Direct methods involve taking measurements of the tool itself to determine wear or breakage of the cutting edge. Direct methods lead to lost production time because of the incapability of detecting failure while the tool is actually cutting. Indirect methods have the potential to alleviate these problems by using in-process measurements to monitor the tool condition. Direct methods may be acceptable for small lot sizes, where frequent access to the tool does not cost much production time. However, an indirect method is generally preferred. Most current research work is concerned with pattern recognition by establishing a correlation between sensor signals and tool wear. Progress in this direction would lead to more universally applicable monitoring systems [59].

Direct methods for tool condition monitoring:
1. Touch trigger probes can be used to check the dimensions of the workpiece. The deviation of the part geometry from the desired dimensions due to tool wear can be compensated for in subsequent cuts until the tool needs replacement.
2. There are several direct methods that rely on the higher reflective properties of the worn surface as compared with those of an unworn surface. The problem with these methods is that precise optics generally do not adapt well to the dirty conditions of the factory floor.
3. Computer vision techniques have also been considered as a direct method of monitoring tool wear [60, 61].

Indirect methods for tool condition monitoring:

1. The most straightforward of the indirect methods are tool/work displacement methods. The location of a machined surface is compared with the surface machined by an unworn tool. This approach is similar to the touch trigger direct method but the direct method's disadvantages are avoided if the measurements are made in-process.

2. Some electrical resistance methods rely on the area of contact between the tool and workpiece. The electrical resistance across the tool/workpiece junction increases as the tool wears. The problem with these methods is that the cyclic temperature variations from intermittent cutting operations will significantly influence the resistance.

3. Some radioactive techniques exist; the preferred method implants a small amount of radioactive material in the flank face of the cutting tool a known distance from the edge [59]. When the material is removed, the tool is discarded. However, there are questions about the safety of these methods, the costs of clean-up and disposal.

4. The most frequently considered measured variables for monitoring tool wear and detecting tool breakage are the cutting forces [62]. Many of these methods utilize an analysis of the frequency components of the force or torque signal [63]. Forces and torques are easily measured in-process. There are several ways of measuring the cutting forces:
   - a. tool-holder dynamometer on lathes,
   - b. table dynamometer on milling and drilling machines, and machining centers,
   - c. dynamometer built into the spindle bearings,
   - d. evaluating the cutting force based on the spindle motor current, voltage, and speed.

5. A substantial amount of research has also been performed to develop methods of using acoustic emission (AE) signals for tool wear monitoring and tool breakage detection [64, 65].

6. Efforts have also been made to correlate the tool temperature with the tool wear [66].

7. Sensor Fusion. Recently several researchers have investigated the possibility of fusing the information from a variety of sensors in order to obtain more accurate and reliable estimates of the tool wear [67].

Workpiece Monitoring

Workpiece monitoring sensors must monitor five items: (1) work identification, (2) workpiece dimensions, (3) mounting position, (4) confirmation of mounting/faulty mounting, and (5) unsatisfactory chucking or loose chuck.

1. Proximity and limit switches and mechanical touch sensors are used for work identification.
2. Workpiece dimensions are measured with contacting (stylus), electrostatic capacitance, ultrasonic and electromagnetic sensors.
3. Mounting position is detected by proximity switches, touch sensors, and air sensors.
4. Mounting confirmation is made with air sensors, air micrometers, pressure sensors, and touch sensors.
5. Chucking is checked with proximity switches, pressure sensors, and limit switches.

Machine Performance Monitoring

The goal of applying sensors for machine performance monitoring is to satisfy requirements in the areas of quality control and manufacturing automation, and to increase the competitiveness of the product. Satisfying these requirements demands that a greater number of sensors be used than that in a conventional setup (e.g. vision tasks, temperature distribution sensing, or tactile sensors). For monitoring machine performance, manufacturers use ammeters for the spindle motor to measure cutting and drive power. Some use strain gauges, load cells, or ammeters for the feed drive motor, and others use electromagnetic sensors [68].

In addition, new and more efficient approaches are needed to process the increased amount of sensory information. The information provided by these approaches will be utilized at higher levels of the manufacturing control hierarchy. The more advanced the sensory information processing is, the more complex the tasks that the control system can perform.

The successful implementation of an in-process control system requires, among other things, the linkage of sensors to the control computer [69]. Particularly for in-process control systems, sensors used to detect the current state of a process must be interfaced directly with a personal computer (Fig. 3.137). Furthermore, if a computer is being used to activate actuators in order to influence a process, these actuators must also be interfaced with the computer system. Interface requirements are not identical for all sensors and actuators. Requirements vary from one element to another, and the characteristics of the computer used must be taken into account. The interfacing requirements, which depend on signal levels, can be divided into three categories:

1. Low voltage input, in the range of millivolts
2. Low current output, in the range 4-20 mA
3. High voltage output, zero to several volts

A number of devices connected to the computer can be used to interface the computer with the sensors. The following devices are used for this interface purpose: (1) analog-to-digital converters (ADCs), (2) digital-to-analog converters (DACs), (3) multiplexers, (4) sample-and-hold circuits, and (5) amplifiers [69].

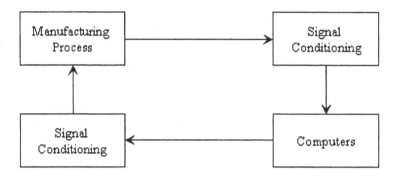

Fig. 3.137. 3.128 Computer Interfacing with a Manufacturing Process

Multiplexers and Demultiplexers

Very often in in-process control systems, data must be acquired from a number of sensors. For most in-process control systems, a processor, having a limited number of input ports, is used. Thus it is not possible to link a number of sensors directly to the processor input. A multiplexer (Fig. 3.138) can therefore be used to share an input port among a number of sensors and convert data from several input lines into one common output line. The multi-channel input end of the analog multiplexer is connected to a number of sensors, and the single output line is connected to a "sample-and-hold" circuit whose analog output is the input to the ADC.

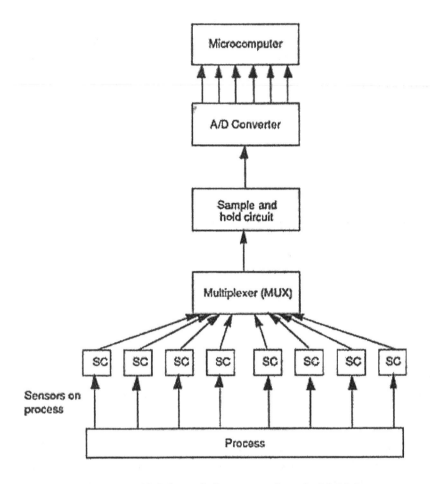

Fig. 3.138. 3.129 Schematic Representation of a Multiplexer

Amplifiers

Most in-process control systems involve the use of differential or instrumentation amplifiers, which amplify the maximum value of the lower-level sensor signals to the full-scale level, required by the analog-to-digital converter. Noise effects are reduced significantly with the use of differential amplifiers. It is possible to use differential or instrumentation amplifiers prior to multiplexing operations in order that the gain of individual amplifiers can be set to suit the level of electrical signal provided by the sensor. However, an arrangement of this type necessitates the use of as many amplifiers as the number of sensors and can therefore be costly. An alternative to this arrangement is the use of programmable gain amplifiers, whose gain can be increased or decreased under software control to suit particular sensors. A single programmable gain amplifier can then be used to properly amplify the signals from all sensors attached to the multiplexer. Important amplifier selection parameters are:

1. The gain range of the amplifier
2. The amplifier settling time
3. The linear or nonlinear nature of the amplifier
4. The time required to cover the entire amplifier range in response to a large step input
5. The common-mode rejection capabilities of the amplifier

An Adaptive Control Example

Real time control of the cutting forces in machining is important for a variety of reasons: better surface finish and part dimensional accuracy, prolonged tool life, and better machine utilization which results in high productivity. Force control for turning and milling is typically accomplished by adjusting the feedrate, which determines the chip thickness. The design of a controller requires a model of the dynamics, relating the command feedrate to the chip thickness, since the success of a controller is dependent upon a system's dynamic response. Furthermore, the feedback controller must account for the cutting force dependence on material properties, tool conditions, and cutting depth, which in general is difficult to accurately predict. Hence adaptive control is an attractive approach for controlling the cutting force. This section considers a controller, which adapts to changes in the cutting conditions and material properties. This adaptive scheme estimates only one parameter and is extremely simple to implement [57].

The adaptive control scheme utilizes a discrete time dynamic model of the cutting process. The model is based on input/output identification and geometric considerations for the chip thickness during the transient. The model accounts for:

1. Carriage motion dynamics
   Least squares identification based on the feedrate command and measured feedrate, provided the following second order discrete time model in order to represent the dynamics of the tool carriage and driver.

$$G_{tc}(z) = \frac{U_a(z)}{U_c(z)} = \frac{0.552z + 0.4529}{z^2 + 0.215z + 0.2466} \qquad (3\text{-}83)$$

where

U_c : command feedrate (volts)

U_a : actual feedrate (cm/sec)

2. Feedrate/chip thickness dynamics

Simple geometric consideration suggests that the chip thickness transient response for a unit step change in the feedrate be characterized by a straight line from the old steady state value to the new one (Fig. 3.139). It takes one full rotation of the spindle before the chip thickness settles down to a new value. A discrete time model to describe the chip thickness transient response is

$$G_{mc}(z) = \frac{f_r(z)}{U_a(z)} = \frac{B(z^4 + z^3 + z^2 + z + 1)}{z^5} \qquad (3\text{-}84)$$

where $B = 0.01$ for the uncut chip thickness/feed, f_r, and actual feedrate, U_a, in cm and cm/sec, respectively. This transfer function depends on the sampling time and the spindle speed.

3. Cutting process

Given the chip thickness, depth of cut, and other cutting conditions, as well as material properties, the cutting force is determined by the following equation.

$$F = K \cdot d \cdot f_r^p \qquad (3\text{-}85)$$

where

F : cutting force

K : constant

d : depth of cut

p : constant exponent

f_r : feed

In order to have a model linear with respect to the chip thickness, the exponent, p, in Equation (3-90) is assumed to be equal to unit. This assumption did not cause any serious difference between the predicted and experimental data.

The output of the force sensor was filtered to remove high frequency noise. The transfer function of the filter was:

$$G_f(z) = \frac{F_f(z)}{F(z)} = \frac{0.454}{z - 0.5464} \qquad (3\text{-}86)$$

where

F : cutting force

F_f : filtered cutting force

Summarizing Equations (3-83) through (3-86), provides a discrete time model of the lathe system for force feedback control (Fig. 3.140).

$$G_p(z) = K_p \frac{(z^4 + z^3 + z^2 + z + 1)(0.522z + 0.4529)}{z^5 (z^2 + 0.215z + 0.2466)(z - 0.5464)}$$

(3-87)

where $K_p = 0.454$ B K d.

All the uncertain aspects of the metal cutting are represented by the gain, K_p, which is included in the model. This aspect of the model is convenient for designing and implementing an adaptive controller.

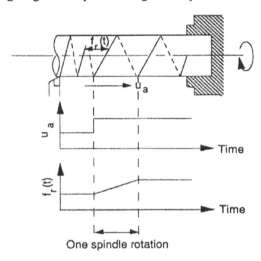

One spindle rotation

Fig. 3.139. Chip Thickness Versus Feedrate

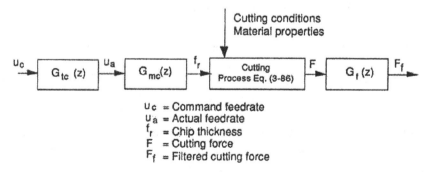

u_c = Command feedrate
u_a = Actual feedrate
f_r = Chip thickness
F = Cutting force
F_f = Filtered cutting force

Fig. 3.140. 3.131 Discrete Time Model for Lathe Metal Cutting Process

The adaptive control approach is based on a PI controller (Fig. 3.141). The fixed gain PI control system provides excellent regulation performance in the tuned cased, and performance deterioration is due to variations of the loop gain. The objective of the adaptive control approach (Fig. 3.142) is to maintain the cutting force, Ff, at a desired value, Fd, by making appropriate adjustments to the command feedrate, Uc while the cutting speed remains constant. The controller consists of two control loops. The first loop is a standard feedback loop utilizing a PI controller; the second loop tunes the gain, $\hat{K}_p$, so that the difference between the cutting force estimated by the model, $\hat{F}_f$, and the measured cutting force, F, is minimized. The process gain, $\hat{K}_p$, which depends on the depth of cut, the tool conditions and the material, is estimated by a recursive least squares identifier. The identification of $\hat{K}_p$ is based on m(k), the command feedrate processed by the known portion of the cutting process dynamics. In the identifier, f(k) represents the adaption gain and 1 is the forgetting factor. The controller parameters are tuned for an assumed nominal process gain Kpo. The controller output is scaled by Kpo/Kp(k-1) to make the overall loop gain unchanged under the process gain variations

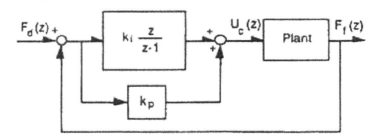

Fig. 3.141. Proportional Plus Integral Control of Cutting Force

In the implementation of adaptive control for the turning process, the cutting force is measured by a piezoelectric crystal force sensor (Fig. 3.143) mounted on the tool carriage (Fig. 3.144). The force sensor output is filtered to eliminate high frequency noise. For force feedback control, the tool feedrate in the axial direction is the manipulated variable. Other quantities that define the cutting condition, such as the depth of cut and material properties, are treated as known or unknown disturbances.

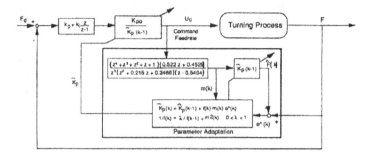

Fig. 3.142. Adaptive Control Scheme Applied to Turning Process

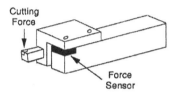

Fig. 3.143. Force Sensor Mounted on Tool Holder

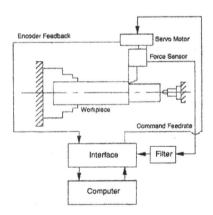

Fig. 3.144. Computer Controlled Lathe

3.4.5 Manufacturing Automation and Computer Integrated Manufacturing (CIM)

In discrete-part manufacturing, computers are essential to the development of NC, robotics, computer-aided design (CAD), computer-aided manufacturing (CAM), and flexible manufacturing systems (FMS). To address problems in the overall corporation computer-integrated manufacturing (CIM) systems are being developed. CIM includes not only manufacturing functions, but also business and other engineering functions. Thus, CIM can be regarded as a concept comprising of three aspects: Automation of manufacturing equipment, integration of systems [70], production planning and control [71].

Computer-Integrated Manufacture (CIM) is concerned with providing computer assistance, control and high level integrated automation at all levels of the manufacturing industries, by linking islands of automation into a distributed processing system. The technology applied in CIM makes intensive use of distributed computer networks, data processing techniques, artificial intelligence (AI) and data base management systems.

A formal definition of CIM is [72]: "A series of interrelated activities and operations involving the design, materials selection, planning, production, quality assurance, management and marketing of discrete consumer and durable goods. CIM is the deliberate integration of automated systems into processes of producing a product. CIM can be considered as the logical organization of individual engineering, production and marketing/support functions into a computer integrated system."

The objective of the CIM environment is to increase productivity by integrating all the phases of the design industrialization production process in the same environment. In this way, the data is introduced once, verified, and then maintained by the system. The software tools, are integrated into a unique environment and used as needed.

In computer-integrated manufacturing, the traditionally separate functions of research and development, design, production, assembly, inspection, and quality control are all linked. Consequently, integration requires that quantitative relationships among product design, materials, manufacturing process and equipment capabilities, and related activities be well understood. Thus, changes in material requirements, product types, and market demand can be accommodated. The goal is to find the optimum method of producing high-quality, low-cost products efficiently [70].

Computer Integrated Manufacturing, in this respect, covers all activities related to the manufacturing business, including:

- Evaluating and developing different product strategies.
- Analyzing markets and generating forecasts.
- Analyzing product/market characteristics and generating concepts of possible manufacturing systems (i.e. FMS cells and FMS systems).
- Designing and analyzing components for machining, inspection, assembly and all other processes relating to the nature of the component

and/or product (i.e. welding, cutting, laser manufacturing, presswork, painting, etc.).

- Evaluating and/or determining batch sizes, manufacturing capacity, scheduling and control strategies relating to the design and fabrication processes involved in the particular product.
- Analysis and feed back of certain selected parameters relating to the manufacturing processes, evaluation of status reports from the DNC (Direct Numerical Control) system (source data monitoring and machine function monitoring in real-time).
- Analyzing system disturbances and economic factors of the total system.

Computer-integrated manufacturing systems consist of subsystems that are integrated into a whole. These subsystems consist of business planning and support, product design, manufacturing process planning, process control, shop floor monitoring systems, and process automation. The subsystems are designed, developed and applied in such a manner that the output of a subsystem serves as the input of another subsystem. Organizationally, these subsystems are usually divided into business planning and business execution functions, respectively. Planning functions include activities such as forecasting, scheduling, material-requirements planning, invoicing and accounting. Execution functions include production and process control, material handling, testing and inspection.

The effectiveness of CIM depends greatly on the presence of a large-scale integrated communication system, involving computers, machines and their controls. Major problems have arisen in factory communication because of the difficulty in interfacing different types of computers, purchased by different company vendors at various times. The trend is strongly towards standardization in order to make communications equipment compatible.

The goal of CIM is to use advanced information processing technology in all areas of the manufacturing industry in order to:

- Make the total process more productive and efficient.
- Increase product reliability.
- Decrease the cost of production and maintenance relating both to the manufacturing system as well as to the product.
- Reduce the number of hazardous jobs and increase the involvement of well educated and competent humans in the manufacturing activity and design.
- Respond to rapid changes in market demand, product modification, and shorter product life cycles.

- Achieve better use of materials, machinery, and personnel, and reduced inventory.
- Achieve better control of production and management of the total manufacturing operation.

FMS

In general, an automated manufacturing system can be described as a collection of mechanical, electrical, and electronic components, coupled together to perform one or more manufacturing tasks. Fixed systems are typically used for the production of high-volume products and are limited to the production of a single part or, at most, a few parts (Fig. 3.145). For a fixed automation system the setup time required to change over from one part to another is usually orders of magnitude greater than the cycle time. Automated transfer lines are typically groups of specially designed machine components that are integrated into a system by way of specially designed parts-transfer and material handling systems.

A flexible manufacturing system (FMS) is a "reprogrammable" manufacturing system capable of producing a variety of products automatically (Fig. 3.145). Transfer lines are able to perform a variety of manufacturing operations automatically. However, altering these systems to accommodate even minor changes in the product is quite difficult. Entire machines might have to be introduced to the system while other machines or components are modified or removed to accommodate small changes in a product.

Table 3.2 shows the relative characteristics of transfer lines and FMS. Note that in FMS the time required for a changeover to a different part is very short.

Production Systems

Fixed (transfer line)   Flexible

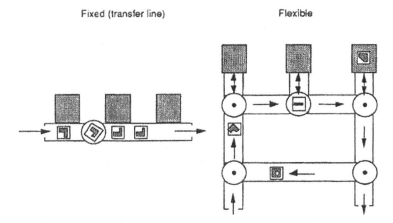

Features:

- Specific machining units
- Identical workpieces
- Constrained sequence
- Cyclic control

- Interchangeable and/or specific machining units
- Various workpieces within a component range
- Usually free component selection

Fig. 3.145. Fixed and Flexible Production Systems

CHARACTERISTIC	TRANSFER LINE	FMS
Types of part made	Generally few	Infinite
Lot size	> 100	1-50
Part changing time	1/2 to 8 hr	1 min
Tool change	Manual	Automatic
Adaptive control	Difficult	Available
Inventory	High	Low
Production during break-down	None	Partial
Efficiency	60-70%	85%
Justification for capital expenditure	Simple	Difficult

Table 3.2. Comparison of the characteristics of transfer lines and flexible manufacturing systems

The quick response to product and market-demand variations is a major attribute of FMS. A variety of FMS technologies are available from machine-tool manufacturers.

Compared with conventional systems, the benefits of FMS are:

- Parts can be produced randomly and in batch sizes as small as one and at lower unit costs.
- Direct labor and inventories are reduced, with major savings over conventional systems.
- Machine utilization is as hl-h as 90 percent, thus improving productivity.
- Shorter lead times are needed for product changes.
- Production is more reliable because the system is self-correcting and product quality is uniform.
- Work-in-progress inventories are reduced.

An FMS employs programmable electronic controls that, in some cases, can be set up for random parts sequences without incurring any set-up time between parts. Numerical control (NC) and robotics have provided reprogramming capabilities at the machine level with minimum setup time. NC machines and robots provide the basic physical building blocks for reprogrammable [1] manufacturing systems (Fig. 3.146).

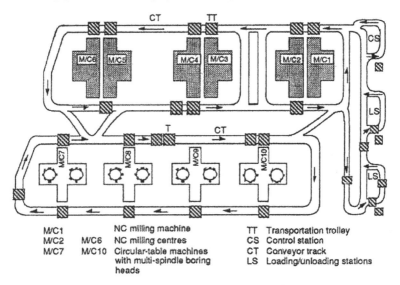

M/C1		NC milling machine	TT	Transportation trolley
M/C2	M/C6	NC milling centres	CS	Control station
M/C7	M/C10	Circular-table machines with multi-spindle boring heads	CT	Conveyor track
			LS	Loading/unloading stations

Fig. 3.146. Layout of a Flexible Manufacturing System

Since the automation must be programmable in order to accommodate a variety of product-processing requirements, easily alterable as well as versatile machines must perform the basic processing. For this reason, CNC turning centers, CNC machining centers, and robotic workstations comprise the majority of equipment in these systems. These machines are not only capable of being easily reprogrammed, but are also capable of accommodating a variety of tooling via a tool changer and tool storage system. It is not unusual for a CNC machining center to contain 60 or more tools (mills, drills, boring tools, etc.), and for a CNC turning center to contain 12 or more tools (right-hand turning tools, left-hand turning tools, boring bars, drills etc.).

Parts must also be moved between processing stations automatically. Several different types of material-handling systems are employed to move these parts from station to station. The selection of the type of material-handling system is a function of several system features. The material-handling system, first, must be able to accommodate the load and bulk of the part and perhaps the part fixture. Large, heavy parts require large, powerful handling systems such as roller conveyors, guided vehicles, or track-driven vehicle systems. The number of machines to be included in the system and the layout of the machines also present another design consideration. If a single material handler is to move parts to all the machines in the system, then the work envelope of the handler must be at least as large as the physical system. A robot is normally only capable of addressing one or two machines and a load-and-unload station. A conveyor or automatic guided vehicle (AGV) system can be expanded to include miles of factory floor. The material-handling system must also be capable of moving parts from one machine to another in a timely manner. Machines in the system will be unproductive if they spend much of their time waiting for parts to be delivered by the material handler. If many parts are included in the system and they require frequent visits to machines, then the material-handling system must be capable of supporting these activities. This can usually be accommodated by using either a very fast handling device or by using several devices in parallel, for example, instead of using a single robot to move parts to all the machines in a system, a robot might only support service to a single machine.

The tooling used in an FMS must be capable of supporting a variety of products or parts. The use of special forming tools in an FMS is not generally employed in practice. The contours obtained by using forming tools can usually be obtained through a contour-control NC system and a standard mill. The standard mill can then be used for a variety of parts rather than to produce a single special contour. An economic analysis of the costs and benefits of any special tooling is necessary to determine the best tooling combination. However, since NC machines have a small number of tools that are accessible, very few special tools should be included.

One of the commonly neglected aspects of an FMS is that of the fixturing used. Work on creating "flexible fixtures" that can be used to support a variety of components has only recently begun. One unique aspect of many FMSs is that the part is also moved about the system in the fixture. Fixtures are made to the same dimensions so that the material-handling system can be specialized to handle a single geometry. Parts are located precisely on the fixture and moved from one station to another on the fixture. Fixtures of this type are usually called pallet fixtures, or pallets. Many of the pallet fixtures, employed today, have standard "T-slots" cut in them, and use standard fixture kits to create the part-locating and holding environment needed for machining.

In recent years, Local-Area-Network (LAN) techniques have become more popular in both shop floor and office automation. LAN consists of two or more connected stations (PCs, servers, computer) in the same limited area, sharing data and peripheral devices, and operating at the speed of 1 Mbps to about 1 Gbps. Lan is much faster than the previously mentioned (point-to-point) communication methods. It also allows many-to-many communication on the same network through the same cable. As the price of implementing local-area networks becomes relatively low, many factories are installing some type(s) of LAN.

The development of computer-networking techniques has enabled a large number of computers to be connected. A local-area network is one that is confined to a 10-km distance.

An ideal LAN has the following characteristics:
- high speed: greater than 100 Mbps
- low cost: easily affordable on a computer and/or machine controller
- high reliability/integrity: low error rates, fault tolerant, reliable
- expandability: easily expandable to install new nodes
- installation flexibility: easy to be installed in an existing   environment
- interface standard: standard interface across a range of computers and controllers

In a local-area network, computers and controllers can communicate with each other. Terminals can also have access to any computer on the network without a physical hardwire. It is beneficial since the same terminal can now access all devices on the shop floor.

A LAN consists of software that controls data handling and error recovery, hardware that generates and receives signals, and media that carry the signal. The software and hardware designs are governed by a set of rules called protocol. This protocol defines the logical, electrical, and physical specifications of the network. In order for devices in a network to communicate with each other, the same protocol must be followed. For a device to send a message across the network, the sender and the receiver must be uniquely identified. The message must be properly sent and error checking must be performed in order to ensure the correctness of the information. Similar to a parcel prepared for the Postal Service, the software has to package and identify the data with appropriate labels. The label contains the address of the destination and information concerning the contents of the message. The package is then converted into the appropriate electric signals and waits to be transmitted.

Unfortunately, at the beginning there was little standardization in the way LANs were implemented. Devices manufactured by different vendors typically could not communicate with each other. A separate computer had to be dedicated to translate data between two networks from two different vendors, and sometimes even from the same vendor. According to a study [73], 50% of the cost on shop floor computers was spent on networks. This cost was far too high to be justifiable. In the early 1980s, GM spearheaded an effort to establish a common standard – Manufacturing Automation Protocol (MAP) – that defined the physical and logical communication standard for manufacturing facilities. MAP has since been adopted as the standard for the shop floor intervendor device data communication. The International Standards Organization (ISO) is adopting MAP as the international standard, but many other communication protocols are still being used.

Further Reading

SME's *Tool and Manufacturing Engineers Handbook* [74], Weck's *Handbook of Machine Tools* [1] together with Kalpakjian's *Manufacturing Engineering & Technology* [4] provide thorough presentations of different machine tool designs and methods for investigating machine tool structures and controls. A thorough introduction in the precision manufacturing is available in Slocum's *Precision Machine Design* [7].

The Working Papers of the MIT Commission on Industrial Productivity [2], particularly the report on the machine tool industry, is a very informative document on the structure of the machine tool industry in the United States and internationally. Relative information for the European machine Tool industry can be accessed in *European committee for Cooperation of the Machine Tools Industry* website [117].

Chang, Wysk and Wang's book *Computer-Aided Manufacturing* [75], together with Weck's *Handbook of Machine Tools* [1], provide good reference materials related to automation and computer-aided manufacturing.

Chryssolouris' *Laser Machining: Theory and Practice* [8] and *LIA Handbook of Laser Materials Processing* [12] are ideal for those interested in reading more about laser technology and laser processing.

Numerical control as well as CAD/CAM/CAE issues are presented in detail in Kunwoo's *Principles of CAD/CAM/CAE Systems* [46] and in Sava and Pusztai's *Computer Numerical Control Programming* [44].

Process control is an area where significant research effort has been devoted in the past few years. Wright's book *Manufacturing Intelligence* [58] discusses very elegantly some of the issues in process control. The works of Hardt [76, 77], Tomizuka [57], Koren [45, 78], Ulsoy [79], Wu [80], Tlusty [81, 82] and a few others provide insight into process control issues for different manufacturing processes. Chryssolouris [83, 84 and 85] and Dornfeld [86] worked most recently in concepts involving sensor fusion or sensor synthesis with the help of neural networks and other artificial intelligence and statistical techniques.

Ranky's book *Computer Integrated Manufacturing* [87] provides sufficient information on the topic while Kusiak's book *Intelligent Manufacturing Systems* [88] deals more with the application of knowledge based systems in manufacturing.

References

1. Weck, M., *Handbook of Machine Tools*, John Wiley & Sons, New York 1984.

2. T*he Working Papers of the MIT Commission on Industrial Productivity*, MIT Press, Cambridge, MA 1989.

3. Lange, K., *Handbook of Metal Forming*, McGraw Hill, New York, 1985.

4. Kalpakjian, S., *Manufacturing Engineering and Technology*, 4th Edition, Prentice Hall, 2000.

5. Wick, C. and R. F. Veilleux, *Tool and Manufacturing Engineers Handbook, Vol. 3 - Materials, Finishing and Coating*, Society of Manufacturing Materials, Michigan, 1985

6. Kawahara, O., S.Audo and H. Khara, "Internal grinding machine with magnetic bearing spindle", *Proc. Jap. Soc. of Grinding Eng. Conf.*, (1997) pp. 39-42.

7. Slocum, A. H., *Precision Machine Design*, SME, Prentice-Hall Inc, 1992.

8. Chryssolouris, G., *Laser Machining: Theory and Practice*, Springer-Verlag, Mechanical Engineering Series, New York, 1991.

9. Basov, N.G., Danilychev, V.A., "Industrial High-Power Lasers. Science and Mankind", *Znanie*, Moscow (1985), pp.261-278.

10.Prokhorov, A.M. (ed.), *Laser Handbook, Vol. 1*, Sov. Radio, Moscow, 1978.

11.Prokhorov, A.M. (ed.), *Laser Handbook Vol. 2*, Sov. Radio, Moscow, 1978.

12.Ready, J.F., *LIA Handbook of Laser Materials Processing*, Laser Institute of America, Magnolia Publishing, Inc., 2001.

13.Grigoryants, A. G., *Basics of Laser Material Processing*, Mir Publishers, Moscow, 1994.

14.Giesen, A. et al, "Einflu, der Optik auf den Bearbeitungsproze", *Opto Elektronik Magazin*, (Vol. 4, No 1, 1988).

15. Weck, M. et al, "Analyse des Verformungsverhaltens optischer Systeme während der Lasermaterialbearbeitung" *Laser und Optoelektronik*, (Vol. 26, No.2 , 1994).

16. Borik, S., "Einflu, optischer Komponenten auf die Fokussierbarkeit", *Laser und Optoelektronik*, (Vol. 26, No.2, 1994).

17. Weck, M. et al, "Betriebsverhalten transmissiver Optiken bei der Laser-Materialbearbeitung", *Laser und Optoelektronik*, (Vol. 26, No.2, 1994).

18. Weck, M. et al, "Laser - A Tool for Turning Centres" *Proceedings of the LANE' 94*, Meisenbach Bamberg, (Vol. I, 1994), pp. 427-437.

19. Tsoukantas, G., K. Salonitis, P. Stavropoulos and G. Chryssolouris, "An overview of 3D Laser Materials' Processing Concepts", *Proceeding of the 3rd GR-I Conference on New Laser Technologies and Applications* (2001)

20. Chryssolouris G., N. Anastasia and P. Sheng, "Three - Dimensional Laser Machining for Flexible Manufacturing", *Proc. Symposium on Intelligent Design and Manufacturing for Prototyping*, ASME Winter Annual Meeting, Atlanta, GA, (December 1991).

21. Chryssolouris G., "3-D Laser Machining: A Perspective", *Proceedings of the LANE '94*, Meisenbach Bamberg, (Vol.1, 1994).

22. Wiedmaier M., E. Meiners, T. Rudlaff, F. Dausinger and H. Hugel, "Integration of Materials Processing with YAG-Lasers in a Turning Center", *Laser Treatment of Materials,* Informationsgesellschaft-Verlag, (1992), pp.559-564

23. Wiedmaier M., E. Meiners, T. Rudlaff, F. Dausinger and H. Hugel, "Integration of Materials Processing with YAG-Lasers in a Turning Center", *Laser Treatment of Materials,* Informationsgesellschaft-Verlag, (1992), pp.559-564

24. Salem B.W., G. Marot, A. Moisan, J. P. Longuemard, "Laser Assisted Turning during Finishing Operation Applied to Hardened Steels and Inconel 718", *Proceedings of the LANE '94*, Meisenbach Bamberg, (Vol.1, 1994).

25. Salonitis, K., G. Tsoukantas, P. Stavropoulos and A. Stournaras, "A critical review of stereolithography process modeling", *Proceedings of the International Conference on Advanced Research in Virtual and Rapid Prototyping-VR@P,* Leiria, Portugal (October 2003), pp. 377-384.

26. Agarwala, M., D. Klosterman, N. Osborne and A. Lightman, "Hard metal tooling via SFF of ceramics and powder metallurgy", *SFF Symposium*, Austin, TX (1999).

27. Jacobson, D.M., "Metal layer object manufacturing", *State of the Art Report of the RAPTIA Thematic Network* (2002).

28. Suh, J.D., Lee, D.G., "Composite machine tool structures for High Speed Milling Machines", *Annals of the CIRP*, (Vol. 51, No.1, 2002), pp. 285-288.

29. Crandall, S.H., N. Dahl, T.J. Lardner, *An Introduction to the Mechanics of Solids*, McGraw-Hill, Inc., New York, 1978.

30. Bathe, K.J., *Finite Element Procedures in Engineering Analysis*, Prentice-Hall, Englewood Cliffs, N.J., 1982.

31. Boothroyd, G. and W.A. Knight, *Fundamentals of Machining and Machine Tools*, 2nd edition, Marcel Dekker, Inc., New York , 1989.

32. Ogata, K., *Modern Control Engineering*, Prentice-Hall, Inc., New Jersey, 1970.

33. Meirovitch, L., *Introduction to Dynamics and Control*, John Wiley & Sons, Inc., New York, 1985.

34. Inasaki, I., B. Karpuschewski and H.-S. Lee, "Grinding Chatter – Origin and Supresión", *Annals of the CIRP*, (Vol. 50, No.2, 2001), pp. 525-534.

35. Govekar, E., A. Baus, J. Gradisek, F. Klocke and I. Grabec, "A new method for chatter detection in grinding", *Annals of the CIRP*, (Vol.51, No.1, 2002), pp. 267-270.

36. Weck, M., P. McKeown, R. Bonse and U. Herbst, "Reduction and compensation of thermal errors in machine tools", *Annals of the CIRP*, (Vol. 44, No. 2, 1995), pp. 589-598.

37. Bryan, J., "International Status of Thermal Error research", *Annals of the CIRP*, (Vol. 39 ,No.2,1990), pp. 645-656.

38. Spur, G., E. Hoffman, Z. Paluncic, K. Bensinger and H. Nymoen, "Thermal behaviour optimization of machine tools", *Annals of the CIRP*, (Vol. 37, No. 1, 1998) pp. 401-405.

39. Tamisier, V., S. Font, M. Lacour, F. Carrere and D. Dumur, "Attenuation of vibrations due to unbalance of an Active Magnetic bearings Milling Electro-Spindle", *Annals of the CIRP*, (Vol.50, No.1, 2001).

40. Weck, M. and U. Wahner, "Linear magnetic bearing and levitation system for machine tools", *Annals of the CIRP*, (Vol. 47, No.1, 1998), pp. 311-314.

41. Koren, Y., *Computer Control of Manufacturing Systems*, McGraw-Hill, Inc., New York, 1983.

42. Acherkan, N., et al., *Machine Tool Design*, U.S.S.R., 1973.

43. Pritschow, G., "A comparison of linear and conventional electromechanical drives", *Annals of the CIRP*, (Vol. 47, No.2, 1998), pp. 541-548.

44. Sava, M. and J. Pusztai, *Computer Numerical Control Programming*, Prentice Hall, 1990.

45. Koren, Y., "The Optimal Locus Approach With Machining Applications," ASME *Journal of Dynamic Systems, Measurement and Control*, (Vol. 111, 1989), pp. 260-267.

46. Kunwoo Lee, *Principles of CAD/CAM/CAE Systems*, Addison-Wesley, 1999.

47. Sutherland, I.E., "SKETCHPAD: A Man-Machine Graphical Communication System," *SJCC 1963*, Spartan Books, Baltimore, (1963) pg. 329 and *MIT Lincoln Lab. Tech. Rep.* (May 1965) pg.296.

48. Amirouche, M. L. F., *Computer-Aided Design and Manufacturing*, Prentice Hall, New Jersey, 1993.

49. Van Houten, F. J. A. M., *PART: A Computer Aided Process Planning System*, PhD Thesis Report, (1991).

50. Knox, C.S., *CAD/CAM Systems-Planning and Implementation*, Marcel-Dekker, New York, 1983.

51. Onosato, M., and K. Iwata., "Development of a Virtual Manufacturing System by Integrating Product Models and Factory Models", *Annals of the CIRP*, (Vol.42, No.1, 1993), pp. 475-478.

52. Aukstakalnis S. and D. Blatner , *Silicon Mirage, The Art and Science of Virtual Reality*, Peachpit-Press, 1992.

53. National Research Council, *Virtual Reality: Scientific and Technological Challenges*, National Academy Press Washington, 1995.

54. Astheimer, P., W. Felger and S. Mueller, "Virtual Design: A Generic VR System for Industrial Applications", *Computers & Graphics*, (Vol. 17, No. 6, 1993), pp. 671-677.

55.Bowyer, A., G. Bayliss; R. Taylor and P. Willis. "A Virtual Factory" *International Journal of Shape Modeling*, (Vol. 2, No. 4, 1996), pp.215-226.

56.Chryssolouris G., D. Mavrikios, D. Mourtzis, K. Pistiolis and D. Fragos. "An Integrated Virtual Manufacturing Environment for Interactive Process Design and Training - The Virtual Machine Shop", *Proceedings of the 32nd CIRP International Seminar on Manufacturing Systems*, (2001) pp. 409-415.

57.Tomizuka, M. and S. Zhang, "Modelling and Conventional Adaptive PI Control of a Lathe Cutting Process", *ASME, Journal of Dynamic Systems, Measurements and Control*, (Vol. 110, 1988), pp. 350-354.

58.Wright, P.K., *Manufacturing Intelligence*, Addison-Wesley, Reading, MA., 1988.

59.Lister, P.M., and G. Barrow, "Tool Condition Monitoring Systems", *Proceedings of the Twenty-sixth International Machine Tool Design and Research Conference*, (1984).

60.Giusti, F., M. Santochi and G. Tantussi, "On-Line Sensing of Flank and Crater Wear of Cutting Tools", *Annals of CIRP*, (1987) pp. 41-44.

61.Lee, Y.H., P. Bandyopadhyay and B. Kaminski, "Cutting Tool Wear Measurement Using Computer Vision", *Proc. of NAMRC*, (1987), pp. 195-212.

62.Altintas, Y., "In-Process Detection of Tool Breakages Using Time Series Monitoring of Cutting Forces", *Int. J. Mach. Tools Manufact.*, (Vol. 28, No. 2, 1988), pp. 157-172.

63.Bandyopadhyay, P., and S.M. Wu, "Signature Analysis of Drilling Dynamics for On-Line Drill Life Monitoring", *Sensors and Control for Manufacturing*, ASME Annual Winter Meeting, (PED-Vol. 18, 1985), pp. 101-110.

64.Chryssolouris, G. and M. Domroese, "Some Aspects of Acoustic Emission Modeling for Machining Control", *Proc. of NAMRC*, (1989), pp. 228-234.

65.Liang, S.Y., and D.A. Dornfeld, "Tool Wear Detection Using Time Series Analysis of Acoustic Emission", *J. Eng. Ind.*, (Vol. 111, 1989), pp. 199-212.

66.Chow, J.G., and P.K. Wright, "On-Line Estimation of Tool/Chip Interface Temperatures for a Turning Operation", *J. Eng. Ind.*, (Vol. 110,
1988), pp. 56-64.

67.Liu, T.I., and S.M. Wu, "On-Line Drill Wear Monitoring" *Sensors and
Controls for Manufacturing*, ASME Annual Winter Meeting, (PED-Vol.
33, 1988), pp. 99-104.

68.Kazuaki, J., *Sensing Technologies for Improving Machine Tool Function*, NMBTA Data Files.

69.Chryssolouris, G.,S.R. Patel, "In-Process Control for Quality Assurance," in *Manufacturing High Technology Handbook*, Tijunelis and
McKee Eds., Marcel Dekker, New York (1987), pp. 609-643.

70.Rolstadas, A., "Architecture for integrating PPC in CIM", *Proceedings
of the IFIP TC5/WG 5.3, Eighth International PROLAMAT Conference,
Man in CIM*, Tokyo, (1992), pp. 187-195.

71.David, S.T. and K., Cheballah "User interface for project management
in the CIM environment", *Proceedings of the IFIP TC5/WG 5.3, Eighth
International PROLAMAT Conference, Man in CIM*, Tokyo, (1992), pp.
525-533.

72.Bunce, P., "Planning for CIM," *The Production Engineer*, (Vol. 64, No.
2, 1985), pg. 21.

73.General Motor's Manufacturing Automation Protocol, *A Communications Network Protocol for Open Systems Interconnection*, Warren, MI.
GM MAP Task Force, 1984.

74.Society of Manufacturing Engineers, *Tool and Manufacturing Engineers Handbook*, McGraw-Hill, New York, 1988.

75. Chang, T.C., R.A. Wysk and H.P. Wang, *Computer Aided Manufacturing*, Prentice-Hall, Englewood Cliffs, N.J., 1991.

76.Hardt, D.E., T. Jenne, M. Domroese and R. Farra, "Real-Time Control
of Twist Deformation Processes", *Annals of the CIRP*, (1987).

77.Hardt, D.E., A. Suzuki and L. Valvani, "Application of Adaptive Control Theory to On-Line GTA Weld Geomery Regulation", *ASME Journal of Dynamic Systems, Measurement and Control*, (Vol. 113, 1991),
pp. 93-103.

78.Ko, T.R., and Y. Koren, "Cutting Force Model for Tool Wear Estimation," *Proc. of NAMRC XVII*, (1989), pp. 166-169.

79.Park, J.J., and A.G. Ulsoy, "On-Line Flank Wear Estimation Using Adaptive Observers", *Automation of Manufacturing Processes*, ASME Winter Annual Meeting, (1990), pp. 11-20.

80.Fassois, S.D., K.F. Eman and S.M. Wu, "A Fast Algorithm for On-Line Machining Process Modeling and Adaptive Control", *ASME Journal of Engineering for Industry* (Vol. 111, May 1989), pp. 133-139

81.Altintas,Y., I. Yellowley and J. Tlusty, "The Detection of Tool Breakage in Milling Operations", *ASME Journal of Engineering for Industry*, (Vol. 110, 1988), pp. 271-277.

82.Tlusty, J., and G.C. Andrews, "A Critical Review of Sensor for Unmanned Machining," *Annals of the CIRP* (1983).

83.Chryssolouris, G., M. Domroese and P. Beaulieu, "Sensor Synthesis for Control of Manufacturing Processes", *Proceedings of the Symposium on Control of Manufacturing Processes*, ASME Winter Annual Meeting, (1990), pp. 67-76.

84.Chryssolouris, G., V. Subramaniam and M. Domroese, "A Game Theory Approach to the Operation of Machining Processes," *ASME Winter Annual Meeting*, (1991).

85.Chryssolouris, G., M. Domroese and P. Beaulieu, "A Statistical Approach to Sensor Synthesis," *Proceedings of NAMRC XIX*, (1991), pp. 333-337.

86.Rangwala, S., and D. Dornfeld, "Integration of Sensors via Neural Networks for Detection of Tool Wear States," *Intelligent and Integrated Manufacturing Analysis and Synthesis*, ASME Winter Annual Meeting, (1987).

87.Ranky, P.G., *Computer Integrated Manufacturing*, Prentice-Hall, Englewood Cliffs, N.J., 1986.

88.Kusiak, A., *Intelligent Manufacturing Systems*, Prentice-Hall, Englewood Cliffs, N.J., 1990.

Other sources

89.Forging & Industrial Equipment, www.whitemachinery.com

90.SMS EUMUCO GmbH, www.sms-eumuco.de

91.www.jelsigrad.com

92.FU SHENG Group, www.fusheng.com

93.Elix LTD, www.elixmec.gr

94.College of Engineering, www.eng.fiu.edu

95.ENCO, www.enco.at

96.RAFAMET, www.rafamet.com

97.www.machinetooldistribution.com

98.RMT Technology, www.rmt.net

99.GOTTOPI, www.gottoppi.com

100.Hardinge Machine Tools, www.hardinge.com

101.DANOBAT, www.danobat.com

102.SAMPUTENSILI, www.samputensili.com

103.TRUMPF GmbH, www.trumpf.com

104.3D Systems, www.3dsystems.com

105.Stratasys Inc., www.stratasys.com

106.www.cubictechnologies.com

107.Electro Optical Systems GmbH, www.eos-gmbh.de

108.Z Corporation, www.zcorp.com

109.ANSYS Inc, www.ansys.com

110.SYSWELD Software, www.sysweld.com

111.MSC software Corp., www.mscsoftware.com

112.ADINA R&D Inc., www.adina.com

113.FAG, www.fag.com

114.Faulhaber Group, www.micromo.com

115.SKF, www.skf.com

116.ROLLVIS S.A., www.rollvis.com

117.European Commitee for Cooperation of Machine Tools Industry: Cecimo, www.cecimo.be

Review Questions

1. Briefly define the term "machine tool." What are two major classes of machine tools? Provide examples of each machine in each of these classes.

2. What are some of the critical criteria to be considered in the design of a machine tool?

3. Which category of deforming machine provides the highest production rate? Which types of deforming processes is this type of machine used for?

4. Which type of forming machine is used most widely? What are some of the reasons it is used so extensively?

5. List and briefly describe the basic components of machines which perform material removal operations.

6. In general which operation, turning or drilling, provides a higher material removal rate? Explain your answer.

7. What difficulties cause the design of machine tools to rely heavily on empirical knowledge and expertise?

8. What is the main goal in the design of a machine tool frame? What types of loading make it difficult to achieve this goal? Briefly describe the sources of these loads.

9. What are the most frequent modes of static loading for machine tool frames? Describe an effective method of increasing the stiffness of the frame with respect to these types of loading without significantly increasing the weight of the structure.

10. Why is it frequently necessary to utilize numerical techniques, such as finite element analysis (FEA), in analyzing machine tool frames? Briefly outline the steps required in applying FEA to the analysis of a machine tool.

11. List and briefly describe the sources of the three basic classifications of machine tool vibrations. What are the detrimental effects of these vibrations?

12. Discuss some of the methods available for reducing the development of chatter.

13. If a machine tool vibrates excessively what are some of the techniques which are available for reducing these vibrations? What are some of the benefits and drawbacks of each of these techniques?

14. What are the contributing factors to the thermal loading of a machine tool? What are some of the techniques for reducing the thermal load?

15. Briefly describe the types of guideways and bearings which are available for use in a machine tool. What are some of the advantages and disadvantages of each of these types of mechanisms?

16. Briefly describe why the stick-slip effect occurs. How should the parameters of a hydrodynamic slide be changed in order to reduce the stick slip effect?

17. Which types of applications would be most appropriate for using the following types of drives: (1) hydraulic, (2) DC motor, (3) stepping motor, (4) AC motor?

18. Discuss the attributes and appropriate applications for the different transmissions used in machine tools.

19. What are the motivations for introducing automation into a manufacturing system?

20. Briefly describe the function of each of the components in a CNC machine tool.

21. Describe the main classifications of control strategies used in machine tools. Use some sketches to help your explanations.

22. What are the reasons for using closed-loop motion control rather than using open-loop control?

23. What is a part program? What are the costs associated with developing a part program?

24. There are four basic part programming methods. Briefly describe the steps used in each of these methods to produce a part program.

25. What are the methods available for NC part program verification? What are the advantages and disadvantages of each of these methods?

26. Why would a manufacturing company want to rely on a CAD system for design rather than more traditional drafting methods?

27. What is the benefit of using a CAD system based on solid modelling rather than wire-frame models?

28. What is the difference between machine and process control?

29. What are the objectives of adaptive control constraint (ACC) and adaptive control optimization (ACO) process control schemes?

30. Briefly describe how process sensors are used for the functions of tool monitoring, workpiece monitoring and machine performance monitoring.

31. There are two basic methods for tool condition monitoring. What are these methods? List some of the tool monitoring techniques that can be classified under each of these methods.

32. What type of device can be used if the number of sensors used to monitor a process is greater than the number of input ports on the processor used to control the process?

33. In a manufacturing plant what functions does a computer-integrated manufacturing (CIM) system include?

34. What are the differences between a fixed automation system and a flexible manufacturing system (FMS)? Under what circumstances would each of these types of systems be most appropriate?

35. Describe some of the material-handling systems which can be used in an FMS.

36. What is a LAN? What functions would a LAN be used for in a manufacturing environment?

4. Process Planning

4.1 Introduction

When the design of a mechanical component is completed, it is usually documented in a drawing or a CAD file, which specifies its geometric features, dimensions, tolerances, etc. In order for the design to be manufactured, a set of instructions is needed regarding the processes, equipment, and/or people to be involved in the manufacturing process. Such instructions are usually documented in an operation sheet or process plan (Fig. 4.1),

OPERATION SHEET (Process plan)				
Part No. —————— Part Name ———— Orig. ———— Checked ————		Material ———————— Changes ———————— Approved ————————		
No.	Operation Description	Machine	Set-up Description	Operate Hr/unit
5	Rough and finish mill 2 mating surfaces	Cinc. Mill (Kender 136)	Gang 6 castings in fixture	
10	Spotface and drill two holes 33/84 in. D; drill 27/84 in. D pipe hole; tap ¼ in. pipe thread	Multispindle drill press	Piece on table Piece in 35" drill jig	
15	Rough and finish bore 6 ¼ in. D; bore 6 ½ in Dx 3/8 in. wide groove	Bullard vert. Boring Mill (kender = 335)	Clamp to eccentric Connector Half (S563-5), then mount both parts in 4-jaw chuck	

Fig. 4.1. Typical Process Plan [1]

which contains data about the part, its material, and other pertinent information. It lists the sequence of operations, including a brief description of every task to be performed, equipment or operator skills, set-up descriptions and/or process parameters to be used during the manufacturing process. Thus, process planning can be defined as *the function, which establishes the sequence of the manufacturing processes to be used in order to convert a part from an initial to a final form, where the process sequence incorporates process description, the parameters for the process and possibly equipment and/or machine tool selection* [2].

Process planning takes into account a number of factors, which influence the selection of the different processes and their operating parameters. Such factors include the *shape* and *size* of the workpiece, the required *tolerances, surface quality,* the *material* the workpiece is made of, and the *quantity* to be made.

Process planning entails on a human the ability to interpret a particular design, and substantial familiarity with manufacturing processes and equipment. It has been estimated that industry may need more people than are actually available to do process planning. Furthermore, it is conceivable that expertise and experience within a company will diminish over time, creating the need to document and capture that expertise. It is often said that two process planners will never come up with the same process plan for the same item, reflecting the fact that often process planning performed by humans is not consistent or optimal with respect to certain performance criteria. All this leads to the need for computerized systems that will allow the process planning function to be performed either totally or partially by a computer, providing the user with optimum process plans in a quick and consistent fashion.

4.1.1 Computer Aided Process Planning

Computer-aided process planning (CAPP) has been investigated for more than 30 years; it can be categorized in two major areas: *variant* process planning, where library retrieval procedures are applied to find standard plans for similar components, and *generative* process planning, where plans are generated automatically for new components without reference to existing plans. The latter system is the most desirable but also the most difficult way of performing computer-aided process planning.

The process planning function bridges the gap between engineering design and manufacturing, and is thus a critical element in integrating activities within manufacturing organizations. Current CAPP systems range from simple editors for manual planning to fully-automated systems for planning a range of products. Some of the specific benefits of CAPP follow:

1. *Improved productivity* – More efficient use of machines, tooling, material and labor. "Best practice" (in the form of optimized process plans)

is documented for consistent application throughout the organization, rather than captured mentally by the process planner.

2. *Lower production cost* – Cost reductions are realized through productivity improvements. Also, the skill level required to produce process plans is less than that required for manual methods.

3. *Consistency* – Computerized methodologies assure consistent application of planning criteria. Also, the number of errors generated during process planning is reduced.

4. *Time savings* – Time savings can range from days to minutes. Lead times are reduced and flexibility is increased due to the ability of reacting quickly to new or changing requirements. The amount of paper work and clerical effort involved with design changes is also reduced.

5. *Rapid integration of new production capabilities* – With the rapid changes in production capabilities, maintaining a competitive advantage requires fast integration of new processes. CAPP allows process plans to be quickly updated to include new process technologies.

There are also several problems associated with automation of the planning process:

1. The designer's intention may not always be obvious to the process planner, who must act on the designer's intentions. Differences in terminology and perspectives separate these two functions.

2. In order to fully automate process planning, the features of a part must be extracted from the product model without human intervention; however, engineering drawings sometimes do not convey all the information about a part. Information may be inaccessible or in a form incompatible with CAPP.

3. One problem source for CAPP systems is the interface between CAD and CAPP, where features are translated into a form recognizable by CAPP. Different CAD systems have different methods of representing dimensions. Translation from CAD to CAPP often requires a human interface.

4. The designer is often unaware of potential manufacturing constraints and may produce a design that is either infeasible or costly to produce.

5. The generation and execution of a production plan may take a long time and may involve several organizations in different geographical locations. Plan-monitoring and improvement may be complex and difficult to automate.

Of the various CAPP methods, *variant process planning* is the easiest to implement. Variant systems allow rapid generation of process plans through comparison of features with other known features in a database.

However, in order to implement variant process planning, products must first be grouped into part families based on feature commonality. If a new part cannot be easily placed into an existing family, then a feasible process plan cannot be generated. Also, as the complexity of feature classifications increase, the number of part families also increases, causing excessive search times during process plan generation. When there are only a few part families with little feature deviation for new designs, variant process planning can be a fast and efficient method for generating new process plans.

Generative process planning (GCAPP) relies on a knowledge base to generate process plans for a new design independent of existing plans. The knowledge base is a set of rules derived from the experience of a human process planner. With generative methods, process plans can be generated for a wide variety of designs with dissimilar features. However, generative methods are difficult to implement in terms of constructing a set of rules, which can encompass all anticipated design features, likely to be encountered. GCAPP systems are difficult to implement due to the large quantity of data and knowledge required to provide even simple manufacturing feedback. Countless scenarios must be represented, and the required memory space comes at a cost of computing ability. The largest single limitation of many current GCAPP systems is their inability to accurately and completely represent the product or part model. Technicians are often called upon to do the dimensioning and tolerancing, because CAPP is neither capable of doing it in a reasonable time frame nor is it simply capable of doing it at all. Commercially, generative process planning methods have been implemented in some specialized applications.

4.2 Basic Concepts of Process Planning

There are a few basic approaches to process planning, including manual and automated methods. The manual methods also include the workbook approach. The automated methods include the commonly used variant approach, the fully automated *generative* approach investigated in research, and the hybrid *semi-generative* method.

4.2.1 Manual Process Planning

The manual approach to process planning begins when a detailed engineering drawing and data on batch size are issued to a production engineer. This information is used to determine the following:

- The manufacturing processes involved
- The machine tools required to execute these processes
- The tools required at each stage of processing
- The fixtures required at each stage of processing
- The number and depth of passes in a machining operation
- The feeds and speeds appropriate to each operation
- The type of finishing process necessary to achieve the specified tolerances and surface quality

As a first step, the production engineer examines the part drawing to identify similarities with previously produced parts. If similarities are recognized, a process plan is manually retrieved for the similar item. The process plan is either used without modifications for identical parts or modified to meet the manufacturing requirements of the new part. Although old process plans are used as references for similar parts, there is still significant duplication of effort due to the lack of efficient information retrieval, comparison, and editing techniques. The manual method may also lead to inconsistency in the final plans because it is unlikely that two process planners will generate identical process plans.

As a part design changes during the product development cycle, the process plan must also change to incorporate new features in the part. As equipment, processes and batch sizes change, the optimum method for manufacturing the part also changes, and these changes must be reflected in current process plans. However, the lack of consistency and the labor intensity of the manual method make rapid incorporation of process changes extremely difficult.

The experience of the process planner plays a significant role in modifying or creating process plans, since the planner selects processes and process variable settings, which have been successfully implemented in similar situations in the past. Since manual process planning is largely subjective, the quality of the process plan is directly related to the skill and experience of the planner.

For these reasons, it is difficult or impossible to achieve consistent, optimized process plans with the conventional manual method. As a consequence, planning and manufacturing costs are increased because of the duplication of effort in the process planning function as well as specification of excessive tooling and material requirements. Production lead times also increase due to redundancies in the planning function.

4.2.2 Workbook Approach

The workbook approach involves cataloging sequences of operations for given families of workpieces. The sequences can be quickly accessed by the production engineer to formulate a process plan. The workbook approach differs from that for the manual process only in the way that the process planner identifies part similarities, using catalog groupings of workpieces rather than past process plans.

The catalog allows for greater consistency of process plans and provides the process planner with greater functionality. However, this method is limited by the number of variables (such as materials, geometry and quality) that can be cataloged. As the variety increases, the number of possible permutations and pages in the workbook, increase exponentially. The workbook approach, like the manual method, is a subjective function, based on the experience of the planner, his/her personal preference, extent of shop knowledge, interpretation of design requirements and many judgmental factors. Both methods require continual re-education of production engineers about the introduction of new processes and the withdrawal of obsolete equipment.

4.2.3 Variant Approach

Variant process planning explores the similarities among components and searches through a database to retrieve the *standard process plan* for the part family to which the component belongs. A standard process plan is a process plan that applies to an entire part family. When a standard plan is retrieved, a certain degree of modification follows in order to accommodate the details of the design. In general, variant process planning has two operational stages [2], a preparatory stage and a production stage:

- The *preparatory* stage involves coding, classifying and grouping existing components into a family matrix, as well as deriving out of this matrix a set of standard plans that can be used and modified later to become process plans for new components.
- The *production* stage of a process planning system involves coding and classifying new components so that the family most closely matching them can be found. The standard plan for the family is retrieved and modified to produce the plan for the new component (Fig. 4.2).

Group Technology

The purpose of Group Technology is to organize the vast amount of manufactured components, which construct a product. Similar to biology, where millions of living organisms are classified into genus and species, or in libraries where taxonomies are used to classify books in stacks, group technology is the method whereby manufactured components are classified into part families. This method has the advantage of providing a tractable database, where information about a part is easily managed, retrieved and implemented in computer algorithms.

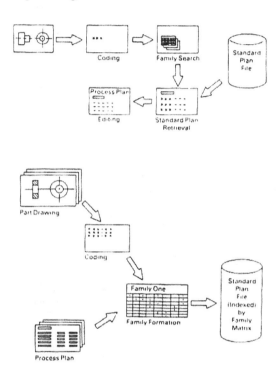

Fig. 4.2. Variant Process Planning Procedures [2]

Coding and Classification

Coding and grouping components is made with the help of a coding system. Coding systems, which are the subject of group technology, involve the application of a matrix, as in Figure 4.3, where a coding system of four digits is presented [1]. The first digit corresponds to primary shape, the second digit corresponds to secondary shape, the third to the auxiliary shape, and the fourth to the initial form of the raw material. The values for these four digits depend upon the particular geometric features of the component, and are systematically presented in the matrix shown in Table 4.1.

This process helps codify components based on their geometric characteristics and features, and, depending on the size of the coding system, other
factors, such as materials. Coding systems can have many digits; some
commercially available codes can have up to 25 digits.

		Digit1	Digit2	Digit3	Digit4
		Primary Shape	Secondary Shape	Auxiliary Shape	Initial Form
0		$\frac{L}{D} = 0.05$	No shape element	No shape element	Round Bar
1		$0.05 < \frac{L}{D} < 3$		No shape element	No shape element	Hexagonal Bar
2	Rotational	$\frac{L}{D} = 3$	Steps with round cross sections	With screw thread	With screw thread	Square Bar
3		$\frac{L}{D} = 2$ with deviation		With functional groove	With functional groove	Sheet
4		$\frac{L}{D} = 2$ with deviation	Rotational cross section	Holes	Drill with pattern	Plate and Slab
5		Flat	Rectangular cross section		Two or more from 2-4	Cast or forged
6	Nonrotational	Long	Rectangular with chamfer	Stepped plane surface	Welded assembly
7		Cubic	Hexagonal Bar	Curved surface	Premachined

Table 4.1 Coding System

The coding system in variant process planning is the first step in establishing a *family formation procedure,* which allows the classification of existing parts and their process plans into families to be used for the creation
of standard process plans. Generally speaking, the family formation procedure follows some rules of similarity. If these rules are strict, they will
lead to the formation of a large number of relatively small families, while
if they are loose, a small number of large families will be formed. For

every component an operation plan is established which entails a number of operation codes (Table. 4.2).

An operation code is usually an alphanumeric expression describing the operations that take place in one set-up, on one machine, for one part [1]. An ordered set of operation codes provides the operation plan. Thus, a step in the family formation procedure is to establish an operation plan for every component that is to be placed within a family [2] (Fig. 4.3).

In Table 4.3, a set of components has been coded, and operation code sequences have been established next to the codes [1].

Operation Code	Operation Plan
01 SAW 01 02 LATHE 02	Cut to size Face end Center drill Drill Ream Bore Turn straight Turn groove Chamfer Cutoff Face Chamfer
03 GRIND 05 04 INSP 06	Grind Inspect dimensions Inspect finish

(a) Operation Plan Code (OP Code) and Operation Plan

01   SAW 01
02   LATHE 02
03   GRIND 05
04   INSP 06

(b) OP Code Sequence

Table 4.2 Preparation Plans and Operation Code

Figure 4.4 illustrates the same information in a different form. A components-operations matrix is established listing the components in columns, and the processes/operation codes in rows [1].

If a component requires a particular process, a check is marked at the intersection of the process row and the component column. If the matrix is relatively small, components can be manually grouped in similar processes. As the number of components and operations increases, though, a more systematic approach to part family formation is needed and can be accomplished with the use of clustering algorithms [1]. The goal of a clustering algorithm is to rearrange the components-operations matrix so that components can be separated in "families" which require the same group of equipment to be manufactured.

A clustering algorithm developed by King [1], uses simple mathematical operations to assign a weight to each component and operation. Components are weighted according to the number of operations assigned to them so that a component that requires many operations is weighted more heavily. The same procedure is applied to the operations or rows of the matrix. More specifically, the weight for each column or component is calculated by assigning "unity" for each required operation and multiplying by 2^i where i is the row number corresponding to each operation. For example, the weight for component A112 in Figure 4.4 is:

$$w_1 = 2^1\left(1\right) + 2^3\left(1\right) + 2^7\left(1\right) + 2^{10}\left(1\right) = 2 + 8 + 128 + 1024 = 1162$$

After each weight has been calculated for each component and operation, the columns and rows of the matrix are ranked in ascending weight order. After the rearrangement of the matrix is final (Fig. 4.5), part families can be readily identified. This example shows two families, one comprising of the parts A123, A120, A131, A432, A451 and A112 that requires operations Lathe01, Grinding05, Lathe02, and Saw01, and the other comprising of A115, A212, A230 and A510 that requires operations Milling02, Grinding06, Drilling01 and Milling05. The steps for a clustering algorithm are:

Step 1. For A calculate the total weight of column w_j:

$$w_j = \sum_{A^i} 2^i M_{ij}$$

Step 2. If w_j is in ascending order, go to step 3. Otherwise, rearrange the columns to make w_j fall in an ascending order.

Step 3. For A_i calculate the total weight of row w_i:

$$w_i = \sum_{A^j} 2^j M_{ij}$$

Step 4. If w_j is in descending order, stop. Otherwise, rearrange the columns to make w_i fall in an ascending order. Go to step 1.

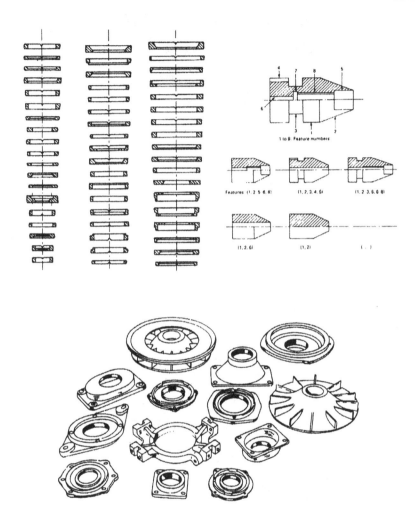

Fig. 4.3. Similar Components [2]

Component	Code	Process	OP code sequence		
A-112	1110	SAW01	LATHE02	GRIND05	INSP06
A-115	6514	MILL02	DRL01	INSP03	
A-120	2110	SAW01	LATHE02	GRIND05	INSP06
A-123	2010	SAW01	LATHE01	INSP06	
A-131	2110	SAW01	LATHE02	INSP06	
A-212	7605	MILL05	INSP03		
A-230	6604	MILL05	INSP03		
A-432	2120	SAW01	LATHE02	INSP06	
A-451	2130	SAW01	LATHE02	INSP06	
A-510	7654	MILL05	DRL01	GRIND06	INSP06
A-511					
A-511					
A-512					
A-550					
A-556					
B-105					
B-107					
B-108					
B-109					
B-115					
B-116					
B-117					
B-118					

Table 4.3 Components Operations Plans

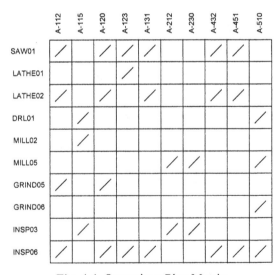

Fig. 4.4. Operations Plan Matrix

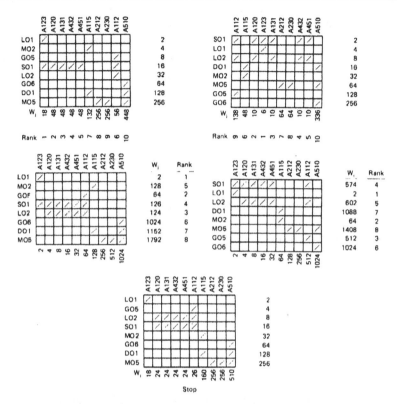

Fig. 4.5. Rank Order Clustering Procedure for Family Formation [1]

A standard plan related to both of these families will contain a standard sequence of operations/operation codes. This plan can be accessed whenever an incoming component is identified, based on its code, as belonging to either of the families; the plan is then modified to become the final plan for the new component.

Often, the process parameters for the particular process must be established. Figure 4.6 shows how the parameters of a face milling process can be determined in a hierarchical fashion. On the first level, a decision is made regarding the kind of material to be used, if this is not already specified by the design; on the second level, the physical properties of the material (expressed, for example, in hardness) are established; on the third level, the depth of cut and cutting speed are chosen; and on the fourth level the cutter diameter is chosen [1].

Current trends toward wide product variety and smaller lot sizes are driving a move to cell-type shop organization (to minimize routing paths), which is essentially based on group technology. Some advantages of Group Technology are:

- Group Technology facilitates the storage and retrieval of existing designs, minimizing duplication of effort. It also promotes standardization of design features (e.g., chamfers, corner radii), leading to standardization of production tools and holding fixtures

- More efficient material handling through cell organization

- Decreased WIP and lead times due to reduction in setup and material handling

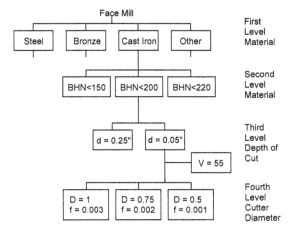

Fig. 4.6. Selection Procedure for Process Parameters

4.2.4 Generative Approach

Generative process planning synthesizes manufacturing information, particularly regarding the capabilities of different manufacturing processes, and creates process plans for new components. An ideal generative process planning system receives information about the design of the part and generates the process plan, including processes to be used and their sequences, without human intervention. Unlike the variant approach, which uses standardized process-grouped family plans, the generative approach is based on defining the *process planning logic* using methods such as:

- Decision trees
- Decision tables
- Artificial intelligence-based approach
- Axiomatic approach

Generative process planning systems are to be rapid and consistent in generating process plans. They should create process plans for entirely new components, unlike variant systems, which always need a standard plan of previously existing components; and they must allow the integration of these activities with the design of a part (upstream) and the creation of tapes for numerically controlled machines, etc. (downstream).

Generative process planning, attempts to imitate the process planner's thinking, by applying the planner's decision-making logic. There are three areas of concern in a generative process planning system:

- *Component definition*, or the representation of the design in a precise manner so that it can be "understood" by the system.
- Identification, capture, and representation of the *knowledge of the process planner,* and the reasoning behind the different decisions made about process selection, process sequence, etc.
- Component definition and planner's logic should be *compatible* within the system.

Most of the generative process planning systems use "built-in" decision logic, which checks condition requirements of the component. Some systems have "canned" process plan fragments, which correspond to particular geometric features that are combined in a final process plan.

In general, generative process planning can be executed either in a *forward* fashion, where planning starts from the initial raw material and proceeds by building up the component using relevant processes, or *backwards*, where planning starts from the final component and proceeds to the raw material shape.

One approach to define components is to use group technology concepts, as discussed in the section on variant process planning systems. Another approach is to describe the component using descriptive language (Fig. 4.7). This approach is more general and allows a variety of components to be described. General language can apply to a large number of components, however, more effort is required to describe a particular component. Less general language may be adequate for only a set of components, but they are easier to apply [1].

Process Planning Logic

Process planning logic determines the kind of processes to be used for the different geometric features of the component by matching process capabilities with design specifications.

Decision Trees

A decision tree is comprised of a root and a set of branches originating from the root. In this way, paths between alternate courses of action are established. Branches are connected to each other by nodes (Fig. 4.8), which contain a logic operation, such as an "and" or "or" statement. When a branch is true, travelling along the branch is allowed until the next node has reached, where another operation is assigned, or an action is executed [1]. A decision tree (Fig. 4.9) can be used as a base for developing a flow chart (Fig. 4.10) and eventually as the code to be used in a generative process plan.

Decision Tables

Decision Tables organize conditions, actions and decision rules in tabular form. Conditions and actions are placed in rows while decision rules are identified in columns. The upper part of the table includes the conditions that must be met in order for the actions (represented in the lower part of the table) to be taken. When all conditions in a decision table are met, a decision is taken [1]. The information content of both approaches is the same, although decision tables have a modular structure, which allows them to be easily modified and written in array format.

Artificial Intelligence (AI) / Expert Systems

Artificial intelligence techniques such as formal logic, for describing components, and expert systems, for codifying human processing knowledge, are also applicable to process planning problems. An expert system can be defined as *a tool, which has the capability of understanding problem-specific knowledge and of using the domain knowledge intelligently to suggest alternative paths of action.* Component definition can be made with methods used for declarative knowledge, or more specifically, *first order predicate calculus (FOPC).* Process selection knowledge is *procedural knowledge*, and as such, it is usually applied to production systems, which consist of production rules in the form of "if then" logic. Declarative knowledge, using first order predicate calculus, is represented by a well-formed formula (WFF), which is an atomic formula including a predicate symbol, a function symbol, and a constant, which can be true or false. To represent, for example, a hole, one can write "Depth hole (x), 2.5" where the depth is the predicate symbol, hole is the function symbol, and the constant is 2.5. With procedural knowledge, rules can be structured to relate processes to certain geometric or other features of the component, and with an adequate inference engine a set of processes and their sequences can be defined for a given component. Moreover several approaches are focused on design and development of process planning systems, capable of generating process plans from CAD drawings [32] [33].

Axiomatic Approach

Axiomatic design has been developed at MIT at the Laboratory of Manufacturing and Productivity [3]. Its intention is to provide a logical framework for designing products and processes. A set of desired characteristics of the design, known as *functional requirements* (FRs), must be defined to establish a *design range*.

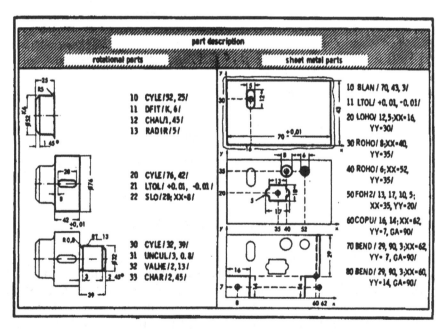

Fig. 4.7. Descriptive Process Plan Language [1]

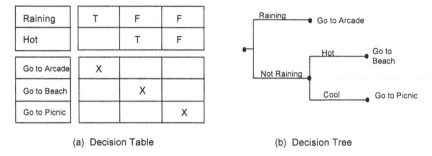

(a) Decision Table   (b) Decision Tree

Fig. 4.8. Decision Table and Decision Tree

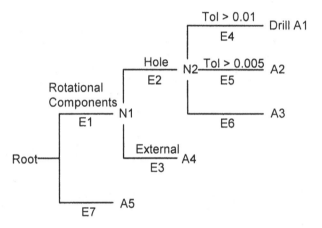

Fig.4.9. Decision Tree for Hole Drilling

A set of *design parameters* (DPs), which may vary and are limited by the nature of the system, must also be identified in order for the *system range* to be defined. A successful solution achieves the best compromise between FRs and DPs, and can be viewed as a mapping of functional requirements from the *functional domain* to design parameters in the *physical domain*. Axiomatic design is used to define and simplify the relation between the FRs and DPs through a set of axioms. The entire framework is based on two axioms:

- Axiom 1 *The Independence Axiom*
  Maintain the independence of functional requirements
- Axiom 2 *The Information Axiom*
  Minimize the information content

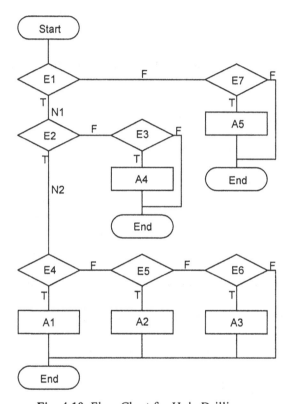

Fig. 4.10. Flow Chart for Hole Drilling

To implement axiomatic design in process planning, the following steps are applied:

1. List all the design (or production) parameters to be evaluated.
2. Divide the surfaces to be produced into surface groups, each of which is to be machined by a single machine.
3. List candidate machines for each surface group.
4. Evaluate all alternatives for the production and machine parameters.
5. Obtain the total information content and select the best machine combination based on the information content.

In a machining operation, the FRs may be surface roughness, dimensional accuracy and cost. The first task is to determine whether these FRs satisfy the first axiom, assuming they are probabilistically independent. Next, the second axiom must be satisfied. The information content of surface roughness can be experimentally determined by measuring the surface roughness given by each machine. In this manner, a lower and upper bound on surface roughness for the system range is defined. The information content, associated with dimensional accuracy, can be determined in a similar manner so that the system's dimensional accuracy range is bounded. The upper and lower bounds of the system range for cost, are determined for each machine by multiplying the maximum and minimum times taken for manufacture by the cost per unit time. Once the information content for each FR has been found, the surfaces needed to be machined by one machine are grouped into surface groups. The information content for each surface group is then calculated. For example, suppose that a particular part has two types of surfaces, which require two operations, one on a lathe and the other on a cylindrical grinder. The information content discussed above is calculated for each machine and is listed in ascending order, starting from the left-most column (Table 4.4). Each column represents a possible path of action, the best being the one, which best satisfies the axioms.

Column 1 in Table 4.4 has the lowest total information content and should be chosen as the best course of action. To verify this method, an experiment was set up to simulate the analytical problem [3]. The experimental results are tabulated in Table 4.5.

Actual machining times for Combinations 1 and 2 had almost the same duration, although machining times for Combinations 3 and 4 were much longer, as predicted by the analytical method.

	Combination			
	1	2	3	4
Surface Group 1 Lathe Order Information	L1 1 1.36	L2 2 1.39	L3 5 1.94	L4 24 4.27
Surface Group 2 Cylindrical Grinder Order Information	G1 1 0.58	G2 2 2.19	G3 3 3.73	G3 3 3.73
Total Order Information	1 1.94	21 3.58	49 5.67	72 8.00

Table 4.4 Information Required to Generate the Surface Groups by Various Machines

	Combination			
	1	2	3	4
Surface Group 1 Lathe Measured Time	L1 1,453	L2 1,405	L3 1,665	L4 2,487
Surface Group 2 Cylindrical Grinder Measured Time	G1 1,158	G2 1,205	G3 1,433	G3 1,481
Total Time	2,661	2,610	3,098	3,948

Table 4.5 Measured Total Machine Time in Seconds

Feature Regognition

It is desirable to design a methodology for transforming a part design into a set of specific process data consistently and without human intervention. Two steps are associated with this technique: part decomposition and feature recognition.

To apply an expert system, a numerical model of the part containing all information about the part geometry, in analytical form, must be created. The features of the part must then be separated based on the combination of geometric shapes (planes, curved surfaces or circles, lines, arcs as seen from different views) used to create them. Once an aggregate of geometric shapes has been found to identify with a certain feature, the feature is removed from the drawing (feature decomposition), and is classified in a group depending on the semantics and degree of similarity of the feature (feature recognition).

Several approaches have been taken to implement such techniques including:

Syntactic Pattern Recognition
State Transition Diagram and Automata
Decomposition Approach
Logic Approach
Graph-Based Approach

Syntactic Pattern Recognition

In syntactic pattern recognition, a picture is decomposed into "semantic primitives" represented by letters. A pattern can be represented by a set or string of letters. This principle is used to recognize and classify pictures in particular categories, similar to formal language processing, where a sentence is analyzed to see whether it is grammatically correct. If the syntax of a picture language agrees with that of the grammar, then the picture is classified as belonging to a particular pattern class [4]. Using these primitives, a simple geometry can be converted into a sequence of letters. For example the outline of the hole shown in Figure 4.11 can be represented by the string "BBCCCBBAAABB."

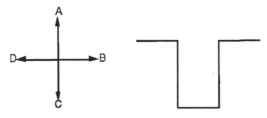

Fig. 4.11. Pattern Primitives and Simple Hole Drawing

State Transition Diagram

In this system, part geometry is described by sweeping operations and uses AND/OR operations for the union of the sweeping volumes. The CIMS/PRO and CIMS/DEC [5] systems use this technique. Here the generating surface is described by ordered pattern primitives, but instead of grammar, the relationships of adjacent pattern primitives are used. Convex adjacency is assigned a "0," and concave adjacency a "1." Consider the slot shown in Figure 4.12(a). The arrows indicate the direction of sweeping and at each transition from one shape to the other a "0" or "1" is assigned. The simple slot shape shown in Figure 4.12(b) can be represented by the string "0110" [4].

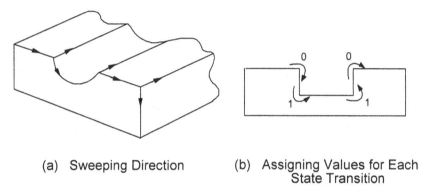

(a) Sweeping Direction   (b) Assigning Values for Each
                                       State Transition

Fig. 4.12. Features of the State Transition Method

The Decomposition Approach

As the name of this method implies, each design is broken into smaller (Fig. 4.13), simple volumes, which can be used to regenerate the original geometry through simple mathematical operations such as addition and subtraction. An attempt is then made to recognize and classify each smaller volume by the semantics of its features. Unfortunately, so far no methods of this type can guarantee the production of usable manufacturing or design features. The principle problem with this method is that the solution is not unique, which is similar to the problems associated with geometry conversion in boundary representation methods.

The Logic Approach

In this method, logic rules are applied to the topological structure of parts to extract and recognize certain features. The topological structure is derived from a boundary representation of the design. The logical rules used are IF/THEN-AND/OR statements. For example:

IF a hole entrance exists
AND the face next to the entrance is cylindrical
AND the face is convex
AND the next adjacent face is plane
AND this face is adjacent only to the cylinder
THEN the entrance face, cylindrical face and plane comprise a cylindrical hole.

The terms "entrance," "adjacent," "convex," "cylindrical," etc. are defined as separate rules in order to be distinguished and extracted from the boundary model. These rules define the part in its predicate form. Since there is a one-to-one mapping between the predicate and the boundary representations, the translation of features from one domain to the other involves searching through all facts until a given rule is proven to be true.

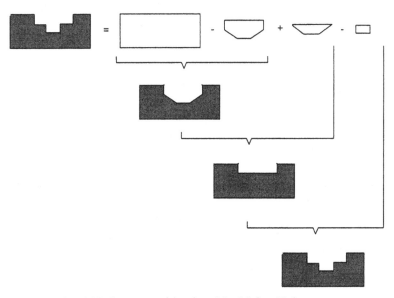

Fig. 4.13. Decomposition into Machining Volumes

The Graph-Based Approach

Typically, expert system-based techniques use a backward chaining procedure for part feature recognition, which is executed by invoking feature primitives one by one and performing an exhaustive search for the presence of the feature. A graph-based part representation that facilitates a heuristic search to uniquely identify part features is the *Attributed Adjacency Graph (AAG)* [6], in which the boundary representation of the part is transformed into a graph. The AAG is defined as a graph $G = (N, A, T)$ where N is the set of nodes, A is the set of arcs, and T is the set of attributes to arcs in A, such that

- for every face f in F, there exists a unique node n in N.
- for every edge e in E, there exists a unique arc a in A, connecting the nodes n_i and n_j, corresponding to faces f_i and f_j, which share the common edge.
- every arc a in A is assigned an attribute t, where $t = 0$ if the faces sharing the edge form a concave angle, and $t = 1$ if the faces sharing the edge form a convex angle.

The AAG (Fig. 4.14) is represented in the computer by a triangular matrix. Each node stores the pointer to the corresponding face information, thus, providing a link to the boundary representation, if desired.

In order for features to be recognized, they must be precisely defined. Minimal sets of necessary conditions can be proposed to classify a few features uniquely [6]. Figures 4.15, 4.16 and 4.17 illustrate generic classifications for slots and pockets. In Figure 4.15, the generic definition for a class of slots is based on the adjacency of F_1 and F_2, the adjacency of F_2 and F_3, the angle between F_1 and F_2 (concave or convex), and the angle between F_2 and F_3. This generalized classification provides a framework for representing the features in a hierarchical manner (Fig. 4.17). The major advantage of this hierarchical organization is the reduction in computational time, required to recognize a feature. Another advantage arises from the notion of inheritance. Since the instance of a particular subclass is also an instance of its superclass, the properties of the superclass can be inherited by the subclass without explicitly being repeated. This provides a more compact representation for subsequent processing.

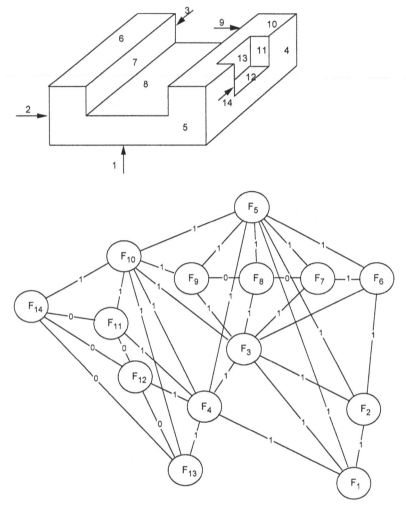

Fig. 4.14. Example of AAG for a Part

The recognition rules for features, at a generic level, are based on the properties of the AAG uniquely to each feature. Figure 4.17 shows some specific instances of generic feature types and their AAGs [4].

As an example of how features can be recognized, all pockets have the following properties in terms of an AAG:

- The graph is cyclic
- There is exactly one node n with number of incident 0 arcs equal to the total number of nodes minus 1
- All other nodes have degree = 3

- The number of 0 arcs is greater than the number of 1 arcs (after deleting node n)

The AAG technique can be extended to allow for features formed by a combination of planar and cylindrical faces [7]. In addition, a Directed Graph Data Structure (DGDS) has been developed to represent the topology of faces and edges of a solid modeler in a more compact structure [7]. The data structure of a DGDS system is compared with that of a solid modeler in Figure 4.18. A knowledge-based heuristic tree search algorithm is then used to determine the sequence of machining operations (process plan). A logical rule-based approach is implemented to generate machinability data (feeds and speeds) using a mathematical modeling system rather than a library retrieval system.

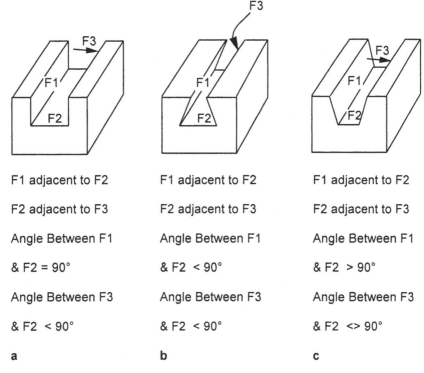

F1 adjacent to F2	F1 adjacent to F2	F1 adjacent to F2
F2 adjacent to F3	F2 adjacent to F3	F2 adjacent to F3
Angle Between F1	Angle Between F1	Angle Between F1
& F2 = 90°	& F2 < 90°	& F2 > 90°
Angle Between F3	Angle Between F3	Angle Between F3
& F2 < 90°	& F2 < 90°	& F2 <> 90°
a	b	c

Fig. 4.15. Different Slot Types and Their Recognition Rules

Modified AAG Technique. In order to account for edges joining cylindrical and planar faces, more arc attribute values were allowed characterizing the edge as straight, cylindrical, round, or hollow (str, cyl, rnd, hol) in addition to concave or convex (cc, cv) (Fig. 4.19) [7]. In order to facilitate cutter path generation, they ensured that all features were composed of three faces: a reference face (where machining begins), a feature face (usually horizontal), and a bottom face (where machining ends). A few feature components were also combined to create recognizable complex features.

Tree Search Algorithm to determine Optimal Cutter Path. Machining precedence is determined by creating a tree structure to relate all faces in a hierarchical structure, where parent nodes have to be finished before the child node can be machined. A "heuristic value", which incorporates the knowledge of an expert planner, is then assigned to every arc (Table 4.6). The optimal path is determined using the hill climbing tree search algorithm (Fig. 4.20).

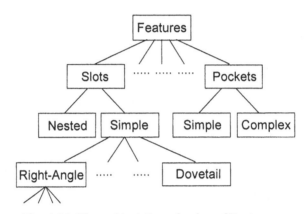

Fig. 4.16. Hierarchical Organization of Features

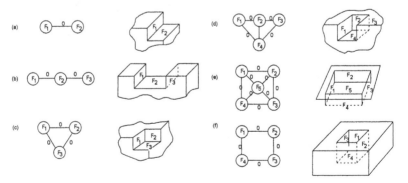

Fig. 4.17. AAG of Feature Instances

Is the same tool used as was used for the parent?	Is the feature face adjacent to one of the parent?	Does the process for the parent machine the reference face?	Heuristic Value
YES	YES	YES	1
YES	YES	NO	2
YES	NO	YES	3
YES	NO	NO	4
NO	YES	YES	5
NO	YES	NO	6
NO	NO	YES	7
NO	NO	NO	8

Table 4.6 Heuristic for Determining Machining Precedence

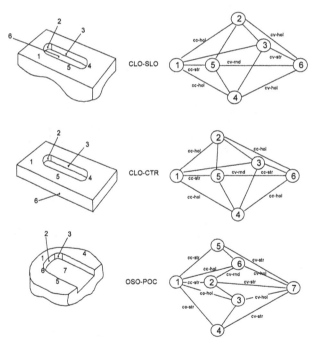

Fig. 4.18. Data Structure in the Solid Modeler Romulus Compared to a DEDS Structure

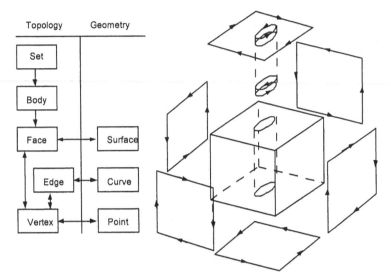

Fig. 4.19. Example of Modified AAG Feature Representations

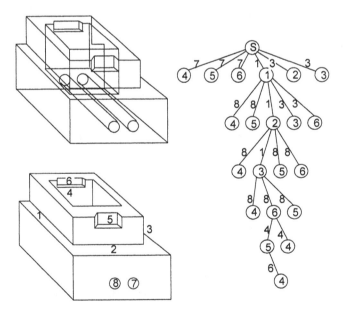

Fig. 4.20. Tree Structure for Machining Precedence

4.2.5 Semi-Generative Approach

Generative forms of process planning, in the true sense of the definition, do not exist in our estimation. However, a number of "semi-generative" systems exist and combine the decision logic of generative systems with that for the modification operations of variant systems. The semi-generative approach can be characterized as an advanced application of variant technology employing generative-type features. Hybrid systems can combine variant and generative features in the following ways:

- Within a given product mix, process plans for some products can be produced, using generative methods, while process plans for the remaining products can be produced using variant planning.
- The variant approach can be used to develop the general process plan and then, the generative approach can be used to modify the standard plan.
- The generative plan can be used to create as much of the process plan as possible, and then the variant approach can be used to fill in the details.
- The user can select either generative or variant modes for planning a part in order to accommodate fast process plan generation or complicated design features.

Some attributes of the process planning methods discussed above are summarized in Tables 4.7 and 4.8.

Characteristic	Manual	Workbook	Variant	Generative
Initial investment cost	Low	Low	High	High
Maintenance cost	Medium	Medium	High	Low
Training cost	High	Medium	Medium	Low
Experience level	High	Medium	Medium	Low
Clerical effort	High	Medium	Medium	Low
Consistency	Low	Medium	Medium	High
Speed	Low	Low	Medium	High
Integration of new capabilities	Low	Low	Medium	High
Accommodation of non-std parts	High	Low	Low	High
Documentation	Low	Medium	High	High
Paper storage	High	High	Low	Low
Ease of retrieval	Low	Low	High	High
Configuration control	Low	Low	High	High
Integration with CIM	Low	Low	Medium	High
Transportability	High	Medium	Low	Low

Table 4.7 Comparison of Process Planning Methods

Process planning steps	Manual	Workbook	Variant	Generative
Examine part drawing	M	M	M	M
Identify "similar-to's"	M	M	A	A
If similar part identified:				
Retrieve plan	M	M	A	
Modify plan	M	M	M	
If no similar part identified:		*	*	
Select material	M			A
Select processes/sequence	M			A
Select machine	M			A
Select required tools	M			A
Identify fixtures	M			A
Determine number & depth of passes	M			A
Determine feeds & speeds	M			A
Select finishing processes	M			A

Table 4.8 Process Planning Steps

4.2.6 Role of Process Planning in Concurrent Engineering

The goal of concurrent (or simultaneous) engineering is to simultaneously release both product and process specifications. To ensure simultaneous release, the design and manufacturing constraints must be considered from the earliest stages of design. When both design and manufacturing professionals can influence the design at the conceptual phase, many difficulties can be avoided downstream. Quality, reliability and low cost can be designed and ultimately be built in. Concurrent engineering is so named because the design and manufacturing constraints must be viewed concurrently while design decisions are made (Fig. 4.21). Specific design and manufacturing constraints should be accessible at all times to ensure this concurrence. In many instances, design and manufacturing engineers cannot continuously consult each other during design; consequently, one function of automated process planning is to model the knowledge of one or both engineers.

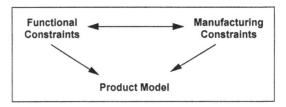

Fig. 4.21. Framework for Concurrent Engineering

Concurrent engineering includes the following levels of feedback on a design:

1. *Manufacturability Analysis* – This is the first level of relationship between product design and manufacturing constraints. In this phase, the process planner examines the design at a feature level in order to determine its technical feasibility, both in terms of engineering and manufacturing constraints. This type of knowledge allows a design engineer to make rapid, informed decisions about the design during the initial phase. Manufacturing feedback early in the design phase, allows for sweeping design changes to improve manufacturability, instead of the incremental changes, which normally occur during sequential product development. In addition, information on constraints of the manufacturing system can be calculated and be fed back to the design engineer for incorporation into the design.

2. *Producibility Analysis* – This determines the ability of a system to produce a part or series of parts, as well as the cost effectiveness of a design. Parts, which are deemed to be unproducible with the current capability constraints, can be reentered into the system with "what if" changes on the constraints. For example, a constraint could be changed to allow a different metal or size of material to be handled. The new constraint would be realized by taking the constraint changes and matching them with machine specifications to determine an acceptable machine system. The "what if" constraints can include actual vendor-supplied machine specifications, which allow the system's procurement to be handled quickly and more accurately with a given part mix. Producibility analysis is also useful in estimating the cost requirements of a particular design. For example, even if a design is manufacturable, it may be lengthy or prohibitively expensive to produce. Trade-offs must be made to decrease the production cost without compromising quality. Producibility analysis effectively allows for early determination of price sensitivity. Since approximately 90% of the total production cost is specified during the design phase, little cost reduction can be accomplished by improving manufacturing methods once a design is complete; therefore, early cost feedback is crucial for competitive designs to be achieved.

3. *Product Changes* – Producibility is performed on the basis of the cost impact of design features on the overall design. Shop floor constraints and processing constraints will suggest changes to the product, changes that include standardization of feature instances within classes and easing of tolerance constraints. Product features can be compared with the standardized features defined by production capability of the shop floor resources. If a product feature is not consistent with the standardized

features, then the standard feature, which is closest to the specified feature, will be suggested as a possible alternate. The process planner can also request larger dimensional tolerances to accommodate process limitations. The request for standard features and larger tolerances is passed back to the design engineer for possible changes.

4. *Vendor Qualification* – A procurer of mechanical parts from multiple vendors may be concerned with the aptitude of the latter to supply quality parts at the quoted prices. A process planner uses models of the vendors' capabilities, based on their shop floor constraints and process limitations to determine which vendor supplies the best quality and cost performance. Other constraints, such as business size, previous performance on bids, and present business nature can also be constraint modeled to assist in determining the best vendor.

5. *Shop Floor Changes* – Producibility analysis results in tooling and fixturing changes to make a part more producible. The process planning function suggests that additions be made to the tooling capabilities of the manufacturing system, resulting in the reduction of shop floor constraints and allowing more shop floor capacity.

Having a fully configured feature-based product model is the first requisite of concurrent engineering. The second is a complete model of factory capabilities and limitations to provide manufacturing feedback. Knowledge of the factory's processes and machine capacities may be modeled in a similar fashion to that of the knowledge-based product model. A product model can be analyzed against the capabilities of a factory in order to determine how producible it is, based on the design requirements of its features. This feedback information shows which parts of the design are outside the limitations of the current factory. The design engineer can use this information to change the product design so as to meet the constraints of the factory. Also, the manufacturing engineer can use the information to upgrade manufacturing capabilities to meet the new design requirements. These models provide vital decision-making information by indicating points of conflict; however, they do not provide any mechanism for conflict resolution. Concurrent engineering environment can be established, by developing a STEP repository, as a core, and then by linking it to various open CAD/CAPP/CAM systems.

The Standard for Exchange of Product Model Data (STEP), ISO 10303, provides a computer sensible format for the communication and storage of product data. STEP includes a number of full International Standards, such as the EXPRESS language for information modeling, the STEP Physical File, which provides neutral format for data exchange and the SDAI, which specifies a computer independent Application Programmer's

Interface (API) for sharing product databases. The STEP format is based on information models, formulated in the EXPRESS language, which describes aspects of a part across its entire life cycle. These information models can form the basis for a standard definition of product data within a concurrent engineering environment.

Several Product Data Management (PDM) systems have been developed and implemented to support concurrent task management. PDM systems maintain control of the data, hold the master data only once in a secure "vault", where their integrity can be assured, and all changes made are monitored, controlled and recorded. However, the implementation of the commercially developed PDM systems frequently faces many unexpected problems, which result in expensive missteps for the enterprises. Typical problems encountered during such efforts, are associated with software integration and conversion. Another point is that PDM is worthless without a business data model and it is required to produce a system that manages the business structure of the whole enterprise. The initial goal, when implementing a PDM system, should be the establishment of a first-class data repository or vault. However, the STEP technology can be used to advance the PDM technology for supporting the concurrent engineering technology.

4.3 Applications

In the past, CAPP research had involved the development of efficient variant systems by using state-of-the-art search techniques and database programs. However, later on, academic research was largely focused on developing fully-generative CAPP systems, which could be directly interfaced with CAD product models to generate process plans. To achieve this capability, researchers have increasingly applied tools used in artificial intelligence, especially expert systems, to CAPP. Expert Systems use a knowledge base to create a process plan with a series of "If-then"-type statements. Since industrial designs are highly complex, expert systems designed for industrial process planning must also be complex, considering variables, such as cost, materials, machine speeds, surface finishes, tooling order, and scheduling on both macro and micro scales. Decision making is performed on a step-by-step basis with consideration for both future and past steps. This is in contrast to the variant method, where a part is presented and the process planner simply finds a similar part and modifies an existing plan.

Some examples of CAPP systems are briefly discussed in the following sections.

4.3.1 Variant Process Planning

The following section describes several process planning systems, which incorporate the variant approach with automated process planning. The basic function of these systems is to analyze a CAD drawing of a part by using decomposition and feature recognition techniques so that the primary features of a part can be identified. A progression in machining operations is then prescribed through a rule-based search of existing data so that possible solutions are found. Many systems use backward chaining logic to generate and to check the feasibility of a particular feature.

CAM-I CAPP [8], *Drilling Operation Planning System (DOPS)* [9], *Expert Computer-Aided Process Planning (EXCAP* [10], *Interactive Computer-Aided Process Planning (ICAPP)* [10] and *Microplan* [11] are some of the early automated process planning applications.

COMPLAN is an automatic process planning and scheduling system, based on non-linear process plans. The system uses production constraints as a means of collaboration between scheduling and process planning. The system architecture contains hierarchical process planning and scheduling modules, workshop resources and corresponding evaluators for each one of them. At all levels of the system, the evaluation modules are provided with feedback information so as to improve the quality of the delivered output. The production constraints used, can be classified into two groups, the group of the general and that of the specific constraints. The system allows automatic, semi-automatic or manual process planning and incorporates options to allow the process planner to control the flow of the search [12] [13].

MCOES (Manufacturing Cell Operator's Expert System) was developed as a design and planning system for short batch production. MCOES uses feature-based part family models for process planning. A variant feature approach was adopted so that varying levels of detail and granularity could be used in the feature model for the part families. The system consists of a design data interface, a generative process plan preparation system and an operative process planning system. The design interface supports feature-based modeling of part families. The generative planner allows manufacturing processes to be described and related to part-family models. The operative planner generates the process plan and the NC code, based on part family descriptions [14].

EXBLIPP (Explanation-based learning approach to intelligent process planning) is an approach to automating the process-planning task. EXBLIPP incorporates a STRIPS-like planning system, an Explanation-Based Learning system (EBL), and a dynamic planner for extending traditional *a priori* planning to allow for execution-time decision-making [15].

Real-Time Computer Aided Process Planner (RTCAPP) is a process planning system, which is capable of generating process plans at each step of the design process, in a timely manner. RTCAPP generates incremental process plans for prismatic parts with internal features, for example, gear housing, bearing block, or cylinder block. RTCAPP evaluates the estimated manufacturing cost of each plan and reports it back to the designer after each design update. The system enables designers to receive manufacturing cost information for their alternative design configurations that meet the desired product functionality requirements, as well as to select the least cost design alternatives [16].

4.3.2 Generative Process Planning

Generative process planning systems have been the focus of research activity in manufacturing. Extensive use of expert systems and search techniques are the main components of such systems. Because of the great degree of complexity of the algorithms and enormous calculation effort, some systems specialize in developing plans for specific types of geometries. A few systems go as far as checking the manufacturability of a part and suggesting changes in the design, if necessary. Finally, developments in AI as well as improvements in computer technology, allow some of these systems to be implemented on mini and micro-computers.

GARI [17], *GCAPPS* and *Printed Wiring Board planner (PWA)* [18], *Hierarchical and Intelligent Manufacturing Automated Process Planner (Hi-Mapp)* [19], *Operation Planning Expert (OPEX)* [20], *and System for Automatic Programming Technology (SAPT)* [21], are some of the early expert generative process planning systems.

PART and *PART-S* are more recent generative process-planning systems. PART is an acronym for *Planning of Activities Resources and Technology*. The difference between the two is that PART is the older system, focused on prismatic parts while PART-S is the newer system, inspired by its predecessor, focused on sheet metal parts. PART and PART-S share the same philosophy and roughly offer the same functionality (apart from specific product and process related functionality). First of all, there is a CAD interface in which a solid model representation from a CAD system can be converted into the internal representation of the modelers, used in PART. If tolerances have not been added to the original model, it is possible to edit tolerances in the tolerance editor and thus, automatic feature recognition can start. The sequence of feature recognition and other activities can be application dependent.

The following activities can be performed: set up selection, machine tool selection, design of jigs and fixtures, determination of machining methods, cutting tool selection, machining operation sequencing, NC output generation and capacity planning. The application area of PART is in the machining of 2.5D prismatic components, focusing on processes, such as milling, drilling, finishing, boring, reaming, etc. The application area of PART-S, on the other hand, is a small batch part manufacturing of sheet metal components, nested in sheets with a thickness between 0.5 and 5 mm. The main processes included in PART-S are laser cutting, nibbling, (special tool) punching, laser welding and air bending [12].

A reference model has been developed that supports integration of design and process planning and enables a simultaneous engineering approach, in the early stages of product development [22]. Moreover, it provides means of creating an integrated product and process modeling procedure for design and process planning activities. The reference model consists of four partial models, such as the activity model, the information model, the technical system model and the model of integrating methods. Using these models, the methodology enables the concurrent processing of design and process planning activities with regard to different components of a product. In order to develop the reference model, it is necessary to clarify the types of interdependence, existing among the elements of the product development process. These interdependencies have been modeled using the EXPRESS modeling method, which was developed for STEP. The reference model can be derived from the connections among the elements of the product development process. In correspondence to the number of elements, the model comprises four partial models: the information model and the technical system model form the core of the reference model. The activity and the method models represent the framework of the reference model. The activities in the activity model are assigned to individual phases.

CyberCut is a web-based design-to-fabrication system consisting of three major components

1. Computer-aided design software, written in Java and embedded in a web page.
2. A computer-aided process planning system with access to a knowledge base, containing the available tools and fixtures, and
3. An open- architecture machine tool controller that can receive the high-level design and planning information and carry out sensor-based machining on a Haas VF- 1 machine tool.

By providing access to the CyberCut CAD interface over the Internet, any engineer with a Word Wide Web browser, becomes a potential user of this on-line rapid prototyping tool. Once fully operational, a remote user will be able to download a CAD file in some specified universal exchange format to the CyberCut server, which will in turn execute the necessary process planning and generate the appropriate CNC code for milling [23].

A generative process planning system for robotic sheet metal bending press brakes has been developed. This process planning system employs a distributed planning architecture. Currently, the system consists of a central operation planner and three specialized domain-specific planners: tooling, grasping and moving. The central operation planner proposes various alternative partial sequences, and each specialized planner evaluates these sequences, based on its objective function. The central operation planner uses state-space search techniques to optimise the operation sequence. The distributed architecture allows the development of an open-architecture environment for making generative process planning and encapsulating the specialized knowledge in specialized planners. This system performs both macro and micro planning. Once a CAD design is given for a new part, the system determines: the operation sequence, the tools and robot grippers needed, the tool layout, the grasp positions, the gage, and the robot motion plans for making the part. These plans are sent to the press-brake controller, which executes them and then returns gauging information back to the planning system for plan improvement. A second plan, which reduces the gauging time, is then formulated by incorporating the reduced uncertainty in the part location [24].

A volume decomposition approach to building an interface, which bridges the gap between a 3D CAD model and an automated process planning system, has been implemented. The volume decomposition approach partitions the whole machining volume, which needs to be removed from the raw material to produce the finished part, into some machinable features. Once the machinable features of a part are found, the machining of the part can be viewed as material removal for a series of machinable features.

The tasks involved in volume decomposition are:

1. extraction of machining volume from the part model,
2. decomposition of the machining volume into machinable features and
3. determination of procedures required for cutting out the machinable features.

By means of a volume decomposition process, the problem of interfacing CAD with a process planning system, can be both solved effectively and efficiently [25].

A robust CAPP system has been developed that can effectively utilise the complete, accurate, computer-interpretable, neutral STEP file for empowering the manufacturing process. This CAPP system designated the *Generative Process Planning Environment* (GPPE), which is a STEP driven process planning system. Its primary functions are the creation of what are called macro process plans, detailed and complete shop floor routings, bills of materials, tooling requirements, and time estimates-from a digital input that is ideally in AP224 STEP compliant format. The GPPE uses feature-based CAPP technology to generate process plans within a data fusion environment [26].

4.3.3 Semi-Generative Process Planning

The *Automatic Machine Programming (AMP)* was originally designed for variant process planning, but subsequent modifications allow the current system to produce process plans generatively. A video library of experts in the process planning field was used to construct the knowledge base, required for planning three-dimensional parts. The system divides its knowledge into two categories: specific and generic. Within the specific categories, additions to the knowledge base can be added as manufacturing process technology changes. The generic portion of knowledge is derived from the researcher's video library and consists of five divisions for task elaboration, task primitives, task purpose, task constraints and deeper expertise, as shown in Table 4.9 [27].

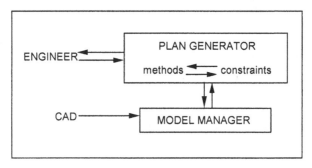

Fig. 4.22. Structure of XMAPP

The *Generative Process Planning System (GENPLAN)* is based largely on CAM-I CAPP and is divided into two modules: Fabrication GENPLAN and Assembly GENPLAN. Similar to variant systems, GENPLAN uses groupings of parts and processes into families. When a new part is introduced into the system, the user has three alternatives:

1. Suitable changes can be made on a standard process plan for the corresponding part family. In this option, the package serves a variant CAPP function.
2. An incomplete process plan can be generated by GENPLAN. The portions of the plan that the computer cannot resolve are left out, and the user completes the plan manually.
3. GENPLAN can start from the beginning and completely create a new plan by using standard process descriptions stored in the computer [8].

CLASS	DESCRIPTION	EXAMPLE
task elaboration	A definition of high level tasks and how to expand each task into a network of sub-tasks	the machine an approach facet, machine each feature accessible from that facet
task primitives	A definition of low level (primitive) actions in a form which may be used for planning output	coutersink the hole
task purpose	The effects and purpose of each action whether at a high or at a low level	dowel location on turnover of the workpiece provides precise registration of the component
task constraints	Preconditions (constraints) to the achievement of an expansion or a primitive action	depth of finishing cut must not exceed 0.1 inches
deeper expertise	Strategies for plan refinement	aggregate all drilling operations to the start of an approach facet sequence

Table 4.9 Process Planning Steps in AMP

The Semi-Intelligent Process Planner (SIPP) is an expert system written in Prolog for machining parts with a knowledge representation, based on frames. There are several frames of information consisting of machine capabilities and machining surfaces. A rule-based control structure is used to manipulate the database in order to generate a process plan. The knowledge base consists of machinable surfaces and capabilities of machining operations. The system uses a "best-first" branch and bound strategy to produce a minimal cost process plan according to user-defined cost objectives [28]. The successor to SIPP is the Semi-Intelligent Process Selector (SIPS), which is written in LISP. SIPS uses a hierarchical knowledge clustering technique, instead of flat frame representation, to store the knowledge base. The knowledge representation is divided into static and problem-solving knowledge. Static knowledge describes geometric features of the part and is stored internally as frames, as knowledge representation is in SIPP. Problem-solving knowledge represents information about operation selection and is stored in a hierarchical fashion with operations clustered into classes. SIPS is being integrated into the AMRF at the National Bureau of Standards [28].

Further Reading

An introduction and brief overview of issues related to process planning are presented in most general manufacturing textbooks. *"Manufacturing Engineering and Technology,"* by Serope Kalpakjian [29], *"Modern Manufacturing Process Engineering,"* by Benjamin Niebel, Alan Draper and Richard Wysk [2], *"Computer Aided Manufacturing,"* by Tien-Chien Chang, Richard Wysk and Hsu-Pin Wang [30], and *"An Introduction to Automated Process Planning Systems,"* by Tien-Chien Chang and Richard Wysk [1], provide definitions and discussion of all existing process planning methods used currently in manufacturing, as well as discussions of process planning logic and decision making strategies.

More information on the role of computers in planning manufacturing processes is presented in *"Computer Integrated Manufacturing Technology and Systems,"* by Urlich Rembold, Christian Blume, and Ruediger Dillmann [31]. An overview of CIM is presented, including topics such as product design and production planning. Specifically for process planning, group technology, coding system methods and direct process plan interfacing with CAD databases are treated.

A more sophisticated and detailed treatment of process planning is presented in *"Expert Process Planning for Manufacturing,"* by Tien-Chien Chang [4]. The book emphasizes computer aided process planning and the use of AI-expert systems in feature recognition, group technology, etc. Process planning logic and its implementation in software are examined, while recent software developments and an example of an expert process planning system are presented.

An application-oriented text that closely follows actual manufacturing plan development is presented in *"Applied Manufacturing Process Planning: With Emphasis on Metal Forming and Machining"*, by Donald H. Nelson and George Schneider. [34].

Finally, more on Axiomatic Design can be found in *"The Principles of Design,"* by Nam P. Suh [3]. Guidelines for decision-making in the manufacturing environment are presented, not only for the "natural" processes, or the physical laws that govern a process, but also for the "creative" processes of design, synthesis and decision-making.

References

1. Chang T.C., Wysk R.A., *An Introduction to Automated Process Planning Systems*, Prentice Hall, Englewood Cliffs, New Jersey, 1985.

2. Niebel, B., A. Draper, and R. Wysk, *Modern Manufacturing Process Engineering*, McGraw-Hill, New York, 1989.

3. Suh, N.P., *The Principles of Design*, Oxford University Press, Oxford, U.K., 1990.

4. Chang, T.C., *Expert Process Planning for Manufacturing*, Addison-Wesley, Reading, MA, 1990.

5. Iwata, K., Y. Kakino, F. Oba, and N. Sugimura, "Development of Non-Part Family Type Computer Aided Production Planning System CIMS/PRO", *Advanced Manufacturing Technology*, ed. P. Blake, North Holland Pub. Co., New York, (1980), pp. 171-184.

6. Joshi, S., and T.C. Chang, "CAD Interface for Automated Process Planning", *Proceedings of the 19th CIRP International Conference on Manufacturing Systems*, Penn. State Univ., (June 1987), pp. 45-53.

7. Lee, Y.C., and K.S. Fu, "Machine Understanding of CSG: Extraction and Unification of Manufacturing Features", *IEEE Computer Graphics and Applications* (1987), pp. 20-32.

8. Ham, I., and C.Y. Liu, "Computer-Aided Process Planning: The Present and the Future", *Annals of the CIRP* (Vol. 37, No.2, 1988), pp. 591-602.

9. Major, F., and W. Grottke, "Knowledge Engineering Within Integrated Process Planning Systems", *Robotics and Computer-Integrated Manufacturing* (Vol. 3, No. 2, 1987), pp. 209-213.

10. Davies, B.J., "Application of Expert Systems in Process Planning", *Annals of the CIRP* (Vol. 35, No. 2, 1986), pp. 451-453.

11. Phillips, R., and V. Arunthavanathan, "An Intelligent Design and Process Planning System", *Proceedings of the 19th CIRP International Conference on Manufacturing Systems*, Penn. State Univ., (June 1987), pp. 17-22.

12. Cay F., Chassapis C., "An IT view on perspectives of computer aided process planning research", *Computers in Industry*, (Vol. 34, 1997), pp. 307-337.

13. Kempenaers J., Pinte J., Detand J, Kruth J-P, "A Collaborative process planning and scheduling system", *Advances in Engineering Software*, (Vol. 25, 1996), pp. 3-8.

14. Mantyla M., "Representation of Process Planning Knowledge for Part Families", *Annals of the CIRP*, (Vol. 42 No.1, 1993), pp. 561-564.

15. S. C. Park and Gervasio M. T., Shaw M. J., DeJong G. F., "Explanation-Based Learning for Intelligent Process Planning", *IEEE Transactions on Systems, Man and Cybernetics,* (Vol. 23, No. 6, 1993), pp. 1597-1616.

16. Park J. Y., "A real-time computer-aided process planning system as a support tool for economic product design", *Journal of Manufacturing Systems*, (Vol.12, No. 2, 1993), pp. 181-193.

17. Descotte, Y., and J.C. Latombe, "GARI: A Problem Solver That Plans How to Machine Mechanical Parts", *Proceedings of the 7th International Joint Conference on Artificial Intelligence*, Vancouver, Canada, (August 1981), pp. 766-772.

18. Gupta, T.B., and K. Ghosh, "Survey of Expert Systems in Manufacturing and Process Planning", *Computers in Industry* (Vol. 11, No. 2, 1989), pp. 195-204.

19. Berenji, H.R., and B. Khoshnevis, "Use of Artificial Intelligence in Automated Process Planning", *Computers in Mechanical Engineering* (Vol. 5, No. 2, 1986), pp. 47-55.

20. Glass, M., "OPEX - An Expert System for CAPP", *6th International Workshop on Expert Systems and Their Applications,* Avignon, France, (April 1986), pp. 141-147.

21. Milacic, V. and M. Kalajdzic, "Logical Structure of Manufacturing Process Design - Fundamentals of an Expert System for Manufacturing Process Planning", *Proceedings of the 16th CIRP International Conference on Manufacturing Systems*, Tokyo, Japan, (July 1984), pp. 93-101.

22. Eversheim, W., W. Bochtler, R. Grabier and W. Kolscheid, "Simultaneous engineering approach to an integrated design and process planning", *European Journal of Operational Research* 100 (1997), pp. 327-337.

23. Brown, S. and P. K. Wrigth, "A progress report on the Manufacturing Analysis Service, an Internet-Based Reference Tool", *Journal of Manufacturing Systems*, (Vol. 17 No.5, 1998), pp. 389-398.

24. Gupta, S. K., et al., "Automated Process Planning for Sheet Metal Bending Operations", *Journal of Manufacturing Systems*, (Vol. 17, No. 5, 1998), pp. 338-360.

25. Lin, A. and S-Y. Lin, "A volume decomposition approach to process planning for prismatic parts with depression and protrusion design features", *Int. Journal of Computer Integrated Manufacturing*, (Vol. 11, No. 6, 1998), pp. 548-563

26. Bradham, J., *STEP-Driven Manufacturing*, SME Blue Book Series, 1998.

27. Willis, D., I.A. Donaldson, J.L. Murray and M.H. Williams, "Knowledge-Based Systems for Process Planning Based on a Solid Modeller", *Computer-Aided Engineering Journal* (Vol. 6, No. 1, 1989), pp. 21-26.

28. Alting, L., "XPLAN - An Expert Process Planning System and Its Further Development", *27th International MASTADOR Conference*, UMIST, UK, (April 1988), pp. 291-297.

29. Kalpakjian, S., *Manufacturing Engineering and Technology*, Addison-Wesley, Reading, MA, 1989.

30. Chang, T.C., R.A. Wysk and H.P. Wang, *Computer Aided Manufacturing*, Prentice Hall, Englewood Cliffs, New Jersey, 1991.

31. Rembold, U., C. Blume and R. Dillmann, *Computer-Integrated Manufacturing Technology and Systems*, Marcel Dekker Inc., New York, 1985.

32. Jiang, B., H. Lau, F.T.S. Chan and H. Jiang, "An automatic process planning system for the quick generation of manufacturing process plans directly from CAD drawings", *Journal of Materials Processing Technology*, (Vol. 87, No.1-3, 1999), pp. 97-106.

33. Choi, J. C., B. M Kim., H. Y. Cho, C. Kim and J. H. Kim, "An integrated CAD system for the blanking of irregular-shaped sheet metal products", *Journal of Materials Processing Technology*, (Vol. 83, No. 1-3 1998), pp. 84-97.

34. Nelson, H. D. and G. Schneider *Applied Manufacturing Process Planning: With Emphasis on Metal Forming and Machining*, Prentice Hall, 2000.

Review Questions

1. What is the purpose of process planning?

2. Why is Computer Aided Process Planning (CAPP) useful in manufacturing? Discuss some of the benefits of CAPP.

3. What are the difficulties associated with CAPP?

4. What approaches to CAPP exist? Discuss their differences and similarities.

5. What are some advantages and disadvantages of manual process planning?

6. What are the steps in variant process planning?

7. How does the "workbook" approach differ from manual process planning?

8. What is the function of "Group Technology"? What advantages can be derived from it?

9. What defines a "part family"? Give an example of a family of parts.

10. What "tools" comprise the "process planning logic"?

11. What is the "Generative Approach" to process planning and how does it differ from "Variant Approach" techniques?

12. List the primary concerns associated with generative process planning.

13. a) What are "Expert Systems" and what advantages do they possess over other   logic structures such as decision tables and decision trees?

    b) List some features of expert systems.

14. How can the "Axiomatic Approach" be applied in decision making?

15. What types of "pattern recognition" techniques are used in CAPP?

16. What is an "Attributed Adjacency Graph"? What are the advantages of this approach?

17. What is the "Semi-Generative Approach" to process planning? How

does it overcome the problems encountered in a fully generative process plan?

18. What is "concurrent engineering"? From what perspective can CAPP aid concurrent engineering?

19. Describe some features of "MCOES" and "PART-S." What are some of their basic differences?

5. The Design of Manufacturing Systems

5.1 Introduction

The manufacture of products in the modern industrial world requires the combined and coordinated efforts of people, machinery, and equipment. Thus, a *manufacturing system* can be defined as a *combination of humans, machinery and equipment that are bound by a common material and information flow.* The materials input to a manufacturing system are raw materials and energy. Information is also input to a manufacturing system, in the form of customer demand for the system's products. The outputs of a manufacturing system can likewise be divided into materials, such as finished goods and scrap, and information, such as measures of system performance.

In general, a manufacturing system design can be conceptualized as the mapping from the performance requirements of a manufacturing system, as expressed by values of certain performance measures, onto suitable values of decision variables, which describe the physical design or the manner of operation of the manufacturing system (Fig. 5.1).

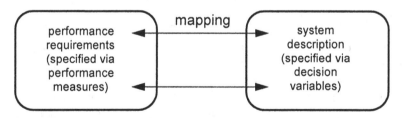

Fig. 5.1. The Manufacturing System Design Problem

A performance measure is a variable whose value quantifies an aspect of the performance of a manufacturing system. Performance measures are either *benefit* measures (the higher the better) or *cost* measures (the lower the better). They can be divided into four categories: time, quality, cost, and flexibility (see Chapter 1). In general, a number of performance measures will be relevant for a given manufacturing system. However, they will differ from one manufacturing system to another.

Given performance requirements, the manufacturing system designer must describe a suitable system design. This design can be captured numerically by specifying the values of an appropriate collection of *decision variables*. Examples of decision variables are the number of each type of machine in a manufacturing system.

Designing manufacturing systems (mapping performance measures onto decision variables) (Fig. 5.1) is a difficult task because

1. Manufacturing systems are large and have many interacting components.
2. Manufacturing systems are dynamic.
3. Manufacturing systems are open systems, which both influence and are influenced by their environment.
4. The relationships between performance measures and decision variables cannot usually be expressed analytically. Well-behaved functions do not apply.
5. Data may be difficult to measure in a harsh processing environment.
6. There are usually multiple performance requirements for a manufacturing system and these may conflict.

Assuming that these difficulties can be surmounted, the next task is the construction of the actual manufacturing system. As this process is very capital-intensive, it places a premium on the cost performance measures of the manufacturing system. No matter how worthy its performance in other aspects may be, a manufacturing system will never be constructed if it is not shown to be financially viable.

Three principal classes of financial measures for the appraisal of manufacturing systems are: *return on investment, payback, and discounted cash flow*. Essentially, *return on investment* (ROI) is the result of dividing the profit generated by an investment by its cost. The result, however, fails to take into account the time phasing of costs and benefits, and completely ignores the time value of money [1].

Payback measures the duration of recouping an investment, in cash terms. It ignores the benefits that accrue beyond the payback period, a factor crucial for the appraisal of manufacturing systems [2]. For example, a manufacturing system with flexible, computer numeric control (CNC) equipment may have a longer payback period than a system composed of dedicated machines. However, the flexible system may be capable of accommodating product design changes easily, resulting in a quicker response to market variations and in a longer useful life.

Discounted cash flow (DCF) measures look at the timing of the cash flows (both revenues and costs) anticipated from an investment, discounting the flows to reflect the time value of money. They call for a more complex analysis than either of the two previous classes of measures do, but do overcome many of the others' shortfalls [3].

Performance measures, based on DCF principles, include *internal rate of return* (IRR) and *net present value* (NPV). The IRR is the discount interest rate at which the present value of the cash inflows (revenues), to be generated by the manufacturing system, is equal to the present value of the cash outflows (costs) necessitated by the system. IRR is defined by the equation

$$C_0^+ + \sum_{p=1}^{N} \frac{C^+(p)}{(1+IRR)} = C_0^- + \sum_{p=1}^{N} \frac{C^-(p)}{(1+IRR)^p}$$

where:

N	$\equiv$	Number of periods for which cash flows will be generated by the investment
C_0^-	$\equiv$	Initial cash outflow
C_0^+	$\equiv$	Initial cash inflow
$C^-(p)$	$\equiv$	Cash outflow in period p
$C^+(p)$	$\equiv$	Cash inflow in period p

The NPV is the difference between the present value of the cash inflows, to be generated by an investment, and the present value of the investment's cash outflows. The cash flows are discounted at an interest rate r called the *opportunity cost of capital*, which is the rate of return offered by a financial security of comparable risk to the investment under consideration. The accepted condition for the acceptance of an investment is that NPV must be greater than zero. In other words, the present value of the investment must be greater than that of a financial security of comparable risk. NPV is defined by the equation

$$NVP = C_0^+ - C_0^- + \sum_{p=1}^{N} \frac{C^+(p) - C^-(p)}{(1+r)^p}$$

Here, the variables have the same definitions as in the previous equation.

Discounted cash flow is widely recognized as the most meaningful accounting methodology for the appraisal of manufacturing investments [1, 2, 3, 4]. However, its application requires having to overcome two difficulties: estimating the complicated cash flows arising from a proposed manufacturing system design, and quantifying the intangible benefits of the design, such as flexibility. Part of the latter difficulty can be addressed by carefully enumerating the possible benefits of a well-designed manufacturing system. Two important benefits are these of improved production organization and improved flexibility. Breaking these further down, the potential benefits of improved production organization include:

- Lower work-in-process inventories
- Increased production rate
- Fewer fixtures, jigs and tooling
- Improved product quality
- Reduction of floor space used
- Reduced labor costs

The potential benefits of improved flexibility include [5]:

- Improved control and status monitoring of machines, tools, and material handling devices
- Improved response time to demand variations
- Improved ability to adjust to machine breakdown
- Improved ability to respond to design or process changes

The application of DCF principles then involves the evaluation of the anticipated cash flows from each of these benefits. This may be a difficult task. In particular, the estimation of cash flows, arising from flexibility, lies beyond the bounds of traditional accounting procedure, and is the subject of continuing research [6, 7].

Another financial method used to measure the risk of a manufacturing system investment is Value-at-Risk (VaR). VaR measures the worst expected loss under normal market conditions, over a specific time interval at a given confidence level [8], which could be very useful when decisions need to be taken as to whether or not new manufacturing equipment should be purchased.

5.1.1 Types of Manufacturing Systems

In general, manufacturing systems are divided into two areas:

1. The area in which materials are processed and individual parts or components are made (the processing area);
2. The area in which, if necessary, individual parts or components are joined together in a subassembly or final product (the assembly area).

In the following paragraphs, methods of structuring the processing area will be described, followed by methods of structuring assembly areas.

In industrial practice, there are five general approaches to structuring the processing area: the *job shop, project shop, cellular system, flow line*, and *continuous system* approaches.

In a *job shop*, machines with the same or similar material processing capabilities are grouped together. The lathes form a turning work center, the milling machines form a milling work center, and so forth. The machines are usually general-purpose machines, which can accommodate a large variety of part types. In this structure, the part or lot of parts moves through the system by visiting the different work centers according to the part's process plan (Fig. 5.2). Material handling must be very flexible in order to accommodate many different part types, therefore, it is usually done by manually controlled implements, such as forklifts and handcarts. Within each work center, a number of machines can be used for a particular operation. This is advantageous for a number of reasons: 1) each operation can be assigned to a machine, which yields the best quality or the best production rate, 2) machines can be evenly loaded, and 3) machine breakdowns can be accommodated easily. These advantages are particularly evident when the job shop's work load includes a large variety of different parts with different process sequences. However, fully exploiting the flexibility of a job shop requires making and implementing complex decisions in real time. For example, when a machine becomes available after finishing an operation, the question arising is which of the possibly many eligible operations it should perform next. Furthermore, lack of decision-making coordination among the work centers can impede the smooth flow of parts from work center to work center. This results in parts spending a long time on the job shop and creates high inventory levels.

In a *project shop* (Fig. 5.3), a product's position remains fixed during manufacturing because of its size and/or weight. Materials, people, and machines are brought to the product as needed. Facilities organized as product shops can be found in the aircraft and shipbuilding industries and in bridge and building construction.

In manufacturing systems organized according to the *cellular* plan (Fig. 5.4), the equipment or machinery is grouped according to the process combinations that occur in families of parts. Each cell contains machines that can produce a certain family of parts. The material flow within the cell may differ for different parts of a part family. Intra-cellular material flow can be performed either automatically or manually. By confining the path of each part to an individual cell, within the manufacturing system (when possible), cellular systems enable each cell to be considered independently for detailed scheduling purposes, thus, greatly simplifying this type of decision. Transportation times and the times spent on the system for the different parts are reduced compared with those of the job shop.

Job Shop

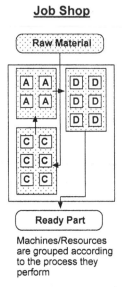

Machines/Resources
are grouped according
to the process they
perform

Fig. 5.2. Schematic of a Job Shop

The fourth way of structuring a manufacturing system is to create a *flow line* in which the machines and other equipment are ordered according to the process sequences of the parts to be manufactured (Fig. 5.5). A typical example is the *transfer line*, which is often used in the automotive industry (Fig. 1.14). A transfer line consists of a sequence of machines, which are typically dedicated to one particular part or to most a few very similar parts. Only one part type is produced at a time. Setups for the production of a new part type, if possible at all, take hours or even days. The machines are linked by automated material handling devices, such as conveyors. The structure of a transfer line presupposes that there is a well-defined, rigid process sequence for the different parts, and that the lot size of each part is high enough to guarantee that the capacity of the equipment will be fully exploited and not wasted on the setups.

Project Shop

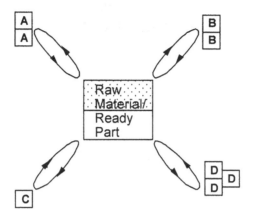

Machines/Resources are
brought to and removed from
stationary part as required

Fig. 5.3. Schematic of a Project Shop

Cellular System

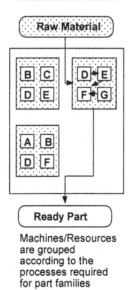

Machines/Resources
are grouped
according to the
processes required
for part families

Fig. 5.4. Schematic of a Cellular Manufacturing System

Flow Line

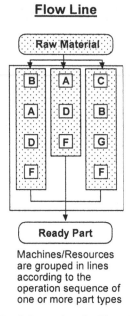

Machines/Resources
are grouped in lines
according to the
operation sequence of
one or more part types

Fig. 5.5. Schematic of a Flow Line

In contrast to the other types of manufacturing systems, which perform the manufacturing of discrete parts, *continuous systems* (Fig. 5.6) produce liquids, gases, or powders. As in a flow line, processes are arranged in the processing sequence of the products. The continuous system is the least flexible of the types of manufacturing systems.

In the actual manufacturing world, these standard system structures often occur in combinations, or with slight changes. The choice of a structure depends on the design of the parts to be manufactured, the lot sizes of the parts, and on market factors, such as the required responsiveness to market changes. In general, job shops and project shops are most suitable for small lot size production whilst flow lines are most suitable for large lot size production, and cellular systems are most suitable for the production of lots of intermediate size (Fig. 5.7). Job shops are most suitable for the low volume production of multiple part types, which are very dissimilar. They possess general-purpose machines and flexible, manual material handling, ideal for this situation. Cellular systems are most suitable for the manufacture, in low-to-medium production volumes and lot sizes, of part types with enough similarity to be clustered into part families. Flow lines are best suited for high volume, high lot size production of a single part type or a few very similar part types. This is a consequence of dedicated machines and material handling equipment (Tbl. 5.1).

Continuous System

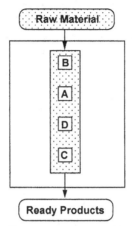

Processes are grouped in lines according to the process sequence of the products

Fig. 5.6. Schematic of a Continuous System

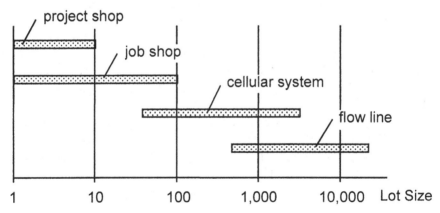

Fig. 5.7. Suitable Manufacturing System Types as a Function of Lot Size

Type of System		
	Job Shop	Cellular	Flow Line
Part Type Similarity	Low	Medium	High
Production Volume	Low	Low-Medium	High

Table 5.1. Best Conditions for the Use of Different Types of Manufacturing Systems

An example of a system, whose structure is a combination of the basic structures discussed above, is the flexible manufacturing system (FMS). A FMS (Fig. 5.8) is a hybrid of a job shop and a cellular manufacturing system. It provides great flexibility in terms of the types of parts and the process sequencing that can be used, a consequence of its high degree of automation. Material and information flow throughout the system is totally automated, and very little human intervention is required.

Another substantial part of a manufacturing company's production facilities is the *assembly* system. Assembly systems can be categorized according to the motion of parts and workplaces (Fig. 5.9). *Stationary part systems* are usually employed for large assemblies, such as airplanes, which are difficult to move around. *Moving part systems* can be divided into stationary workplace systems, in which parts are brought to stationary workplaces, and moving workplace systems, in which the workplaces move along with the parts. Assembly systems with stationary parts tend to have higher floor area requirements. They also tend to have more work than moving part systems at each workplace. These traits are due mainly to the fact that the parts assembled in stationary part systems are usually large. Moving part systems are generally more expensive because complicated material handling equipment is required to move parts quickly from one workplace to another (Tbl. 5.2).

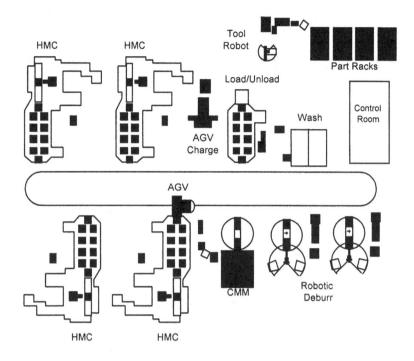

Fig. 5.8. A Flexible Manufacturing System

Stationary Parts		Moving	
Stationary Workplace	Moving Workplace	Stationary Workplace	Moving Workplace

Fig. 5.9. Types of Assembly Systems

The assembly operations at each workplace are usually short in duration and are highly repetitive when moving part-stationary workplace systems. Moving part-moving workplace systems on the other hand, allows for workers to work on each assembly for a longer period of time, thus, making the assembly work less repetitive.

	Stationary Parts		Moving Parts	
	Stationary Workplace	Moving Workplace	Stationary Workplace	Moving Workplace
Area Requirement	high	high	low	medium
Work Content at Each Workplace	high	medium	low	medium
Cost of System	low	medium	high	high

Table 5.2. Characteristics of Different Assembly Systems

5.1.2 Academic Versus Industrial Perspectives

There are a number of approaches to the difficult endeavor of designing manufacturing systems. In the *academic* literature, the overall manufacturing system design problem is usually decomposed into sub-problems of manageable complexity, which are then treated separately (Fig. 5.10). These problems are simplified and abstracted with the aid of assumptions. Many approaches attempt to find optimal solutions to the simplified problems. However, even the simplified problems are usually non-polynomial-hard (NP-hard), meaning that the time required to find the optimal solution, increases exponentially as the problem size increases linearly.

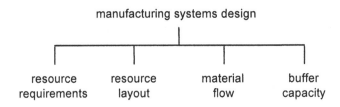

Fig. 5.10. Sub-problems in the Design of Manufacturing Systems

One sub-problem is the resource requirements problem. For this problem, the task is to determine the appropriate quantity of each type of production resource (for example, machines or pallets) in a manufacturing system. The objective is usually cost-based, such as the maximization of investment efficiency, or time-based, such as the maximization of production rate.

The resource layout problem is the problem of locating a set of resources in a constrained floor space. The objective is typically to minimize some combination of material handling cost, travel time, and resource relocation cost.

In material flow problems, the objective is to determine the configuration of a material handling system so that some combination of flexibility, cost, production rate, and reliability of the manufacturing system is maximized.

The buffer capacity problem is concerned with the allocation of work, in process or storage capacity, in a manufacturing system. Adequate levels of work in process, maximize machine utilization and production rate, but add to floor space and inventory holding costs. The goal is to find an optimum trade-off between these conflicting benefits and costs.

In *industrial* practice, *trial and error* remains the most frequently used design approach. It proceeds as follows:

1. Guess a suitable manufacturing system design (guess values for an appropriate collection of decision variables).
2. Evaluate the performance measures of the system. If they satisfy the performance requirements, then stop the design process. Otherwise, return to Step 1.

Clearly, when designing a large and complicated manufacturing system, many iterations of these steps may be required. The success of the trial and error approach relies heavily on the skill of the designer or "guesser" in Step 1. Intuition and rules of thumb derived from experience often help.

General guidelines for the design process do exist, however [9]. The first step in these guidelines is the definition of system objectives. Objectives differ from company to company, but a survey of industrial planners [10] has found that *economic objectives,* such as the return on investment, tend to be emphasized most, followed by the *efficient use of resources* and of the *system flexibility.*

The second step is the development of detailed system requirements and constraints. Among other tasks, machine types, which are capable of performing the processes for the set of parts to be manufactured, are determined; available floor space and existing equipment, which can be incorporated into the new system, are also determined.

Based on the requirements and constraints, a number of alternative manufacturing systems are developed and then evaluated based on some predefined scenarios. The basis for these scenarios is a document, such as that of a long-range business plan, which contains forecasts of demand patterns for the products to be manufactured by the system.

Despite the existence of accepted design guidelines, a proposed manufacturing system design is rarely implemented in its entirety. This is because a number of scenarios can drastically change the requirements of a manufacturing system from the time the design is approved until the first part is produced [11]:

1. Product development announces a new product line and several new options for the existing line of products.
2. Marketing reacts to the new product line and to the last quarter's statistics and makes major adjustments to the production forecasts.
3. Manufacturing engineering specifies new processes as well as new equipment to perform those processes that greatly enhance manufacturing throughput.
4. The manufacturing planning department unveils plans to utilize a new material. Processing methods will change, but at this time it is not clear how they will change.
5. Package engineering determines that the present practices have been responsible for product damage. A new method is introduced.

Manufacturing system design is therefore viewed as a continuous, cyclical activity, involving the definition of the system's objectives, the development of detailed system requirements and constraints, and the implementation of the design. In general, the higher the technology of the products produced and the lower the investment in the manufacturing system, the greater will be the rate at which this cycle is traversed – that is, the greater the rate of manufacturing system change [11].

From the above discussion, we may conclude the following about the industrial manufacturing system design problem:

1. The objectives of a manufacturing system are often neither well defined at the time a manufacturing system should be created, nor are they subject to change. Design flexibility is therefore very important.
2. Data regarding manufacturing resources, such as machines and material handling systems are imprecise, especially if the manufacturing process is new.

This vagueness of the inputs to the manufacturing system design process makes the optimization in a quantitative fashion difficult. This vagueness tends to render futile efforts to hone solutions to some mathematical optimum.

5.2 Methods and Tools

The fundamental activity in design is decision making: the design of a manufacturing system is the process of deciding the values of the decision variables of the manufacturing system. In this section, some methods and tools, which are useful for making design decisions, will be described. Emphasis will be given on the underlying ideas; discussion of specific applications in the area of manufacturing system design, will be deferred to Section 5.3.

The methods and tools for the design of manufacturing systems fall into three broad categories: *operations research*, *artificial intelligence*, and *simulation*. The divisions among these categories are fuzzy. For example, simulation is often considered to be an operations research tool, while the mathematical programs of operations research are solved using search methods, studied in artificial intelligence. These categories are an organization convenience; in truth, they overlap to a considerable extent.

5.2.1 Operations Research

Mathematical Programming

Mathematical programming is a family of techniques for optimizing (minimizing or maximizing) a given algebraic *objective function* of a number of *decision variables* [12]. The decision variables may either be independent of one another, or they may be related through *constraints*.

Mathematical programming solves problems of the form

$$\text{Minimize or maximize } f(x_1, x_2, ..., x_n) \tag{5-1}$$

Subject to the constraints:
$$g_1(x_1, x_2, ..., x_n) \lozenge b_1$$
$$g_2(x_1, x_2, ..., x_n) \lozenge b_2$$
$$\bullet$$
$$\bullet$$
$$\bullet$$
$$g_m(x_1, x_2, ..., x_n) \lozenge b_m$$

where the symbol $\lozenge$ stands for one of the relations $\leq$, $=$, or $\geq$ (not necessarily the same relation for each constraint). Such a problem statement is called a *mathematical program*. With some ingenuity, it is possible to state many problems in the design and operation of manufacturing systems in the above form, so that standard algorithms for solving mathematical programs may be applied.

When referring to a mathematical program, a *point* means a set of values for the decision variables x_1, x_2, ..., x_n of the program. The *feasible region* is the set of all points that satisfy the program's constraints.

The goal in mathematical programming is to find the *optimal solution*, the point in the feasible region, which minimizes the objective function (in the case of a minimization problem) or maximizes the objective function (in the case of a maximization problem).

Linear Programming

A mathematical program is called a *linear program* if the objective function $f(x_1, x_2, ..., x_n)$ and each constraint function $g_i(x_1, x_2, ..., x_n)$ are linear in their arguments [9]. A linear program is therefore of the form:

$$\text{Minimize or maximize } c_1x_1 + c_2x_2 + ... + c_nx_n \qquad (5\text{-}2)$$

$$\text{Subject to:}$$
$$a_{11}x_1 + a_{12}x_2 + ... + a_{1n}x_n \lozenge b_1$$
$$a_{21}x_1 + a_{22}x_2 + ... + a_{2n}x_n \lozenge b_2$$
$$\bullet$$
$$\bullet$$
$$\bullet$$
$$a_{m1}x_1 + a_{m2}x_2 + ... + a_{mn}x_n \lozenge b_m$$

Again the symbol $\lozenge$ stands for one of the relations $\leq$, $=$, or $\geq$. Any mathematical program that is not a linear program is *nonlinear*. The formulation and the solution of nonlinear programs are the subject of *nonlinear programming*.

There are a number of assumptions, which are made in the formulation and solution of a linear program [12]:

– *The Proportionality Assumption.* The contribution from each decision variable to the objective function and to the left hand side of each constraint, is proportional to the value of the decision variable.

- *The Additivity Assumption.* The contribution from each decision variable to the objective function and to the left hand side of each constraint, is independent of the values of the other decision variables.
- *The Divisibility Assumption.* Decision variables may assume any real value.
- *The Certainty Assumption.* All coefficients c_i and a_{ij} and all right hand side constants b_i in the constraints are known with certainty.

Solving Linear Programs

A linear program is in *standard form* if all constraints are modeled as equalities and all decision variables are restricted to being non-negative. It is important that they be familiar with the standard form because most linear programming methods require that problems be stated in this form. In standard form, any decision variable (say, x_1) which is not constrained to be non-negative is replaced by the difference of two new variables (say, x_2-x_3) which are so constrained. A constraint of the form $a_{i1}x_1 + a_{i2}x_2 + \ldots + a_{ij}x_3 + \ldots + a_{in}x_n \leq b_i$ may be stated as an equality by adding a *slack variable* to the left hand side. For example, the constraint

$$3x_1 + 5x_2 - 8x_3 \leq 100{,}000$$

may be restated as

$$3x_1 + 5x_2 - 8x_3 + s_1 = 100{,}000$$

A constraint of the form $a_{i1}x_1 + a_{i2}x_2 + \ldots + a_{ij}x_3 + \ldots + a_{in}x_n \geq b_i$ may be stated as an equality by subtracting an *excess variable* (also called a surplus variable) from the left hand side. For example, the constraint

$$8x_1 - 7x_2 + 2x_3 \geq 50{,}000$$

may be restated as

$$8x_1 - 7x_2 + 2x_3 - e_1 = 50{,}000$$

The solution of a linear program requires an initial feasible and non-negative solution, called a *basic feasible solution* [13]. A method to generate such a solution requires that a new variable, called an *artificial variable*, be added to the left-hand side of each constraint equation that does not contain a slack variable. Each constraint will then contain either one slack variable or one artificial variable. The basic feasible solution is then obtained by setting each slack variable and each artificial variable equal to the right hand side of the constraint in which it appears, and by setting all other variables to zero.

For example, the constraints of the linear program

$$\text{Maximize: } 6x_1 + 8x_2$$

$$\text{Subject to:}$$
$$x_1 + 2x_2 \leq 9$$
$$3x_1 - x_2 \geq 7$$
$$x_1 + 4x_2 = 10$$

may be cast into standard form by adding a slack variable s_3 to the left hand side of the first constraint, an excess variable e_4 and an artificial variable v_5 to the left hand side of the second constraint, and an artificial variable v_6 to the left hand side of the third constraint:

$$x_1 + 2x_2 + s_3 = 9$$
$$3x_1 - x_2 \quad - e_4 + v_5 = 7$$
$$x_1 + 4x_2 \quad\quad\quad + v_6 = 10$$

The initial basic feasible solution obtained is then: $s_3 = 9$, $v_5 = 7$, $v_6 = 10$, and $x_1 = x_2 = e_4 = 0$. Note, however, that the point $x_1 = x_2 = 0$ does not satisfy the original constraints. This is normal, and does not present a problem. The price of transforming the original problem into the standard form, required in order to apply standard solution procedures, has been the expansion of the linear program from two variables (x_1, x_2) to six variables $(x_1, x_2, s_3, e_4, v_5, v_6)$.

While slack and excess variables do nothing to alter the meaning of the original constraints, such is not the case with artificial variables. A solution of the expanded problem will satisfy the constraints of the original problem only if the value of all artificial variables is zero. To ensure that this will be the case, the objective function is modified in a way which makes the assignment of any non-zero value to an artificial variable very sub-optimal. In minimization problems, artificial variables are incorporated into the objective function with a very large positive coefficient $+M$; in maximization problems, a very large negative coefficient $-M$ is used. For example, the objective function of the example above would become:

$$6x_1 + 8x_2 - Mv_5 - Mv_6$$

If either v_5 or v_6 were to be non-zero, the objective function value would likely be lowered enough by the terms with the $-M$ coefficients to make it sub-optimal.

In matrix notation [13], the standard form of a linear program is

$$\text{Optimize: } z = c^T x$$

$$\text{Subject to: } Ax = b \qquad\qquad (5\text{-}3)$$

$$\text{with: } x \geq 0$$

where x is the vector of unknown variables, consisting of all of the original decision variables x_1, x_2, ..., x_n; all slack variables s_i; all excess variables e_i; and all artificial variables v_i. The vector c contains the objective function coefficients, the matrix A contains the constraint coefficients, and the vector b contains the right hand side values of the constraints. We further define x_0 to be a vector of the unknown slack and artificial variables (in the order in which they appear in the constraints), and c_0 to be a vector containing the objective function coefficients of the variables in x_0. The initial basic feasible solution required for the solution of this linear program is then given by $x_0 = b$, where it is understood that all variables not in x_0 (the original decision variables and the excess variables) are set to zero.

The most widely applied method for the solution of linear programs is the *simplex method*, which solves linear programs in standard form (Eq. 5-3) for which an initial basic feasible solution is known [10]. Starting with this solution, the method locates successively other basic feasible solutions, having better values of the objective, until the optimal solution is obtained.

It is convenient to describe the simplex method, in terms of a shorthand display of variables, variable values, and coefficients called a *simplex tableau* (Fig. 5.11). As shown, the tableau applies to maximization problems only. For minimization problems, the tableau is written with the signs in the last row reversed – that is, the terms in the last row should be $(\mathbf{c}^T - \mathbf{c}_0^T\mathbf{A})$ and $-\mathbf{c}_0^T\mathbf{b}$ respectively.

	$\mathbf{x}^T$	
$\mathbf{x}_0$	$\mathbf{A}$	$\mathbf{b}$
	$\mathbf{c}_0^T\mathbf{A} - \mathbf{c}^T$	$\mathbf{c}_0^T\mathbf{b}$

Fig. 5.11. The Simplex Tableau for Maximization Problems

The steps of the simplex method will be described below. Examples will be presented to demonstrate the effect of each step on the numbers of an actual problem. The example problem [10] to be solved is:

$$\text{maximize: } z = x_1 + 9x_2 + x_3$$

$$\text{subject to: } x_1 + 2x_2 + 3x_3 \le 9$$
$$3x_1 + 2x_2 + 2x_3 \le 15$$

with: all variables nonnegative

This program can be put into a standard form by introducing slack variables s_1 and s_2 in the first and second constraint inequalities, respectively. These then become

$$x_1 + 2x_2 + 3x_3 + s_1 = 9$$
$$3x_1 + 2x_2 + 2x_3 + s_2 = 15$$

We may then define

$$\mathbf{x} \equiv [x_1, x_2, x_3, s_1, s_2]^T \qquad \mathbf{c} \equiv [1, 9, 1, 0, 0]^T$$

$$\mathbf{A} \equiv \begin{bmatrix} 1 & 2 & 3 & 1 & 0 \\ 3 & 2 & 2 & 0 & 1 \end{bmatrix}, \quad \mathbf{b} \equiv \begin{bmatrix} 9 \\ 15 \end{bmatrix}, \quad \mathbf{x}_0 \equiv \begin{bmatrix} s_1 \\ s_2 \end{bmatrix}, \quad \mathbf{c}_0 \equiv \begin{bmatrix} 0 \\ 0 \end{bmatrix}$$

The initial simplex tableau (Fig. 5.11) then becomes

	x_1	x_2	x_3	s_1	s_2	
s_1	1	2	3	1	0	9
s_2	3	2	2	0	1	15
	-1	-9	-1	0	0	0

Let the rows of this tableau be numbered 1, ..., 4, from top to bottom. Let the columns of this tableau be numbered 1, ..., 7, from left to right. The steps of the simplex method can then be stated as follows.

Step 1   Locate the most negative number in the bottom row of the simplex tableau, excluding the last column, and call the column in which this number appears the *work column*. If more than one candidate for most negative number exists, choose one.

The most negative number is -9 (circled below). Since it belongs to column 3, this is the work column.

work column

	x_1	x_2	x_3	s_1	s_2	
s_1	1	2	3	1	0	9
s_2	3	2	2	0	1	15
	-1	(-9)	-1	0	0	0

Step 2   Form ratios by dividing each *positive* number in the work column, excluding the last row, into the element in the same row and last column. Designate the element in the work column that yields the *smallest* ratio as the *pivot element*. If more than one element yields the same smallest ratio, choose one. If no element in the work column is positive, the program has no solution.

The ratios to be formed are 9/2 = 4.5 and 15/2 = 7.5. Since 4.5 is the smallest ratio, the 2 in row 2, column 3 (circled below) is designated as the pivot element.

pivot element

	x_1	x_2	x_3	s_1	s_2	
s_1	1	(2)	3	1	0	9
s_2	3	2	2	0	1	15
	-1	-9	-1	0	0	0

Step 3   Use elementary row operations to convert the pivot element into 1 and then to reduce all *other* elements in the work column to 0. (An elementary row operation can be either the multiplication of a row by a constant or the addition of two rows.)

To convert the pivot element to 1, we use the row operation *row 2′* = 1/2 ∞ *row 2*. This gives *row 2′* = [1/2 1 3/2 1/2 0 | 9/2]. To convert the other elements in the work column into 0, we use *row 3′* = *row 3 − row 2*, giving *row 3′* = [2 0 -1 -1 1 | 6]; and *row 4′* = *row 4 + 9/2 ∞ row 1*, giving *row 4′* = [7/2 0 25/2 9/2 0 | 81/2].

pivot element

	x_1	x_2	x_3	s_1	s_2	
s_1	1/2	(1)	3/2	1/2	0	9/2
s_2	2	0	-1	-1	1	6
	7/2	0	25/2	9/2	0	81/2

Step 4   Replace the variable in the pivot row and the first column by the variable in the first row and pivot column. This new first column is the current set of basic variables.

The variable in the pivot row (row 2) and first column is the slack variable s_1. The variable in the first row and pivot column (column 3) is x_2. Therefore, replace s_1 by x_2.

	x_1	x_2	x_3	s_1	s_2	
x_2	1/2	1	3/2	1/2	0	9/2
s_2	2	0	-1	-1	1	6
	7/2	0	25/2	9/2	0	81/2

new basic variables

Step 5   Repeat Steps 1 through 4 until there are no negative numbers in the last row, excluding the last column.

There are already no negative numbers in the last row, so Steps 1 through 4 do not have to be repeated.

Step 6   The optimal solution is obtained by assigning to each variable in the first column that value in the corresponding row and in the last column. All other variables are assigned the value zero. The associated z^*, the optimal value of the objective function, is the number in the last row and last column for a maximization program, but is the *negative* of this number for a minimization program.

The final basic feasible solution has been found. The variables in the first column are x_2 and s_2. From the last column, the optimal values of these variables are $x_2^* = 9/2$ and $s_2^* = 6$. For the other variables, the optimal values are $x_1^* = x_3^* = s_1^* = 0$. Discarding the slack variables, the optimal solution to the example problem is then $x_1^* = 0$, $x_2^* = 9/2$, and $x_3^* = 0$. The optimal objective function value of 81/2 can be found either at the bottom-right corner of the tableau, or by substituting the optimal solution into the objective function.

When a linear program involves = or ≥ constraints, its solution requires the introduction of artificial variables. The effect of this is to introduce into the simplex tableau some very high numbers derived from the -M and +M objective function coefficients of the artificial variables. In this case, application of the simplex method as given, would result in large round off errors when row operations, involving both very small and very large numbers, are performed. Fortunately, the basic simplex method can be modified to overcome this problem; the resulting algorithm is called the *two-phase method*. Details of this method can be obtained from operations research references [9, 10].

Goal Programming

Goal programming is a specific application of linear programming to the making of decisions which involve trade-offs among multiple goals [12]. The goals are modeled as constraints in the linear program.

An example will show how a goal program can be structured to solve a practical problem. Suppose that a company wants to determine the quantities to be purchased from two types of machining centers. The machining centers are to be used in the manufacture of three different part types. The rate (in parts/hour) at which each machining center can manufacture each part type is shown below:

	Part A	Part B	Part C
Machining Center 1	10	7	3
Machining Center 2	5	5	8

Production rate goals for each part have been set on the basis of demand forecasts:

Goal 1. Production rate for Part A should be at least 90 parts/hour.
Goal 2. Production rate for Part B should be at least 80 parts/hour.
Goal 3. Production rate for Part C should be at least 75 parts/hour.

Suppose that for every part/hour by which the company falls short of a production rate goal, the expected annual costs due to lost sales for the company are:

$1,000,000 for Part A
$500,000 for Part B
$300,000 for Part C

Let

x_1 = number of units of Machining Center 1 to be purchased
x_2 = number of units of Machining Center 2 to be purchased

The decision to be made is to find the values of x_1 and x_2. Each unit of Machining Center 1 costs \$400,000, while each unit of Machining Center 2 costs \$250,000. The total budget for the purchase of machining centers is \$2,000,000.

In order to formulate the above problem as a goal program, we require the following *deviational variables*:

d_i^+ = amount by which we numerically oversatisfy (exceed) the ith goal

d_i^- = amount by which we numerically undersatisfy the ith goal

For each goal, only one of the two deviational variables can be non-zero.

The suitable goal program for this problem is:

Minimize $1,000d_1^- + 500d_2^- + 300d_3^-$ (Costs due to lost sales [\$1,000])

Subject to:

$10x_1 + 5x_2 + d_1^+ - d_1^- = 90$ (Part A production rate goal)

$7x_1 + 5x_2 + d_2^+ - d_2^- = 80$ (Part B production rate goal)

$3x_1 + 8x_2 + d_3^+ - d_3^- = 75$ (Part C production rate goal)

$400x_1 + 250x_2 \leq 2,000$ (Budget constraint [\$1,000])

$x_1, x_2, d_1^+, d_1^-, d_2^+, d_2^-, d_3^+, d_3^- \geq 0$

From this formulation we note the main characteristics [12] of goal programming:

- Each goal is stated as an equality constraint through the introduction of two deviational variables. Although the constraint must be satisfied by the solution, the same does not necessarily apply to the goal. This is possible because, in the final solution, one of the deviational variables in the constraint may be non-zero. It is thus, permissible to satisfy only some goals but not others, or, for that matter, it is possible to satisfy no goals at all. This will often be the case when goals conflict. The solution will come closest to satisfying the most important goals, and will undersatisfy or oversatisfy less important goals.
- The objective function, is a function of the deviational variables, which expresses the penalty of deviating from the stated goals.

- The coefficients of the objective function express the relative importance of each goal. The larger the coefficients of a goal's deviational variables, the smaller those deviational variables will be in the solution (so as to avoid a large penalty), and the more closely that goal will be met.

In many cases, the precise relative importance of the goals is not known, but the ranking of goals from the most important to the least important ones is known. In these cases, the coefficients may be set so that the coefficient of the most important goal is much greater than that of the second most important goal, and the coefficient of the second most important goal is much greater than that of the next most important goal, and so forth. This technique is known as *pre-emptive goal programming*.

Integer Programming

An integer program is a linear one with the additional constraint that the decision variables x_1, x_2, ..., x_n must be integers. Very often, the coefficients c_i of the objective function and the coefficients a_{ij} and the constants b_i of the constraints are also constrained to be integral.

Solving Integer Programs

In certain situations, a reasonably good solution to an integer program can be obtained by ignoring the integer requirement and by solving the resulting linear program. The resulting solution is known as the *first approximation* to the solution of the original integer program. If the first approximation happens to be integral, then it is also the optimal solution to the original integer program. The *second approximation* is obtained by rounding the components of the first approximation to the nearest feasible integers. This approximation is most satisfactory when the first approximation contains large numbers, whilst it can be inaccurate when the numbers are small. For example, if the optimal solution to a linear program stated that a bearing plant should manufacture 5,000,000.5 ball bearings per year, then manufacturing 5,000,000 ball bearings per year would be acceptable. On the other hand, if a linear program prescribed the purchase of 1.4 transfer lines at a cost of $20 million each, then rounding to 1 transfer line would play a large role in the outcome, and solution algorithms which provide explicitly integral solutions should be employed.

One such algorithm is the *branch and bound* algorithm. It will be illustrated below with the aid of an example [13]. The example problem to be solved is:

$$\text{maximize:} \quad z = 10x_1 + x_2 \qquad (5\text{-}4)$$

subject to: $2x_1 + 5x_2 \leq 11$

with: x_1 and x_2 nonnegative and integral

If the first approximation to this integer program contains a variable that is not integral, say x_j^*, then $i_1 < x_j^* < i_2$, where i_1 and i_2 are consecutive, nonnegative integers. Two new integer programs can then be created by augmenting the original integer program with either the constraint $x_j \leq i_1$ or the constraint $x_j \geq i_2$. This process, called *branching*, has the effect of shrinking the feasible region in a way that eliminates from further consideration the current nonintegral solution for x_j but still preserves all possible integral solutions to the original problem [13].

For Program 5-4, consider the associated linear program obtained by deleting the integer requirement. By graphing the feasible region defined by the constraints $2x_1 + 5x_2 \leq 11$ and $x_1, x_2 \geq 0$, and then superimposing contours of constant z, the first approximation (Fig. 5.12) is readily found to be $x_1^* = 5.5$, $x_2^* = 0$, with $z^* = 55$. Alternatively, the simplex method could be applied to finding the same solution; however, for a two-variable problem the graphical method yields more insight.

Since $5 < x_1^* < 6$, branching creates the two new integer programs

$$\text{maximize:} \quad z = 10x_1 + x_2 \qquad (5\text{-}5)$$

subject to: $2x_1 + 5x_2 \leq 11$

$\phantom{\text{subject to:}}\quad x_1 \leq 5$

with: x_1 and x_2 nonnegative and integral

$$\text{maximize:} \quad z = 10x_1 + x_2 \qquad (5\text{-}6)$$

subject to: : $2x_1 + 5x_2 \leq 11$

$\phantom{\text{subject to: :}}\quad x_1 \geq 6$

with: x_1 and x_2 nonnegative and integral

For the two integer programs created by the branching process, first approximations are obtained by ignoring the integer requirements once more and solving the resulting linear programs. If either the first approximation is still nonintegral, then the integer program, which has given rise to that first approximation, becomes a candidate for further branching.

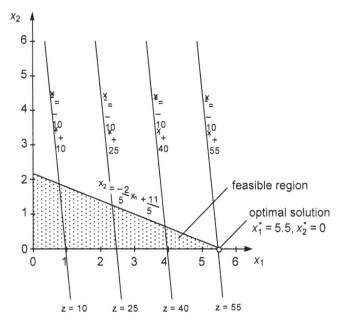

Fig. 5.12. Graphical Method for Obtaining the First Approximation to Program 5-4

Using graphical methods, we find that Program 5-5 has the first approximation $x_1^* = 5$, $x_2^* = 0.2$, with $z^* = 50.2$, while Program 5-6 has no feasible solution. Thus, Program 5-5 is a candidate for further branching. Since $0 < x_2^* < 1$, we augment Program 5-5 with either $x_2 \leq 0$ or $x_2 \geq 1$, and obtain the two new programs

$$\text{maximize:}\qquad z = 10x_1 + x_2 \qquad\qquad (5\text{-}7)$$

$$\text{subject to:}\qquad 2x_1 + 5x_2 \leq 11$$

$$x_1 \qquad\quad \leq 5$$

$$x_2 \leq 0$$

with: x_1 and x_2 nonnegative and integral

maximize: $z = 10x_1 + x_2$ (5-8)

subject to: $2x_1 + 5x_2 \leq 11$

                    $x_1 \qquad \leq 5$

                    $x_2 \qquad \geq 1$

with: x_1 and x_2 nonnegative and integral

With the integer requirements ignored, the solution to Program 5-7 is $x_1^* = 5$, $x_2^* = 0$, with $z^* = 50$, while the solution to Program 5-8 is $x_1^* = 3$, $x_2^* = 1$, with $z^* = 31$. Since both these first approximations are integral, no further branching is required.

We are now ready to discuss the bounding aspect of branch and bound. Assume that the objective function is to be maximized. Branching continues until an integral first approximation (which is thus an integral solution) is obtained. The value of the objective for this first integral solution becomes a lower bound for the problem, and all programs whose first approximations, integral or not, yield values of the objective function smaller than the lower bound ,are discarded.

In our example, Program 5-7 possesses an integral solution with $z^* = 50$; hence 50 becomes a lower bound for the problem. Program 5-8 has a solution with $z^* = 31$. Since 31 is less than the lower bound 50, Program 5-8 is eliminated from further consideration, *and would have been so eliminated even if its first approximation had been nonintegral.*

Branching continues from those programs having nonintegral first approximations that give values of the objective function greater than those of the lower bound. If, in the process, a new integral solution is uncovered having a value of the objective function greater than the current lower bound, then this value of the objective function becomes the new lower bound. The program that yielded the old lower bound is eliminated, as are all programs, whose first approximations give values of the objective function smaller than the new lower bound. The branching process continues until there are no programs with nonintegral first approximations remaining under consideration. At this point, the current lower-bound solution is the optimal solution to the original integer program.

If the objective function is to be *minimized,* the procedure remains the same, except that upper bounds are used. Thus, the value of the first integral solution becomes an upper bound for the problem, and programs are eliminated when their first approximate z-values are greater than the current upper bound.

One always branches from that program, which appears most nearly optimal. When there are a number of candidates for further branching, one chooses the candidate with the largest z-value, if the objective function is to be maximized, or the candidate with the smallest z-value, if the objective function is to be minimized.

Additional constraints are added one at a time. If a first approximation involves more than one nonintegral variable, the new constraints are imposed on that variable, which is furthest from being an integer; i.e., that variable whose fractional part is closest to 0.5. In case of a tie, the solver arbitrarily chooses one of the variables.

Finally, it is possible for an integer program or an associated linear program to have more than one optimal solution (i.e., different sets decision variable values which yield the same optimal value of the objective function). In such cases, the accepted convention is to arbitrarily designate one of the solutions as optimal and disregard the rest.

The application of branch and bound to the example problem [13] can be drawn in a schematic diagram or tree (Fig. 5.13):

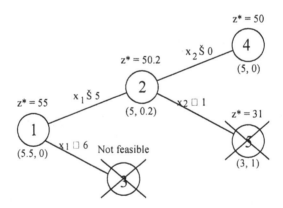

Fig. 5.13. Application of Branch and Bound to Program 5-4

The original integer program, here Program 5-4, is designated by a circled 1, and all other programs formed through branching are designated in the order of their creation by circled successive integers. Thus, Programs 5-5 through 5-8 are designated by circles 2 through 5, respectively. The first approximate solution to each program is written by the circle designating the program. Each circle (program) is then connected by a line to the circle (program) which generated it via the branching process. The new constraint that defined the branch is written above the line. Finally, a large cross is drawn through a circle if the corresponding program has been eliminated from further consideration. Hence, Branch 3 was eliminated because it was not feasible; Branch 5 was eliminated by bounding. Since there are no nonintegral branches left to consider, the tree indicates that Program 5-4 is solved with $x_1^* = 5$, $x_2^* = 0$, and $z^* = 50$.

Dynamic Programming

Dynamic programming is a method for solving problems that can be viewed as *multistage decision processes*. A multistage decision process is a process that can be separated into a number of sequential steps, or stages, which may be completed in one or more ways. The options for completing the stages are called *decisions*. A *policy* is a sequence of decisions, one for each stage of the process. The condition of the process at a given stage is called the *state* at that stage; each decision effects a transition from the current state to a state associated with the next stage. Many multistage decision processes have *returns* (costs or benefits) associated with each decision, and these returns may vary with both the stage and state of the process. The objective in analyzing such processes is to determine an optimal policy, one that results in the best total return [13].

To make the above description less abstract, consider an investor, who begins the day with D dollars and enough time to make n investments by the end of the day.. The expected monetary return of each investment (to be paid out on some given day in the future), is a function of the amount of money invested; the greater the amount, the greater the return is. The objective is to obtain the greatest expected total monetary return for the n investments, without investing more than D dollars.

Suppose that the investor has $D = \$4,000$ to invest and $n = 3$ investment opportunities. Each opportunity requires deposits in $1,000 amounts; the investor may allocate all the money to just one opportunity or split the money between them. The expected returns are tabulated as follows [10].

	Dollars Invested				
	0	1,000	2,000	3,000	4,000
Return from Opportunity 1	0	2,000	5,000	6,000	7,000
Return from Opportunity 2	0	1,000	3,000	6,000	7,000
Return from Opportunity 3	0	1,000	4,000	5,000	8,000

Letting $f_i(x)$ (i = 1, 2, 3) denote the return (in thousand-dollar units) from opportunity i when x units of money are invested in it, we can rewrite the returns table (Tbl. 5.3).

Defining x_i (i = 1, 2, 3) as the number of units of money invested in opportunity i, we can formulate the objective as

maximize: $z = f_1(x_1) + f_2(x_2) + f_3(x_3)$

Since the investor has only 4 units of money to invest,

$$x_1 + x_2 + x_3 \leq 4$$

	Units Invested				
	0	1	2	3	4
$f_1(x)$	0	2	5	6	7
$f_2(x)$	0	1	3	6	7
$f_3(x)$	0	1	4	5	8

Table 5.3. Returns Table for the Investment Problem

In this problem, the process of determining the amount to be invested in each opportunity for maximizing the total return, is a three-stage decision process. Consideration of opportunity i constitutes Stage i (i = 1, 2, 3). The state of the process at Stage i is the amount of funds still available for investment at Stage i. For Stage 1, the beginning of the process, there are 4 units of money available; hence the state is 4. For stages 2 and 3, the states can be 0, 1, 2, 3 and 4, depending on the allocations (decisions) at previous stages. The decision at Stage i is represented by the variable x_i; the possible values of x_i are the integers from 0 to the state at Stage i, inclusive.

Dynamic programming, the procedure by which multistage decision processes may be optimized, is based on Bellman's principle of optimality: an optimal policy has the property that, regardless of the decisions taken to enter a particular state in a particular stage, the remaining decisions must constitute an optimal policy for leaving that state.

To implement this principle, begin with the last stage of an n-stage process and determine for each state the best policy for leaving that state and complete the process, assuming that all preceding stages have been completed. Then move backwards through the process, stage by stage. At each stage determine the best policy for leaving each state and complete the process, assuming that all preceding stages have been completed and make use of the results already obtained for the succeeding stage. In doing so, the entries of Table 5.4 [13] will be calculated, where

u $\equiv$ the state variable, whose values specify the states

$m_j(u)$ $\equiv$ optimum return from completing the process beginning at Stage j in state u

$d_j(u)$ $\equiv$ decision taken at Stage j that achieves $m_j(u)$

The entries corresponding to the last stage of the process, $m_n(u)$ and $d_n(u)$, are generally straightforward to be computed. The remaining entries are obtained recursively; that is, the entries for the jth stage (j = 1, 2, ... , n-1) are determined as functions of the entries for the $(j+1)^{st}$ stage. The recursion formula is problem dependant, and must be obtained anew for each different type of multistage process.

In order to determine an optimal policy for the investor, we begin by considering the last stage of the process, Stage 3, under the assumption that all previous stages, Stage 1 and 2, have been completed. That is, allocations to investments 1 and 2 have been made (although, at this time, we do not know what they are), and we are to complete the process by allocating units of money to investment 3. Since we do not know how many units were allocated to the first two investments, we do not know how many units are available for investment 3; we must therefore consider all possibilities. There will be either 0, 1, 2, 3, or 4 units available.

No matter how many units of money are available at Stage 3, it is clear from the definition of $f_3(x)$ (Tbl. 5.3), that the best way of completing the process is to allocate all the available units to investment 3. This can be verified via a formal calculation:

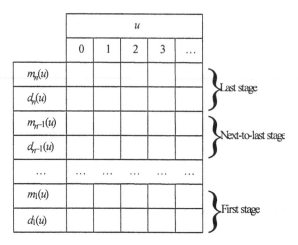

Table 5.4. A Dynamic Programming Table

$m_3(4)$ = max $\{f_3(0), f_3(1), f_3(2), f_3(3), f_3(4)\}$
           = max $\{0, 1, 4, 5, 8\} = 8$ with $d_3(4) = 4$
$m_3(3)$ = max $\{f_3(0), f_3(1), f_3(2), f_3(3)\}$
           = max $\{0, 1, 4, 5\} = 5$ with $d_3(3) = 3$
$m_3(2)$ = max $\{f_3(0), f_3(1), f_3(2)\}$
           = max $\{0, 1, 4\} = 4$ with $d_3(2) = 2$
$m_3(1)$ = max $\{f_3(0), f_3(1)\}$ = max $\{0, 1\} = 1$ with $d_3(1) = 1$
$m_3(0)$ = max $\{f_3(0)\}$ = max $\{0\} = 0$ with $d_3(0) = 0$

These results give us the first two rows in the solution table:

	U				
	0	1	2	3	4
$m_3(u)$	0	1	4	5	8
$d_3(u)$	0	1	2	3	4

Having completed Stage 3, we next consider Stage 2 under the assumption that Stage 1 has been completed (although, at this time, we do not know how). Since we do not know the number of units allocated to investment 1, we do not know how many units are available for investment 2; we must therefore consider all possibilities.

One possibility is that all 4 units are available at Stage 2, which presupposes that no units were allocated to investment 1. Now, all or some of these 4 units can be allocated to investment 2, with the remainder available for Stage 3. If x of these 4 units are allocated to investment 2, the return is $f_2(x)$, and the remaining 4-x units are available for Stage 3. However, we have already found the best continuation from Stage 3 when 4-x units are at hand; namely, $m_3(4-x)$. The total return, therefore is $f_2(x) + m_3(4-x)$; and the value of x (x = 0, 1, 2, 3, 4) that maximizes this total return represents the optimal decision at Stage 2 with 4 units available. This may be formally stated as follows:

$m_2(4)$ $= \max\{f_2(0)+m_3(4-0), f_2(1)+m_3(4-1), f_2(2)+m_3(4-2), f_2(3)+m_3(4-3), f_2(4)+m_3(4-4)\}$
$= \max\{0+8, 1+5, 3+4, 6+1, 7+0\} = 8$ with $d_2(4) = 0$

Similarly treating the other possibilities at Stage 2, we obtain:

$m_2(3)$ $= \max\{f_2(0)+m_3(3-0), f_2(1)+m_3(3-1), f_2(2)+m_3(3-2), f_2(3)+m_3(3-3)\}$
$= \max\{0+5, 1+4, 3+1, 6+0\} = 6$ with $d_2(4) = 3$
$m_2(2)$ $= \max\{f_2(0)+m_3(2-0), f_2(1)+m_3(2-1), f_2(2)+m_3(2-2)\}$
$= \max\{0+4, 1+1, 3+0\} = 4$ with $d_2(2) = 0$
$m_2(1)$ $= \max\{f_2(0)+m_3(1-0), f_2(1)+m_3(1-1)\}$
$= \max\{0+1, 1+0\} = 1$ with $d_2(1) = 0$ (breaking the tie arbitrarily)
$m_2(0)$ $= \max\{f_2(0)+m_3(0-0) \} = \max\{0+0\} = 0$ with $d_2(0) = 0$

Collecting the calculations for Stage 2, we obtain the third and fourth rows of the solution table:

	U				
	0	1	2	3	4
$m_3(u)$	0	1	4	5	8
$d_3(u)$	0	1	2	3	4
$m_2(u)$	0	1	4	6	8
$d_2(u)$	0	1	0	3	0

Having completed Stage 2, we now turn to Stage 1. There is only one state associated with this stage, $u = 4$.

$m_1(4)$ $= \max\{f_1(0)+m_2(4-0), f_1(1)+m_2(4-1), f_1(2)+m_2(4-2), f_1(3)+m_2(4-3), f_1(4)+m_2(4-4)\}$
$= \max\{0+8, 2+6, 5+4, 6+1, 7+0\} = 9$ with $d_1(4) = 2$
With these data we complete the solution table:

	u				
	0	1	2	3	4
$m_3(u)$	0	1	4	5	8
$d_3(u)$	0	1	2	3	4
$m_2(u)$	0	1	4	6	8
$d_2(u)$	0	1	0	3	0
$m_1(u)$	...	...	...	...	9

The maximum return that can be realized from this three-stage investment program beginning with 4 units is $m_1(4) = 9$ units. To achieve this return, allocate $d_1(4) = 2$ units to investment 1, leaving 4-2 = 2 units for Stage 2. But $d_2(2) = 0$, indicating that no units should be expended at this stage if only 2 units are available. Thus, 2 units remain for Stage 3. Since $d_3(2) = 2$, both units should be allocated to investment 3. The optimal policy, therefore, is to allocate 2 units to investment 1, 0 units to investment 2, and 2 units to investment 3.

A number of key characteristics of dynamic programming are demonstrated by the above example:

- Dynamic programming is particularly well suited to multistage decision processes in which each decision pays off separately, independently of previous decisions. In the above example, the expected monetary return of each investment i depends only on the amount x_i dedicated to the investment. It does not depend on the amounts dedicated to other investments.

- Dynamic programming works backwards, from the nth stage to the first stage. For this process to occur, it must be possible to express the optimum returns of each stage j in terms of the optimum returns of the succeeding stage, $j+1$. Specifically, $m_j(u)$, the optimum return from completing the process beginning at Stage j in state u (Tbl. 5.4), must have a recursive definition of the form

$$m_j(u) = f(m_{j+1}(v_1), m_{j+1}(v_2), \ldots, m_{j+1}(v_s))$$

where $v_1, v_2, \ldots, v_s$ are the possible states at Stage $j+1$, given that the state in Stage j is u.

Queueing Theory

A queueing process consists of customers arriving at a service facility, then waiting in a line (*queue*) if all servers are busy, eventually receiving service, and finally departing from the facility [13,14]. A *queueing system* [13] is a set of customers, a set of servers, and an order whereby customers arrive and are processed? (Fig. 5.14). The *state* of the system is the number of customers in the facility. *Queueing theory* is the study of the behavior of queueing systems through the formulation of analytical models.

There are two ways that a queueing system can represent a manufacturing system. One way is to let the servers represent resources (e.g., machines) while the customers represent the parts that have to be operated on by the resources. A "dual" representation is to view the customers as resources and the servers as the fixed number of parts in the manufacturing system. The latter representation is suitable for certain flexible manufacturing systems, in which parts circulate about the system on a fixed number of pallets. (A pallet is a mobile fixture.)

Queueing systems are characterized by five components: the arrival pattern of customers, the service pattern, the number of servers, the capacity of the facility to hold customers, and the order in which customers are served.

The *arrival pattern* of customers is usually specified by the *interarrival time*, the time between successive customer arrivals at the service facility. It may be determined (i.e. known exactly), or it may be a random variable, whose probability distribution is presumed known. It may depend on the number of customers already in the system, or it may be state-independent. Also of interest is whether customers arrive singly or in batches and whether balking or reneging is permitted. *Balking* occurs when an arriving customer refuses to enter the service facility because the queue is too long. *Reneging* occurs when a customer already in the queue leaves the queue and the facility because the wait is too long. Unless stated to the contrary, the standard assumption is that all customers arrive singly and that neither balking nor reneging occurs.

The *service pattern* is usually specified by the *service time*, the time required by one server to serve one customer. The service time may be deterministic, or it may be a random variable, whose probability distribution is presumed known. It may depend on the number of customers already in the facility, or it may be state-dependent. Also of interest is whether a customer is attended completely by one server, or the customer requires a sequence of servers.

The *system capacity* is the maximum number of customers, both those in service and those in the queue(s), permitted in the service facility at the same time. Whenever a customer arrives at a facility that is full, the arriving customer is denied entrance to the facility. Such a customer is not allowed to wait outside the facility (since that effectively increases the capacity) but is forced to leave without receiving service. A system that has no limit on the number of customers permitted inside the facility has infinite capacity; a system with a limit has finite capacity.

The *queue discipline* is the order in which customers are served. This can be on a first-in, first-out (FIFO) basis (i.e. service in order of arrival), a last-in, first-out (LIFO) basis (i.e., the customer who arrives last is the next served), a random basis, or a priority basis.

Kendall's notation for specifying a queue's characteristics is v/w/x/y/z, where v indicates the arrival pattern, w denotes the service pattern, x signifies the number of available servers, y represents the system's capacity, and z designates the queue discipline. Various notations [13] are used for v, w, and z (Tbl. 5.5). If y (system capacity) or z (queue discipline) is not specified, it is taken to be ∞ or FIFO, respectively.

For example, an M/D/2/5/LIFO system has exponentially distributed interarrival times, deterministic service times, two servers, and a limit of five customers allowed into the service facility at any time, with the last customer to arrive being the next customer to go into service. A D/D/1 system has both deterministic interarrival times and deterministic service times, and only one server. Since system capacity and queue discipline are not specified, they are assumed to be infinite and FIFO, respectively.

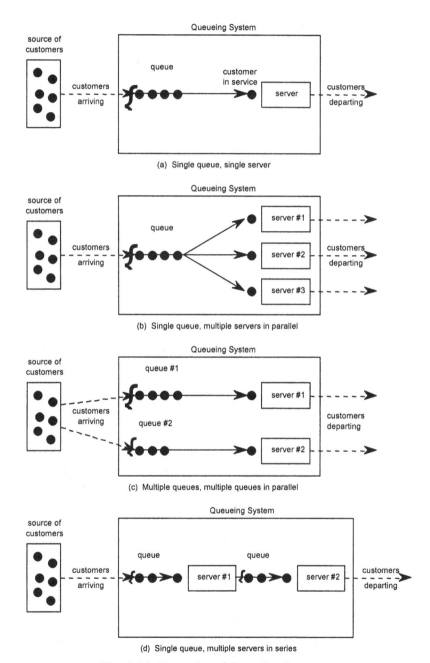

Fig. 5.14. Examples of Queueing Systems

Queue Characteristics	Symbol	Meaning
Interarrival time (v) or Service time (w)	D	Deterministic
	M	Exponentially distributed
	E$_k$	Erlang-type-k ($k=1,2,...$) distributed
	G	Any other distribution
Queue discipline (z)	FIFO	First in, first out
	LIFO	Last in, first out
	SIRO	Service in random order
	PRI	Priority ordering
	GD	Any other specialized ordering

Table 5.5. Queueing Model Notation

M/M/1 Systems

An M/M/1 system is a queueing system having exponentially distributed interarrival times, with parameter λ; exponentially distributed service times, with parameter μ; one server; no limit on the system capacity; and a queue discipline of first come, first served. The constant λ is the average customer arrival rate; the constant μ is the average service rate of customers. Both are in units of customers per unit time. The expected interarrival time and the expected time to serve one customer are $1/\lambda$ and $1/\mu$, respectively.

Since exponentially distributed interarrival times with mean $1/\lambda$ are equivalent, over a time interval τ, to a Poisson-distributed arrival pattern with mean $\lambda\tau$, M/M/1 systems are often referred to as single-server, infinite-capacity, queueing systems having Poisson input and exponential service times.

The *steady-state probabilities* for a queueing system are

$$p_n \equiv \lim_{t\to\infty} p_n(t) \quad (n=0,1,2...)$$

if the limits exist, where $p_n(t)$ is the probability that the system has exactly n customers at time t. For an M/M/1 system, we define the *utilization factor* (or *traffic intensity*) as

$$\rho \equiv \frac{\lambda}{\mu}$$

i.e., ρ is the expected number of arrivals per mean service time. If $\rho < 1$, then steady-state probabilities exist and are given by

$$p_n = \rho^n(1-\rho)$$

If $\rho > 1$, the arrivals come at a faster rate than the server can accommodate: the expected queue length increases without limit and a steady state does not occur. A similar situation prevails when $\rho = 1$.

For a queueing system in steady state, the measures of greatest interest are:

$L \equiv$ the average number of customers in the system
$Lq \equiv$ the average length of the queue
$W \equiv$ the average time a customer spends in the system
$Wq \equiv$ the average time a customer spends (or waits) in the queue
$W(t) \equiv$ the probability that a customer spends more than t units of time in the system
$Wq(t) \equiv$ the probability that a customer spends more than t units of time in the queue

The first four of these measures are related to many queueing systems by

$$W = W_q + \frac{1}{\mu} \tag{5-9}$$

and by *Little's formulas*

$$L = \overline{\lambda} W \tag{5-10}$$

$$L_q = \overline{\lambda} W_q \tag{5-11}$$

The waiting-time equation, (Eq. 5-9), holds whenever (as in an M/M/1 system) there is a single expected service time, $1/\mu$, for all customers. Little's formulas are valid for quite general systems, provided that $\bar{\lambda}$ denotes the average arrival rate of customers *into* the service facility.

For an M/M/1 system, $\bar{\lambda} = \lambda$, and the six measures are explicitly:

$$L = \frac{\rho}{1 - \rho}$$

$$L_q = \frac{\rho^2}{1 - \rho}$$

$$W = \frac{1}{\mu - \lambda}$$

$$W_q = \frac{\rho}{\mu - \lambda}$$

$$W(t) = e^{-t/W} , (t \geq 0)$$

$$W_q(t) = \rho \cdot e^{-t/W} , (t \geq 0)$$

The M/M/1 queue has exponentially distributed interarrival times and service times. An exponentially distributed random variable has the property that "most of" its values are smaller than the mean value. If T has an exponential distribution, with parameter β, the mean value of T is $1/\beta$. Then,

$$P(T \leq 1/\beta) = 1 - e^{-1} \approx 0.632$$

$$P(T \leq 1/2\beta) = 1 - e^{-1/2} \approx 0.393$$

where $P(T \leq 1/\beta)$ is the probability that T is less than or equal to the mean $1/\beta$, and $P(T \leq 1/2\beta)$ is the probability that T is less than or equal to $1/2\beta$. Thus, we might say that 63 percent of the values are smaller than the mean, and, of *those* values, some 63 percent are smaller than half the mean.

The above observation implies that an M/M/1 queue models facilities with a preponderance of interarrival times that are less than the average, with a few that are very long. The net result is that a number of customers arrive in a short period of time, thereby creating a queue, followed eventually by a long interval during which no new customer arrives, allowing the server to reduce the size of the queue. Also, the exponentially distributed service times of an M/M/1 system make it a good model for facilities with a preponderance of shorter-than-average servicings, combined with a few long ones. This would be the situation, for example, at banks where a majority of customers make simple deposits, requiring very little teller time, but a few have more complicated transactions that consume a lot of time.

Exponential distributions also possess the Markovian (or *memoryless*) property:

$$P(T \le a+b \mid T> a) = P(T \le b)$$

Here, a and b are arbitrary constants; $P (T \le a + b \mid T > a)$ is the probability that T is less than or equal to $a+b$, *given* that T is greater than a; and $P (T \le b)$ is the probability that T is less than or equal to b. When T measures interarrival times, the implication is that the time at the next arrival is independent of the time since the last arrival. For service times, the implication is that the time required to complete service on a customer cannot be predicted by knowing (i.e., is independent of) the time the customer has already been in service.

State-Dependent Queueing Systems

In contrast to an M/M/1 system, the number of customer arrivals, in many queueing situations, does not constitute a strict Poisson process, with constant parameter λ; instead, it seems to follow a Poisson-like process in which λ varies according to the number of customers in the system. It may also be the case that departures from the system do not occur at a constant mean rate μ, as they would for an M/M/1 system; rather, the distribution for μ also varies according to the number of customers in the system (the system state). For such queueing processes, $\lambda_n \Delta t$ and $\mu_n \Delta t$ are, respectively, the expected numbers of arrivals and departures in a small time interval Δt, if the system is in a state n at the beginning of the interval. The steady-state probabilities for these processes are found to satisfy

$$p_n = \frac{\lambda_{n-1}}{\mu_n} p_{n-1} \quad \text{or} \quad p_n = \frac{\lambda_{n-1}\lambda_{n-2}\cdots\lambda_0}{\mu_n\mu_{n-1}\cdots\mu_1} p_0 \tag{5-12}$$

in which p_0 is determined by the condition that the sum of all the probabilities be unity. This sum converges provided the λ's are not too large with respect to the μ's. In particular, the existence of a steady state is assured if

$$\frac{\lambda_{n-1}}{\mu_n} < 1$$

for all large n.

Little's formulas (Eqs. 5-10 and 5-11) hold for state-dependent queueing systems if

$$\bar{\lambda} = \sum_{n=0}^{\infty} \lambda_n p_n$$

is the average arrival rate of customers *into* the service facility.

In any queueing system, the expected number of customers in the system is

$$L = \sum_{n=0}^{\infty} n p_n \tag{5-13}$$

and the expected number of customers in the queue is

$$L_q = \sum_{n=0}^{\infty} \left[\max\{n - s_n, 0\}\right] p_n \tag{5-14}$$

where s_n is the number of servers available in state n. If it is possible to evaluate L and L_q, then, knowing $\bar{\lambda}$, we can at once find W (the average time a customer spends on the system) and W_q (the average time a customer spends in the queue) from Little's formulas (Eqs. 5-10 and 5-11).

M/M/s Systems

An M/M/*s* system is a queueing process having a Poisson arrival pattern; *s* servers, with *s* independent, identically distributed, exponential service times (which do not depend on the state of the system); infinite capacity; and a FIFO queue discipline.   The arrival pattern being state-independent, $\lambda_n = \lambda$ for all *n*.   The service times associated with each server are also state-independent; but since the number of servers that actually attend customers (i.e., are not idle) *does* depend on the number of customers in the system, the effective time it takes the *system* to process customers through the service facility is state-dependent.   In particular, if $1/\mu$ is the mean service time for one server to handle one customer, then the mean rate of service completions when there are *n* customers in the system is

$$\mu_n = \begin{cases} n\mu & (n = 0, 1, \cdots, s) \\ s\mu & (n = s+1, s+2, \cdots) \end{cases}$$

Steady-state conditions prevail whenever

$$\rho \equiv \frac{\lambda}{s\mu} < 1$$

The steady-state probabilities are given by Equation 5-12 as

$$p_0 = \left[\frac{s^s \rho^{s+1}}{s!(1-\rho)} + \sum_{n=0}^{s} \frac{(s\rho)^n}{n!} \right]^{-1}$$

and

$$p_n = \begin{cases} \dfrac{(s\rho)^n}{n!} p_0 & (n = 1, \cdots, s) \\[2ex] \dfrac{s^s \rho^n}{s!} p_0 & (n = s+1, s+2, \cdots) \end{cases}$$

With p_0 (the probability that in steady state there will be no customers in the system) given, the average queue length may be calculated as

$$Lq = \frac{s^s \rho^{s+1} p_0}{s!(1-\rho)^2}$$

Once L_q is determined, W_q (the average time that a customer waits in a queue), W (the average time that a customer spends in the system), and L (the average number of customers in the system) are obtained from Equations 5-11, 5-9, and 5-10, respectively, with $\overline{\lambda} = \lambda$. Equation 5-9 applies here, because, regardless of the state of the system, the expected service time for each customer has the fixed value $1/\mu$. Furthermore,

$$W(t) = e^{-\mu t}\left\{1 + \frac{(s\rho)^s p_0[1 - e^{-\mu t(s-1-s\rho)}]}{s!(1-\rho)(s-1-s\rho)}\right\} \qquad (t \geq 0)$$

$$W_q(t) = \frac{(s\rho)^s p_0}{s!(1-\rho)} e^{-s\mu t(1-\rho)} \qquad (t \geq 0)$$

M/M/1/K Systems

An M/M/1/K system can accommodate a maximum of K customers in the service facility at the same time. Customers arriving at the facility when it is full, are denied entrance and are not permitted to wait outside the facility for entrance at a later time. If λ designates the mean arrival rate of customers to the service facility, then the mean arrival rate *into* the facility when the facility is in state n is

$$\lambda_n = \begin{cases} \lambda & (n = 0, 1, \cdots, K\text{-}1) \\ 0 & (n = K, K+1, \cdots) \end{cases}$$

A steady state is always attained, whatever the value of $\rho + \lambda/\mu$. The steady-state probabilities are given by $p_n = 0$ when n is greater than K, and, for $n = 0, 1, ..., K$,

$$p_n = \begin{cases} \dfrac{\rho^n(1-\rho)}{1-\rho^{K+1}} & (\rho \neq 1) \\[4mm] \dfrac{1}{K+1} & (\rho = 1) \end{cases}$$

The measures of effectiveness can be calculated beginning with the average number of customers in the system,

$$L = \begin{cases} \dfrac{\rho}{1-\rho} - \dfrac{(K+1)\rho^{K+1}}{1-\rho^{K+1}} & (\rho \neq 1) \\[4mm] \dfrac{K}{2} & (\rho = 1) \end{cases}$$

with W (the average time that a customer spends on the system), W_q (the average time that a customer waits in a queue), and L_q (the average queue length) obtained from Equations 5-10, 5-9, and 5-11, respectively. For the calculation of these measures,

$$\overline{\lambda} = \lambda\,(1 - p_K) \tag{5-15}$$

M/M/s/K Systems

An M/M/s/K system is a finite-capacity system with s servers having independent, identically distributed exponential service times (which do not depend on the state of the system). Since the capacity of the system must be at least as large as the number of servers, $s \leq K$. For such a system,

$$\lambda_n = \begin{cases} \lambda & (n = 0,1,\ldots,K-1) \\ 0 & (n = K, K+1,\ldots) \end{cases} \qquad \mu_n = \begin{cases} n\mu & (n = 0,1,\ldots,s) \\ s\mu & (n = s+1, s+2,\ldots) \end{cases}$$

Steady-state probabilities exist for all values of $\rho + \lambda/\mu$, and are given by Equation 5-12 as

$$
p_0 = \begin{cases} \left[\dfrac{s^s p^{s+1}\left(1 - p^{K-s}\right)}{s!(1 - p)} + \displaystyle\sum_{n=0}^{s} \dfrac{(sp)^n}{n!} \right]^{-1} & (p \neq 1)\,(p = 1) \\[2em] \left[\dfrac{s^s}{s!}(K - s) + \displaystyle\sum_{n=0}^{s} \dfrac{s^n}{n!} \right]^{-1} \end{cases}
$$

and

$$
p_n = \begin{cases} \dfrac{(s\rho)^n}{n!}\, p_0 & (n = 1, 2, \cdots, s) \\[1.5em] \dfrac{s^s \rho^n}{s!}\, p_0 & (n = s+1, \cdots, K) \\[1.5em] 0 & (n = K+1,\ K+2, \cdots) \end{cases}
$$

The measures of effectiveness can be calculated beginning with the average length of each queue,

$$
L_q = \frac{s^s \rho^{s+1}}{s!\,(1 - \rho)^2}\left[1 - \rho^{K-s} - (1 - \rho)(K - s)\,\rho^{K-s}\right] p_0
$$

with W_q (the average time that a customer waits in a queue), W (the average time that a customer spends on the system), and L (the average number of customers in the system) obtained from Equations 5-11, 5-9, and 5-10, respectively; $\bar{\lambda}$ is again given by Equation 5-15. An M/M/1/K system is a special M/M/s/K system with $s = 1$.

To see the type of analysis that is possible with the queueing theory, consider a machine within a factory attended by an operator, who also functions as the job setter when the machine is not too busy [13]. Parts arrive at the machine according to a Poisson process, at a mean rate of 30 per hour. The time required for the operator to set up the machine and process the part is exponentially distributed, with a mean of 2 min. Whenever there are three or more parts waiting for the machine (including the part in process), a second operator in the factory is instructed to assist the first operator in setting up the jobs. When the operators work together, the service time for a part (including setup and processing) remains exponentially distributed, but with a mean of 1 min. Determine (a) the average number of parts at the machine at the same time, (b) the length of time a part should expect to spend at the machine, and (c) the length of time a part should expect to wait on line before being processed.

Throughout the process the arrival rate remains state-independent at λ_n $= \lambda = 30$ h^{-1}. The service times, however, are state-dependent. When there are fewer than three parts at the machine, the mean service time is 2 min; hence the mean service rate is 30 h^{-1}. When there are three or more parts at the machine, the mean service time is 1 min; hence the mean service rate increases to 60 h^{-1}. Thus,

$$\mu_n = \begin{cases} 30 \text{ h}^{-1} & (n = 1, 2) \\ 60 \text{ h}^{-1} & (n = 3, 4, \cdots) \end{cases}$$

Note that, when a new arrival changes the state of the system from 2 to 3, the part in process is instantly subject to the new exponential distribution (the "memoryless" property).

It follows from Equation 5-12 that

$$p_1 = \frac{\lambda_0}{\mu_1} p_0 = \frac{30}{30} p_0 = p_0 \qquad p_2 = \frac{\lambda_1}{\mu_2} p_1 = \frac{30}{30}(p_0) = p_0$$

$$p_3 = \frac{\lambda_2}{\mu_3} p_2 = \frac{30}{60}(p_0) = \frac{1}{2} p_0 \qquad p_4 = \frac{\lambda_3}{\mu_4} p_3 = \frac{30}{30}\left(\frac{1}{2} p_0\right) = \left(\frac{1}{2}\right)^2 p_0$$

and, in general,

$$p_n = \left(\frac{1}{2}\right)^{n-2} p_0 \qquad (n \geq 2)$$

To find p_0, we use the fact that the sum of the steady state probabilities must be 1,

$$1 = \sum_{n=0}^{\infty} p_n = p_0 + p_1 + \sum_{n=2}^{\infty} p_n = 2p_0 + \sum_{n=2}^{\infty} \left(\frac{1}{2}\right)^{n-2} p_o$$

$$= 2p_0 p_0 \sum_{j=0}^{\infty} \left(\frac{1}{2}\right)^j = 2p_0 + 2p_0 = 4p_0$$

obtaining $p_0 = 1/4$. Therefore,

$$p_n = \begin{cases} \left(\frac{1}{4}\right) & (n = 0,1) \\ \left(\frac{1}{2}\right)^n & (n = 2,3,\dots) \end{cases}$$

(a)   From Equation 5-13,

$$L = \sum_{n=0}^{\infty} np_n = 0 \cdot \frac{1}{4} + 1 \cdot \frac{1}{4} + S, \quad \text{where } S = \sum_{n=2}^{\infty} n\left(\frac{1}{2}\right)^n$$

To calculate S, note that

$$\sum_{n=2}^{\infty} x^n = \frac{x^2}{1-x}, |x| < 1 \qquad \text{(sum of geometric series)}$$

$$\sum_{n=2}^{\infty} x^{n+1} = \frac{x^3}{1-x}, |x| < 1 \qquad \text{(sum of geometric series)}$$

$$\sum_{n=2}^{\infty} nx^n = \sum_{n=2}^{\infty} (n+1)x^n - \sum_{n=2}^{\infty} x^n = \frac{d}{dx}\left(\sum_{n=2}^{\infty} x^{n+1}\right) - \sum_{n=2}^{\infty} x^n$$

$$= \frac{d}{dx}\left(\frac{x^3}{1-x}\right) - \frac{x^2}{1-x} = \frac{-x^3 + 2x^2}{(1-x)^2}, |x| < 1$$

Substituting $x = 1/2$ into the above expression, we obtain $S = 1.50$. Finally,

$L = 0.25 + S = 1.75$ parts on average at the machine at the same time

(b)   Since $\overline{\lambda} = \lambda = 30$ h^{-1}, we obtain from Little's formula (Eq. 5-10)

$$W = \frac{L}{\lambda} = \frac{1.75}{30} = 0.05833 \text{ hours} = 3.5 \text{ minutes}$$

(c) Because the two operators work together, the number of servers is state-independent at $s_n = 1$. Then, from Equation 5-14,

$$L_q = \sum_{n=1}^{\infty} (n-1)p_n$$

From Equation 5-13,

$$L = \sum_{n=1}^{\infty} np_n$$

therefore

$$L - L_q = \sum_{n=1}^{\infty} p_n = 1 - p_0, \text{ or}$$
$$L_q = L - (1 - p_0) = 1.75 - 0.75 = 1.00 \text{ part}$$

and, from Little's formula (Eq. 5-11),

$$W_q = \frac{L_q}{\lambda} = \frac{1.00}{30} = 0.0333 \text{ hours} = 2 \text{ minutes}$$

Observe that the average service time *per part* is

$$W - W_q = 1.5 \text{ minutes}$$

5.2.2 Artificial Intelligence

The field of artificial intelligence may be defined as the study of ideas that enable computers to be intelligent [15]. Its main goals are to make computers more useful, and to understand the principles that make intelligence possible. Since, the first of these goals is the most relevant to designing manufacturing systems, two artificial intelligence tools, which make computers more useful, namely *search* and *rule-based systems*; will be described in the following sections.

Search

If we view a manufacturing system design as a set of values for, say, *n* decision variables, then any feasible design can be viewed as a point in a *n*-dimensional design space. A sensible design *process* must begin at an initial design point. From this point, a designer seeks to explore the design space, moving from point to point (design to design), evaluating each point as it arises. A sophisticated designer uses the information from previous evaluations (i.e. the performance measures of the previously explored designs) to determine the path of future moves through the design space. In the end, the designer, having traced a path through the design space, will arrive at the final design point, either because the optimal design point (according to some performance measure) will have been reached, or because some limit on computational effort will have been exceeded. In either case, one hopes, the final design point to be superior to the initial design point, and describes a manufacturing system that meets the stated performance requirements.

The body of *heuristics* or intuitively "reasonable" rules, which the designer can use to establish a path (hopefully short) through the design space, is called *search*. On an abstract level, search methods find solutions by exploring paths. What distinguishes one search method from another is that of the heuristics, which decides how the exploring is to be done.

Suppose we want to find some path through a net of cities [15] connected by highways (Fig. 5.15). The path is to begin at city S, the starting point, and it is to end at city G, the final goal.

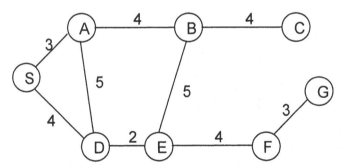

Fig. 5.15. A Basic Search Problem. A Path is to be found from the Start Node, S, to the Goal Node, G.

Finding a path involves two kinds of effort:

- First, there is the effort expended in *finding* either some path or the shortest path.
- And, second, there is the effort actually expended in *traversing* the path.

Should it be necessary to go from S to G often, then it is a lot worth finding a really good path. On the other hand, if only *one* trip is required, and if the net is hard to force a way through, then it is proper to be content as soon as *some* path is found, even though better ones could be found with more work. For the moment, we will consider only the problem of finding *one* path.

The most obvious way to find a solution is to devise a bookkeeping scheme that allows orderly exploration of all possible paths. It is useful to note that the bookkeeping scheme must not allow itself to cycle in the net. It would be senseless to go through a sequence like S-A-D-S-A-D over and over again. With cyclic paths terminated, nets are equivalent to trees. The tree shown in Figure 5.16 [15] is made from the net in Figure 5.15 by following each possible path outward from the net's starting point until it runs into a place already visited.

In nets, the connection between nodes is called *links*, and in trees, the connections are called *branches*. Branches directly connect *parents* with *children*. The node at the top of a tree, the one with no parent, is called the *root node* (Node S in Figure 5.16). The nodes at the bottom, the ones with no children, are called *terminal nodes*. (In Figure 5.16, accumulated distances are drawn beside these nodes.) One node is the *ancestor* of another, a *descendant*, if there is a chain of branches between the two. If the number of children is always the same for every node that has children, that number is said to be the *branching factor*. Drawing in the children of a node is called *expanding* the node. Nodes are said to be open until they are expanded, whereupon they become *closed*.

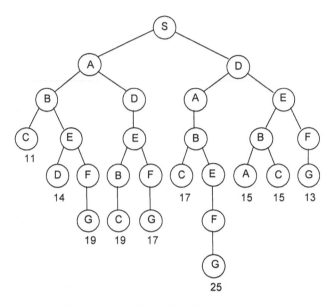

Fig. 5.16. A Tree Made From a Net

If no node in a net is to be visited twice, there can be no more than *n* levels in the corresponding tree, where *n* is the total number of nodes, eight in the map traversal example. In the example, the goal is reached at the end of four distinct paths, each of which has a total path length given by adding up a few distances.

Depth-First Search

Given that one path is as good as any other, one no-fuss idea is to pick an alternative at every node visited and work forward from that alternative. Other alternatives at the same level are ignored completely as long as there is hope of reaching the destination using the original choice. This is the essence of the *depth-first search*. Using the convention that the alternatives are tried in left-to-right order, the first action in working on the situation in Figure 5.16 is a headlong dash to the bottom of the tree along the leftmost branches.

But since a headlong dash leads to terminal node C, without encountering G, the next step is to back up to the nearest ancestor node with an unexplored alternative. The nearest such node is B. The remaining alternative at B is better, bringing eventual success through E in spite of another dead end at D. Figure 5.17 [15] shows the nodes encountered.

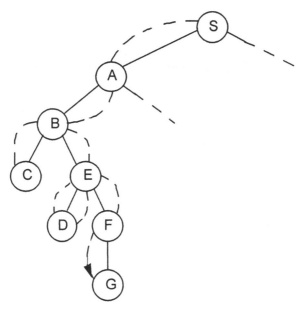

Fig. 5.17. Depth First Search

If the path through E had not worked out, then the procedure would have moved still further back up the tree, seeking another viable decision point to move forward from. On reaching A, the movement would go down again, reaching destination through D.

The depth-first search procedure can be formally stated as follows:

1. Form a one-element queue consisting of the root node.
2. Until the queue is empty or the goal has been reached,
   2a. If the first element is the goal node, do nothing.
   2b. If the first element is not the goal node, remove the first element from the queue and add the first element's children, if any, to the *front* of the queue.
3. If the goal node has been found, announce success; otherwise announce failure.

Hill Climbing: Depth-First Search with Quality Measurements

Search efficiency may improve spectacularly if there is some way of ordering choices so that the most promising ones are explored first. In many situations, simple measurements can be made to determine a reasonable ordering.

To move through a tree of paths using hill climbing, proceed as in depth-first search, but order the choices according to some heuristic measure of remaining distance. Figure 5.18 [15] shows what happens when hill climbing is used on the map-traversal problem (Figs. 5.15 and 5.16), using as-the-crow-flies distance, to order choices.

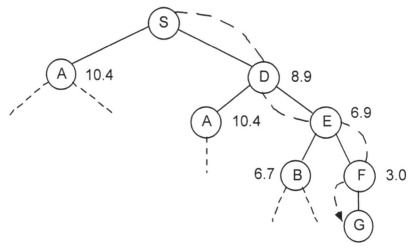

Fig. 5.18. Hill Climbing

From a procedural point of view, hill climbing differs from depth-first search in only one detail, that of the added, italicized part:

1. Form a one-element queue consisting of the root node.
2. Until the queue is empty or the goal has been reached,
   2a. If the first element is the goal node, do nothing.
   2b. If the first element is not the goal node, remove the first element from the queue, *sort the first element's children, if any, by estimated remaining distance,* and add the first element's children, if any, to the front of the queue.
3. If the goal node has been found, announce success; otherwise announce failure.

A form of hill climbing is also used in parameter optimization. To move through a space of parameter values using parameter-oriented hill climbing, take one step in each of a fixed set of directions, move to the best alternative found, and repeat until reaching a point that is better than all of the surrounding points reached by the one-step probes. In applying parameter optimization to manufacturing system design, the parameters would be decision variables, whose values specify the design of a manu-facturing system.

Although simple, parameter optimization via hill climbing has a number of limitations:

- The *foothill problem* occurs whenever there are secondary peaks. An optimal point is found, but it is local, not global.
- The *plateau problem* comes up when there is mostly a flat area separat-ing peaks. The local improvement operation breaks down completely. For all but a small number of positions, all standard-step probes leave the quality measurement unchanged.
- The *ridge problem* is more subtle. Suppose that the current point lies on what seems like a knife edge, running generally from northeast to southwest. Standard one-step probes in the north, south, east and west directions imply that all surrounding points are worse than the current point, even though the current point is not at any sort of maximum, local or global. Increasing the number of directions used for the probing steps may help.

In general, the foothill, plateau, and ridge problems are greatly exacer-bated as the number of parameter dimensions increases.

Breadth-First Search

When depth-first search and hill climbing are bad choices, *breadth-first search* may be useful. Breadth-first search looks for the goal node among all nodes, at a given level, before using the children of those nodes to push on. In the situation shown in Figure 5.19 [15], node D would be checked just after A. The procedure would then move on, level by level, discover-ing G on the fourth level down from the root level.

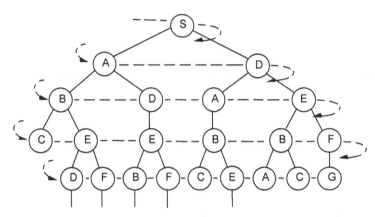

Fig. 5.19. Breadth-First Search

Like hill-climbing, a procedure for breadth-first search, resembles that of the one for depth-first search, differing only in the place where new elements are added to the queue.

1. Form a one-element queue consisting of the root node.
2. Until the queue is empty or the goal has been reached,
   2a. If the first element is the goal node, do nothing.
   2b. If the first element is not the goal node, remove the first element from the queue and add the first element's children, if any, to the *back* of the queue.
3. If the goal node has been found, announce success; otherwise announce failure.

Breadth-first search will work even in trees that are infinitely deep or effectively infinite. On the other hand, breadth-first search is wasteful when all paths leading to the destination node, are at more or less the same depth.

Beam Search

Beam search is like breadth-first search because it progresses level by level. Unlike breadth-first search, however, beam search only moves downward from the best w nodes at each level. The other nodes are ignored.

Consequently, the number of nodes explored remains manageable, even if there is a great deal of branching and the search is deep. If beam search of width w is used in a tree with branching factor b, there will be only wb

nodes under consideration at any depth, not would there be the explosive number if breadth-first search was used. Figure 5.20 [15] illustrates how beam search would handle the map-traversal problem (Figs. 5.15 and 5.16). The numbers beside the nodes are straight-line distances to the goal node.

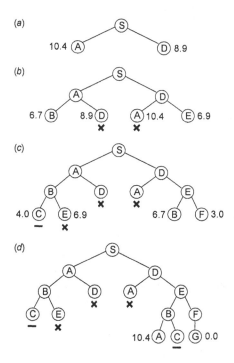

Fig. 5.20. Application of Beam Search ($w = 2$) to the Map Traversal Problem

Best-First Search

When forward motion is blocked, hill climbing demands forward motion from the last choice through the seemingly best child node. In *best-first search*, forward motion is so far from the best open node, no matter what its position is in the partially developed tree.

Selection of a Search Strategy for Finding One Path

There are many ways of finding *one* path, each with advantages, among them the following:

• Depth-first search is good when blind alleys do not get too deep.

- Breadth-first search is good when the number of alternatives at the choice points is not too large.
- Hill climbing is good when there is a natural measure of goal distance and a good choice is likely to be among the good-looking choices at each choice point.
- Beam search is good when there is a natural measure of goal distance and a good path is likely to be among the good-looking choices at each choice point.
- Best-first search is good when there is a natural measure of goal distance and a good path may look bad at shallow levels.

Branch and Bound Search

One way to find *optimal* paths with less work is by using *branch and bound search*. An application of branch and bound search that we have already encountered is the solution of integer programs. Here, the search procedure will be presented in a more generic fashion. The basic idea is simple. Suppose an optimal solution is desired for the net [15] shown in Figure 5.21(*a*). The numbers beneath the nodes in the trees are accumulated distances. Looking only at the first level, in Figure 5.21(*b*), the distance from S to node A is clearly less than the distance to B. Following A to the destination at the next level reveals that the total path length is 4, as shown in Figure 5.21(*c*). This means though that there is no point in calculating the path length for the alternative path through node B since at B the incomplete path's length is already 5 and hence, longer than the path for the known solution through A.

More generally, the branch and bound scheme works as follows: during search there are many incomplete paths contending for further consideration. The shortest one is extended one level, creating as many new incomplete paths as there are branches. These new paths are then considered along with the remaining old ones, and again, the shortest path is extended. This is repeated until the destination has been reached along some path. Since the shortest path was always chosen for extension, the path first reaching the destination is certain to be optimal. However, the last step in reaching the destination may be long enough to make the supposed solution longer than one or more incomplete paths. It might be that only a tiny step would extend one of the incomplete paths to the solution point. Thus, instead of terminating when a path is found, terminate when the shortest incomplete path is longer than the shortest complete path.

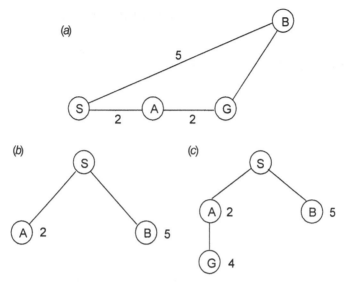

Fig. 5.21. Branch and Bound: B is Not Expanded Because the Partial Path At B is Already Longer Than the Complete Path Through A

Here, then, is the branch and bound search procedure with the proper terminating condition:

1. Form a queue of partial paths. Let the initial queue consist of the zero-length, zero-step path from the root node to nowhere.
2. Until the queue is empty or the goal has been reached, determine if the first path in the queue reaches the goal node.
   2a. If the first path reaches the goal node, do nothing.
   2b. If the first path does not reach the goal node:
      2b1.   Remove the first path from the queue.
      2b2.   Form new paths from the removed path by extending one step.
      2b3.   Add the new paths to the queue.
      2b4.   Sort the queue by the cost accumulated so far, with least-cost paths in front.
3. If the goal node has been found, announce success; otherwise announce failure.

Figure 5.22 [15] illustrates the exploration sequence for the branch and bound method as applied to the map traversal problem (Figs. 5.15 and 5.16). In the first step, A and D are identified as the children of the only active node, S. The partial path distance of A is 3 and that of D is 4; A therefore, becomes the active node. Then B and D are generated from A with partial path distances of 7 and 8. Now the first encountered D, with a partial path distance of 4, becomes the active node, leading to the generation of partial paths to A and E. At this point, there are four partial paths, with the path S-D-E being the shortest.

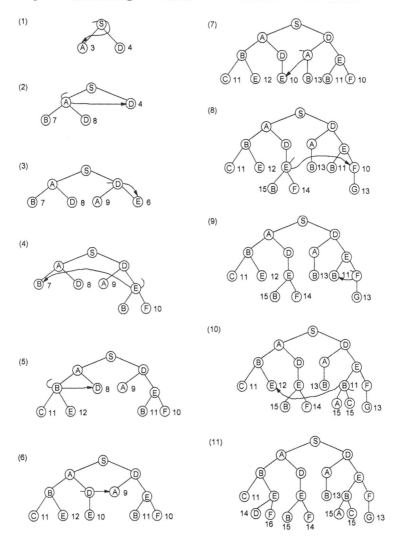

Fig. 5.22. Branch and Bound Search Applied to the Map Traversal Problem

After the seventh step, partial paths S-A-D-E and S-D-E-F are the shortest partial paths. Expanding S-A-D-E leads to partial paths terminating at B and F. Expanding S-D-E-F-, along the right side of the tree, leads to the complete path S-D-E-F-G, with total distance of 13. This is the shortest path, but to be absolutely sure, it is necessary that two partial paths be extended, S-A-B-E, with a partial path distance of 12, and S-D-E-B, with a partial path distance of 11. There is no need to extend the path S-D-A-B, since its partial path distance of 13 is equal to that of the complete path.

Methods of Improving Branch and Bound Search

In some cases, branch and bound search can be improved greatly by using guesses about distances remaining as well as guesses about facts as to the distances already accumulated. After all, if a guess about a distance remaining is good, then the distance guessed, added to the definitely known distance already traveled, should be a good estimate of the total path length, e (total path length):

$$e(\text{total path length}) = d(\text{already traveled}) + e(\text{distance remaining})$$

where d(already traveled) is the known distance already traveled and where e(distance remaining) is an estimate of the distance remaining.

Surely, it makes sense to work hardest on developing the path with the shortest estimated path length, until the estimate changes upward enough to make some other path be the one with the shortest estimated path length. After all, if the guesses were perfect, this approach would keep us on the optimal path at all times. In general, however, guesses are not perfect, and a bad overestimate, somewhere along the true optimal path, may cause us to wander off that optimal path permanently.

But note that *underestimates* cannot cause the right path to be overlooked. An underestimate of the distance remaining yields an underestimate of total path length, u(total path length):

$$u(\text{total path length}) = d(\text{already traveled}) + u(\text{distance remaining})$$

where d(already traveled) is the known distance already traveled and where u(distance remaining) is an *under*estimate of the distance remaining.

Now if a total path is found by extending the path with the smallest underestimate repeatedly, no further work needs be done once all incomplete path distance estimates are longer than some complete path distance. This is true because the real distance along a completed path cannot be shorter than an underestimate of the distance. If all estimates of the remaining distance can be guaranteed to be underestimates, there can be no bungle.

Another way to improve the basic branch and bound search is presented in Figure 5.23 [15]. The root node, S, has been expanded, producing A and D. For the moment we use no underestimates for the remaining path length. Since the path from S to A is shorter than that from S to D, A has also been expanded, leaving three paths: S-A-B, S-A-D-, and S-D. Thus, the path S-D will be the next path extended, since it is the partial path with the shortest length.

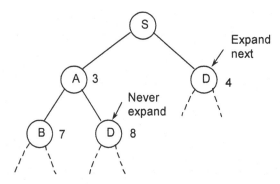

Fig. 5.23. The Dynamic Programming Principle

However, what about path S-A-D? Will it ever make sense to extend it? Clearly, it will not. Since there is one path to D with length 4, it does not make sense to work with another path to D with length 8. The path S-A-D should be forgotten forever as it cannot produce a winner. This illustrates a general truth. Assume that the path from a starting point, S, to an intermediate point, I, does not influence the choice of paths for traveling from I to a goal point, G. Then the minimum distance from S to G, through I, is the sum of the minimum distance from S to I and the minimum distance from I to G. Consequently, the *dynamic-programming principle* holds that when looking for the best path from S to G, all paths from S to any intermediate node, I, other than the minimum-length path from S to I, can be ignored.

A^*: Improved Branch-and-Bound Search

The A^* procedure is branch-and-bound search, with an estimate of the remaining distance, combined with the dynamic-programming principle. If the estimate of the remaining distance, is a lower bound on the actual distance, then A^* produces optimal solutions. Generally, the estimate may be assumed to be a lower bound estimate, unless specifically stated otherwise, implying that A^*'s solutions are normally optimal.

To perform A^* search:

1. Form a queue of partial paths. Let the initial queue consist of the zero-length, zero-step path from the root node to nowhere.
2. Until the queue is empty or the goal has been reached, determine if the first path in the queue reaches the goal node.
   2a. If the first path reaches the goal node, do nothing.
   2b. If the first path does not reach the goal node:
      2b1.   Remove the first path from the queue.
      2b2.   Form new paths from the removed path by extending one step.
      2b3.   Add the new paths to the queue.
      2b4.   Sort the queue by the *sum of* cost accumulated so far *and a lower-bound estimate of the cost remaining*, with least-cost paths in front.
      2b5.   *If two or more paths reach a common node, delete all those paths except for the one that reaches the common node with the minimum cost.*
3. If the goal node has been found, announce success; otherwise announce failure.

Selection of a Search Strategy for Finding an Optimal Path

There are many ways of searching for optimal paths, each with advantages among them, the following:

- Branch and bound search is good when the tree is big and bad paths turn distinctly bad quickly.
- Branch and bound search with a guess is good when there is a good lower-bound estimate of the distance remaining to the goal.
- Dynamic programming is good when many paths reach common nodes.
- The A^* procedure is good when both branch and bound search, with a guess and dynamic programming, are good.

Rule-Based Systems

Rule-based systems (also often referred to as *expert systems*) are built around *rules,* which consist of an *if* part and a *then* part [15] (i.e., if condition 1, condition 2, ... are true, then take action 1, action 2, ...)

A rule-based system consists of two major components: a *rule-base* and an *inference engine*. The rule-base is a collection of rules, which captures human expertise or reasoning in a particular problem domain. The inference engine is a piece of software, which invokes the rules in the rule-base to solve problems.

In most rule-bases, the rules are interrelated in that implementing an action in the *then* part of one rule, may cause a condition in the *if* part of another rule to become true. These interrelationships may be drawn [15] in the form of an inference net (Fig. 5.24). In an inference net, each rule is represented by a labeled semi-circle. A fact is represented by a rectangle; a rectangle which precedes a rule that represents a fact, required to establish the truth or falsehood of a condition in the rule's *if* part, while a rectangle which follows a rule represents a fact, which is established by performing the actions in the rule's *then* part. Open rectangles represent raw facts; solid rectangles represent deducible facts.

There are two ways in which the inference engine of a rule base can operate to solve problems. These are referred to as *forward chaining* and *backward chaining*. In *forward chaining*, the inference engine answers the question, "Which actions should be taken?" It works forward (left to right) through the inference net, taking the following approach:

1. Until a problem is solved or the no rule's *if* parts are satisfied by the current situation:
   1.1. Collect rules whose *if* parts are satisfied. If more than one rule's *if* parts are satisfied, use a conflict-resolution procedure to eliminate all but one.
   1.2. Perform the actions in the rule's *then* part.

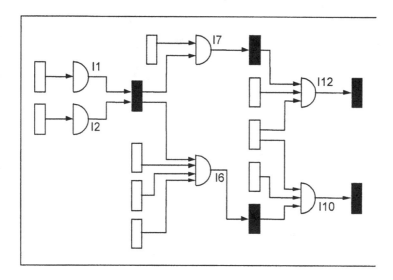

Fig. 5.24. An Example of an Inference Net

When all the conditions in a rule are satisfied by the current situation, the rule is said to *be triggered*. When the actions in the rule are performed, the rule is said to *be fired*. Triggering does not always mean firing, because the conditions of several rules may be satisfied simultaneously, triggering them all, making it necessary for a conflict-resolution procedure to decide which rule actually fires.

Here are some possibilities:

- *Specificity ordering.* Suppose the conditions of one triggering rule are a superset of the conditions of another triggering rule. Use the rule with the superset on the ground that is more specialized to the current situation.
- *Rule ordering.* Arrange all rules in one long priority list. The triggering rule appearing earliest in the list has the highest priority. The others are ignored.
- *Data ordering.* Arrange all possible aspects of the situation in one long priority list. The triggering rule, having the highest priority condition, has the highest priority.
- *Size ordering.* Assign the highest priority to the triggering rule with the toughest requirements, where toughest means the longest list of constraining conditions.

- *Recency ordering.* Consider the most recently used rule to have the highest priority, or consider the least recent to have the highest priority, at the designer's whim.
- *Context limiting.* Reduce the likelihood of conflict by separating the rules into groups, only some of which are active at any time. Have a procedure that activates and deactivates groups.

Of course, having a set of possibilities to choose among does not mean that a science of conflict resolution exists. Selection of a strategy is done ad hoc, for the most part.

In the *backward chaining* mode of operation, the inference engine answers the question, "Should a given action be taken?" It works backward (right to left) through the inference net, starting from a rule, whose *then* part includes the action in question. To determine whether the action should be taken, the inference engine attempts to determine if the rule should be fired. (The action should be taken if the rule should be fired.) If a fact necessary to make this determination can be established by the actions of another rule, the inference engine attempts to determine if *that* rule should be fired, and so forth. To obtain facts, which cannot be supplied by other rules, the inference engine prompts the user of the rule-based system. Once the facts are obtained, the inference engine either deduces whether the action should be taken, or informs the user of not having enough facts to make the deduction.

An inference engine can run forward or backward, but which is better? The question is decided for the purpose of reasoning and by the shape of the search space. Certainly, if the goal is to discover all that can be deduced from a given set of facts, then the inference engine should run forward. On the other hand, if the purpose is to verify or deny one particular conclusion, then the inference engine should run backward, because many conclusions, irrelevant to the target conclusion, can usually come out of an initial given set of facts. If these facts are fed to a forward-chaining inference engine, then much work may be wasted on developing a combinatorial nightmare.

Determining Answer Reliability via Certainty Factors

Rule-based systems, particularly those used for identification, as in the above example, usually work in domains where conclusions are rarely certain. Thus rule-based system developers often build some sort of certainty-computing procedure on top of the basic inference engine. Generally, certainty-computing procedures associate a number between 0 and 1 with each fact. This number, called a *certainty factor*, is intended to reflect how certain the fact is, with 0 indicating a fact that is definitely false and 1 indicating a fact that is definitely true.

Since the calculation of certainty factors is of practical importance, let us look at the existing procedures. Understand, however, that none of these procedures is completely satisfactory.

Note that any procedure for computing certainty factors must embody answers in three questions. First, how are the certainties associated with a rule's *if* part to be combined into the rule's overall *input certainty*? Second, how does the rule itself translate input certainty into *output certainty*? And third, how is a fact's certainty determined when the *then* parts of several rules argue for it, requiring the computation of a *multiply argued certainty*?

A simple, ad hoc procedure answers the question this way:

- The minimum certainty [15] associated with each rule's *if* parts becomes the certainty of the rule's overall input (Fig. 5.25(*a*)). This is analogous to the idea that a chain is only as strong as its weakest link.
- Each rule description includes the amounts to an attenuation factor, which, like the certainty factors, range from 0 to 1. To compute a rule's output certainty [15], the input certainty is multiplied by the attenuation factor (Fig. 5.25(*b*)).
- With several rules supporting some particular fact, the overall certainty [15] attributed to that fact is the maximum certainty, proposed by the supporting rules (Fig. 5.25(*c*)). Thus, the strength of a fact is affected only by the strongest of the supporting rules.

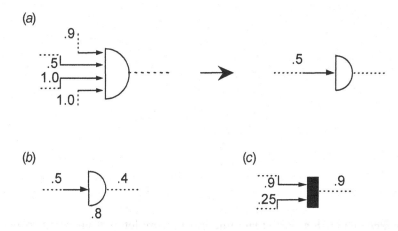

Fig. 5.25. An Ad Hoc Method of Calculating the Certainty of Deduced Facts

Another way to pass certainties through an inference net is based on an analogy between *certainties* and *probabilities*:

- The certainty of a rule's overall input is the product of the certainties associated with the rules *if* parts.
- The certainty of a rule's output is given by a single-valued function having input certainty on one axis and output certainty on the other.
- The certainty of a fact supported by several rules is determined by transforming certainties into related measurements called *certainty ratios*, then pumping the certainty ratios through a simple formula, and finally transforming the result back into certainty.

The simplest of these ideas is that of using the product of *if* part certainties to get the overall certainty of input. The idea is derived directly from the notion that the probability of a joint event is the product of the probabilities of the participating event, as long as those participating events have no influence on one another. In a coin toss, for example, the probability of turning up two heads in a row is the square of the probability of turning up one head on one toss. To summarize the computation is expressed in algebraic notation as a formula involving the certainties of the *if* parts, c_i.

$$c_1 \times \ldots \times c_n$$

A problem with this idea has to do with the prerequisite of independence among the contributing events. In the coin toss example, the events are said to be independent because they do not exert influence on one another. The *if* parts of a rule are often dependent events, and consequently, combination by multiplication is not an operation that computes a probability from probabilities. What we are computing is analogous to a probability, but the combination formula is not justified by the theory of probability.

Once an input certainty is obtained, the next step is to compute the output certainty from that input certainty. To do this, a human expert is asked to construct a function [15] relating input to output, such as the function shown in Figure 5.26(*a*). As drawn, the function indicates that the output certainty is .8 when the input certainty is absolutely certain; the output certainty is 0 when the input certainty is 0; and in between, the output certainty lies on a straight line between the points (0, 0) and (1.0, .8).

In fact, using this particular function is completely equivalent to using the attenuation-factor method with an attenuation equal to .8. Using the function [15] of Figure 5.26(*b*) is not equivalent to using an attenuation factor, though, because the line does not go through the origin. Evidently, the output probability goes to .2 when the input probability goes to 0. This means that the *if* parts need not be true in order for the *then* part to be true.

Unlike the examples [15] of Figure 5.26(*a*) and 5.26(*b*), most functions have two straight-line segments, rather than one. The reason is that the relation between input and output should reflect not only end-point considerations, but also before-analysis estimates. A before-analysis estimate, called an *a priori value*, is a statement as to what the certainty is in absence of any knowledge about the particular case in hand.

The use of a priori values [15] is illustrated in Figure 5.26(*c*) in which the input-output line jogs through the a priori certainty of the input, .5, and the a priori certainty of the output, also .5, coincidentally. Using a two-segment function to relate input to output certainties is a capitulation to necessity. Strictly speaking, according to the classical theory of probability, all the points evoked from a human should lie on a straight line. People evidently, are not always well-modeled by elegant mathematics.

Finally, to calculate multiply-argued certainties, certainty ratios are used. Certainty, c, and a certainty ratio, r, are related as follows:

$$r = \frac{c}{1-c} \qquad\qquad c = \frac{r}{r+1}$$

Note that a certainty of .5 corresponds to a neutral certainty ratio of 1.

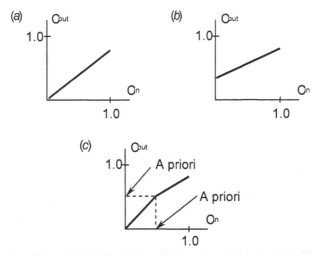

Fig. 5.26. Use of Probability-Based Certainty Factors

After certainties are transformed into certainty ratios, the certainty ratio of a multiply-argued conclusion is given by the following formula:

$$r_0 \times \frac{r_1}{r_0} \times \cdots \times \frac{r_n}{r_0}$$

where r_0 is the certainty ratio corresponding to the a priori certainty of the *if* part, and the r_i are the certainty ratios corresponding to the certainties read from input-output functions of the contributing rules. Note that the formula reduces to the product of certainty ratios in the special case, where the a priori certainty of the conclusion is .5, corresponding to an a priori certainty ratio of 1. This formula is an illegitimate adaptation of a legitimate probability formula, as is the formula for combining the *if* part certainties into input certainties.

Transforming certainties into certainty ratios to compute the certainty of multiply-argued conclusions, illustrates a powerful idea: if a problem is hard when expressed in one representation, try to transform the problem into another representation that makes the problem easy. Going to certainty ratios is like going to logarithms to do multiplication or to Fourier transforms to make convolution.

Although rule-based systems can answer to simple questions as to how they reach their conclusions and why they ask questions, they still lack many of the characteristics of human experts. Some of the distinguishing factors are:

- Basic rule-based systems do not learn.
- They do not look at problems from different perspectives.
- They do not know how and when to break their own rules.
- They do not have access to the reasoning behind their rules.

Neural Networks

Over the past thirty years, neural networks have been studied with the purpose of capturing in a "black box" the general relationship between variables, which are difficult or impossible to relate to analytically. A neural network is a powerful data modeling tool, capable of capturing and representing complex input/output relationships. The motivation for the development of neural network technology, stemmed from the desire to develop an artificial system that could perform "intelligent" tasks similar to those performed by the human brain. A neural network "learns" the relationship between variables of interest simply by means of exposure to examples of the relationship [16, 17, 18]. Typically, neural networks are simulated in software, although hardware implementations are starting to become more common.

There are two types of neural network applications (Fig. 5.27): *recognition* and *generalization* [19]. The procedure for using a neural network in each case consists of two phases: a *training* phase and a *use* phase. For both types of applications, the *training* phase consists of "training" a neural network by exposing it to a list of input-output pairs $\{I_1\text{-}O_1, I_2\text{-}O_2, ..., I_n\text{-}O_n\}$. In the *use* phase of *recognition* applications, the trained network is given an input I_k, $1 \leq k \leq n$ that has been corrupted by noise, and is expected to provide the corresponding output O_k, in spite of the noise. Handwritten character recognition is an example of a recognition application of neural networks. In the *use* phase of *generalization* applications, the trained network is given an entirely new input I_{n+1}, different from any of its training inputs $I_1, I_2, ..., I_n$, and is expected to predict the appropriate output O_{n+1} based on the internal model of the input-output relationship that it has developed via exposure to the training pairs $\{I_1\text{-}O_1, I_2\text{-}O_2, ..., I_n\text{-}O_n\}$. The prediction of a company's bond ratings, from the values of its key financial ratios, is an example of a generalization application of neural networks [19].

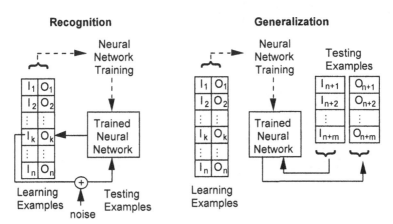

Fig. 5.27. The Two Types of Neural Network Applications: Recognition and Generalization

Structurally, a neural network consists of many nonlinear computational elements, called nodes. A node takes one or more input values, combines them into a single value and then transforms this value into an output value. The nodes are densely interconnected via directed links. A link takes the output value of a node, transforms it and then submits the resulting value as an input value to another node. Via links, the output of any node becomes the input to many other nodes. A neural network takes input numbers and maps them onto output numbers.

Existing research on the use of neural networks has favored a type of network structure called a multi-layer perceptron [19,20]. Multi-layer perceptrons (MLPs) are feedforward neural networks trained with the standard backpropagation algorithm. They are supervised networks so they require a desired response in order to be trained. They learn how to transform input data into a desired response; therefore, they are widely used for pattern classification. They can virtually approximate a wide range of input-output maps. They have been shown to approximate the performance of optimal statistical classifiers in difficult problems. In a multi-layer perceptron, the nodes are arranged in layers: an input layer, an output layer, and between the two, a number of so-called hidden layers. The input nodes receive the values of the input variables, which are then propagated concurrently through the network, layer by layer. At the output layer, the nodes output the values of the output variables. The number of layers and nodes can be selected arbitrarily. A multi-layer perceptron's structure may be specified using notation of the form $L_1-L_2-...-L_n$, where L_1 denotes the number of nodes in the input layer, L_n denotes the number of nodes in the output layer, and $L_2, ..., L_{n-1}$ denote the number of nodes in each of the hidden layers (Fig. 5.28).

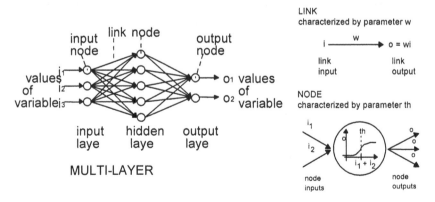

Fig. 5.28. A 3-5-2 Multi-layer Perceptron and Associated Terminology

Nodes and links can be thought of as value transformation functions. The transformation behavior of a node is controlled by a parameter *th*. Likewise, the transformation behavior of a link is controlled by a parameter *w*. The mapping from input numbers to output numbers is thus fixed by the settings of the *th* and *w* parameters. Training is simply the procedure of adjusting the settings of these parameters until the network maps all example input values onto the corresponding example output values (within a certain tolerance). Adjustments are performed after the presentation to the network of every *n*; usually, *n* = 1. Once the total number of examples is exhausted, the examples are presented again, starting from the first one. Training is thus, an iterative presentation-adjustment-presentation-adjustment process. The most established training algorithm, backpropagation, is based on hill climbing [21]. Backpropagation is the basis for training a supervised neural network. At the core of all backpropagation methods, is an application of the chain rule for ordered partial derivatives to calculate the sensitivity that a cost function has, with respect to the internal states and weights of a network. In other words, the term backpropagation is used to imply a backward pass of error to each internal node within the network, which is then used to calculate weight gradients for that node. Learning progresses by alternately propagating forward the activations and propagating backward the instantaneous errors.

During the last fifteen years a number of commercially or freely available software tools have been developed to support the application of neural networks in a broad spectrum of data-intensive applications, such as process modeling and control, machine diagnostics, portfolio management, quality control, voice recognition, financial forecasting and others.

5.2.3 Computer Simulation

Computer simulation is the generic name for a class of computer software, which simulates the operation of a manufacturing system. Conceptually, the inputs of a computer simulator are decision variables, which specify the design (e.g., machine processing and failure rates, machine layout), the workload (e.g., arrivals of raw materials over time, part routings), and the operational policy (e.g., "first come, first served") of a manufacturing system. The simulator assembles these data into a *model* of the manufacturing system, including the rules as to how the components of the system interact with each other. The user of the simulator specifies the initial *state* of the manufacturing system (e.g., the number and types of parts initially in inventory at various points in the system). Starting from this initial state, the simulator follows the operation of the model over time, tracking *events* such as parts movement, machine breakdowns, machine setups. etc. over time. At the conclusion of the simulation, the output provided by the simulator is a set of statistical performance measures (e.g. the average number of parts in the system over time) by which the manufacturing system may be evaluated (Fig. 5.29).

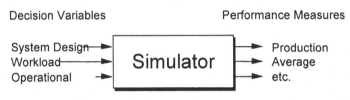

Fig. 5.29. Computer Simulation

Simulation is an analysis tool because it determines only the performance of a *given* manufacturing system design. When used for the design of manufacturing systems, simulation must interface with an external element, capable of prescribing or synthesizing new designs. Often, the external element is a person who creates a number of intuitively feasible alternative designs, and then evaluates them with a simulator. Or it may be historical (or reference data) provided by other software tools [22] On the basis of the simulation outputs, either the best alternative design is chosen, or new and possibly improved alternative designs are proposed. Simulation is thus, often used in the trial and error design approach, described in Section 5.1.3.

Mechanics of Simulation

Most simulation software programs model a manufacturing system as it evolves over time by a representation in which the variables that track the system's state (the state variables) change instantaneously at separate points in time [23]. These points in time, are the ones at which an event occurs, where an event is defined as an instantaneous occurrence that may change the state of the system. A model of this type is called a *discrete event simulation model.*

Because of the dynamic nature of discrete event simulation models, the current value of simulated time must be tracked as the simulation proceeds, and a mechanism to advance simulated time from one value to another is needed. The variable, in a simulation model that gives the current value of simulated time, is called the *simulation clock.* As for the simulation time advance mechanism, the most widely used approach is called the *event-driven* approach.

With the event-driven approach, the simulation clock is initialized to zero and the times of occurrence of future events are determined. The simulation clock is then advanced to the time of occurrence of the *most imminent* (first) of these future events, at which point the state of the system is updated to account for the fact that an event has occurred, and knowledge of the times of occurrence of future events is also updated. Then the simulation clock is advanced to the time of the (new) most imminent event, the state of the system is updated, and future event times are determined, etc. This process of advancing the simulation clock from one event time to another is continued until eventually some prespecified stopping condition is satisfied. Since all state changes occur only at event times for a discrete event simulation model, periods of inactivity are skipped over by jumping the clock from event time to event time. Successive jumps of the simulation clock are generally unequal in size.

All discrete event-driven simulation models share the following components [23]:

- *System state:* The collection of state variables necessary to describe the system at a particular time.
- *System clock:* A variable giving the current value of simulated time.
- *Event list:* A list containing the next time when each type of event will occur.
- *Statistical counters:* Variables used for storing statistical information about system performance.
- *Initialization routine:* A subprogram to initialize the simulation model at time zero.
- *Timing routine:* A subprogram that determines the next event from the event list and then advances the simulation clock to the time when that event is to occur.
- *Event routine:* A subprogram that updates the system state when a particular type of event occurs (there is one event routine for each event type).
- *Library routines:* A set of subprograms used to generate samples from probability distributions that were determined as part of the simulation model.
- *Report generator:* A subprogram that computes estimates (from the statistical counters) of the desired measures of performance and produces a report when the simulation ends.
- *Main program:* A subprogram that invokes the timing routine to determine the next event and then transfers control to the corresponding

event routine to update the system state appropriately. The main program may also check for termination and invoke the report generator when the simulation is over.

The logical relationships (flow of control) among these components are as follows. The simulation begins at time 0 with the main program invoking the initialization routine, where the simulation clock is set to zero, the system state and the statistical counters are initialized, and the event list is initialized. After control has been returned to the main program, it invokes the timing routine to determine which type of event is most imminent. If an event of type i is the next to occur, the simulation clock is advanced to the time that event type i will next occur and control is returned to the main program. Then the main program invokes event routine i, where typically three types of activities occur: (1) the system state is updated to account for the fact that an event of type i has occurred; (2) information about system performance is gathered by updating the statistical counters; and (3) the times of occurrence of future events are generated and this information is added to the event list. It is often necessary to generate random observations from probability distributions in order to determine these future event times; such a generated observation is called a *random variate*. After all processing has been completed, either in event routine i or in the main program, a check is typically made to determine (relative to some stopping condition) whether the simulation should be terminated at that point. If it is time to terminate the simulation, the report generator is invoked from the main program to compute estimates (from the statistical counters) of the desired measures of performance and to produce a report. If it is not time for termination, control is passed back to the main program and the main program–timing routine–main program–event routine–termination check cycle is repeated [23] until the stopping condition is eventually satisfied (Fig. 5.30).

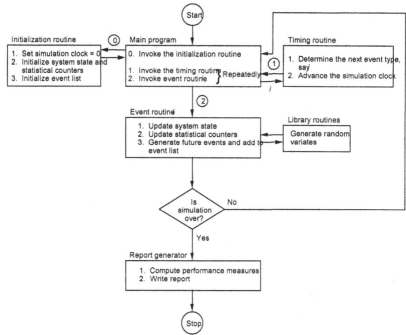

Fig. 5.30. Flow of Control in a Discrete Event-Driven Simulation Program

Features of Simulation Software

Since manufacturing systems are complex and differ widely, an important feature of simulation software is *modeling capability*. Provided a simulator lacks in ability to model a specific aspect of a manufacturing system (e.g., a particular operational policy); that aspect must be approximated with a construct, supported by the simulator. This leads to questionable precision in the final model. It is therefore desirable for a simulator to have elements incorporating customizable attributes (e.g., part number, due date, etc.) which can be accessed and changed as required.

Beyond modeling capability, *modeling ease* becomes the major issue. Simulation models are usually constructed via specialized simulation languages, which resemble high-level programming languages, such as FORTRAN. The major advantage of a good simulation language is modeling flexibility, whereas the major disadvantage is that programming expertise is required [24]. Consequently, they are difficult to code and debug, and may be difficult to comprehend. Alternatively, *icon-based simulation software* can be used to build simulation models. Using this software, a model is constructed by defining the attributes of graphical icons, which represent parts, machines, and other equipment. The graphical displays greatly simplify model coding and debugging. Many such systems even have animation capability: every time the state of the simulation changes, a corresponding change appears in the graphical representation. Animation has become a widely accepted part of the simulation of manufacturing systems because it is well suited for communicating the dynamic behavior of a simulation model. This, in turn, significantly increases the model's credibility. Besides being a useful communications device, animation can be useful for debugging, validating and improving models. It can also be used to train manufacturing personnel in the operation of a system. The major disadvantage of simulators is that they are not as flexible as simulation languages are, since they do not allow full programming. The vendors of the major manufacturing-oriented simulators have introduced programming into their simulation software either by the use of "programming-like" constructs (e.g., setting values for attributes or global variables, if-then else logic, etc.) or by calling external routines, written in a general-purpose programming language. Animations of three simulations of manufacturing systems are illustrated in Figure 5.31.

Manufacturing processes exhibit both deterministic flows and statistical fluctuation. Sources of randomness include processing times, machine operating times, machine repair times, and so on, and they require the estimation of the probability distributions of the variables, and not just their means. Therefore, *statistical capabilities* represent an important area of functionality for simulation software. These capabilities can be divided into a number of major categories:

- *Distributions*. Simulation software should contain a wide variety of standard distributions (including exponential, gamma and triangular) to facilitate the modeling of distributions, based on observed shop-floor data.
- *Random-number generators*. Simulators should be equipped with multiple stream random-number generators.
- *Independent replications*. When random factors are implemented into models, results are also somewhat random. To check the range of possible results, a number of runs through the same simulation starting from the same point but with different streams of random numbers are required. To accommodate this, simulators should have commands, which automatically repeat the simulation and track the statistics separately for each run.
- *Warm-up period and confidence interval*. To determine the statistical accuracy of simulation results, a user should be able to specify a warm-

up period at the end of which output statistics are reset to zero and confidence intervals for desired performance measures, such as mean daily throughput, are constructed.

The features discussed so far have been related to the ability of a simulator to faithfully reproduce the behavior of a manufacturing system. However, even if a simulator is successful in this aspect, it will not be a valuable tool, unless it can effectively document the performance of the modeled system and it can provide a range of standard file format inputs/outputs. For this reason, *standard reports* for commonly tracked performance statistics, such as utilization, queue size and throughput should be readily available. *Custom reports,* which are specific to the particular simulation, should be easy to construct. These include reports that track non-standard groups of statistics and which require special formats (such as those prepared for management use) while *graphical information*, including, for example, histograms and time plots of important variables, can effectively summarize large quantities of data for easier comprehension. Reports in Hypertext Markup Language (HTML)[25], for reporting in web, is usually available, as well. Apart from enhanced reporting capabilities, a simulation software should be easily integrated with IT tools available in a company. For this reason, they are equipped with interfaces to exchange data with different types of applications, such as CAD tools, PDM, office tools, either in specific data format or using standards, such as eXtended Markup Language (XML)[25]

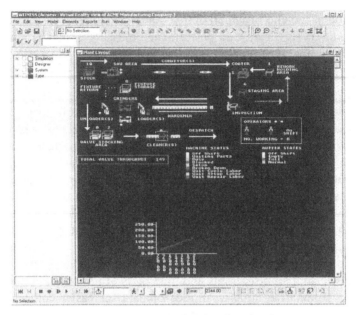

Fig. 5.31. An Example of Animation [23

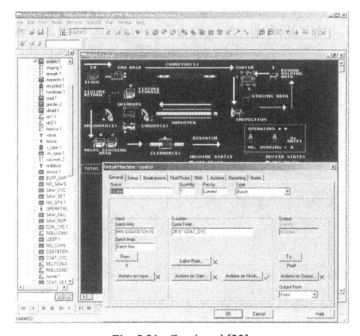

Fig. 5.31. Continued [23]

Design of Simulation Experiments

Simulation can be used to map a set of decision variables onto a set of performance measures. However, the design process is the *inverse* of the simulation process: it maps desired performance measures onto decision variables. That is, given the performance requirements of the system, the task is to find a system design (values of the system's decision variables) that will achieve the required performance.

The optimum decision variables are usually deduced after running multiple simulations. If this search for optimum decision variables is not properly organized, it can become extremely tedious, resembling the proverbial search for the needle in the haystack.

Use of statistically designed experiments (SDE) can minimize simulation effort. SDE can be defined as the process of formulating a plan to gather the desired information at minimal cost, enabling the modeler to draw valid inferences.

The SDE method presupposes that each of the decision variables has several discrete, legal states. If a decision variable is continuous, it must be discretized. Given a problem with n decision variables, each with m discrete, legal states, the number of potential manufacturing system designs is n^m. Conceptually, it is a simple matter to simulate all n^m design possibilities and then select the design with the best simulation performance measures. Practically, however, n^m is usually a very large number (e.g., $20^5 = 3.2$ million); it is not computationally feasible to perform all n^m simulations. The ultimate aim of SDE is to be able to deduce the best of the n^m designs by simulating only a very small fraction of those designs.

The first and most crucial step in the statistical design approach is the formation of an orthogonal array. An orthogonal array is a matrix in which the rows represent simulation experiments, the columns represent decision variables, and the elements of the matrix contain the states of the decision variables. The rows of the matrix thus, prescribe the decision variable states to be used in each simulation experiment. Orthogonality implies that if the states of each decision variable are numbered consecutively from 1, each state number will occur the same number of times in each column, and each combination of two state numbers (e.g., 1-1, 1-2, 2-1) will occur an equal number of times in any pair of columns. (In the SDE literature, decision variables and their state numbers are referred to as *factors* and *levels*, respectively.) The property of orthogonality provides 'balanced' experiments in which equal exposure is given to the effect of each decision variable on the simulation performance measures.

In the second step, simulation runs are performed using the decision variables as specified by the rows. For each simulation run (for each row of the orthogonal array), an objective function which measures the 'goodness' of the simulation performance measures, is calculated.

Next, an array of the average objective function values for each factor at each level is formulated. The optimum state of each decision variable is deduced from the array by choosing the level of each factor that has the best mean objective function value.

An analysis of variance (ANOVA) can then be used on the orthogonal array to determine which of the decision variables affect the value of the objective function the most [26].

As a concrete example of how SDE works, consider a design problem containing three factors, each with three levels. The implementation of SDE would occur as follows.

Step 1. Formulate an orthogonal array.

Expt. No.	Factor 1	Factor 2	Factor 3	Obj. Func.
1	1	1	1	R1
2	2	2	2	R2
3	3	3	3	R3
4	1	3	2	R4
5	2	1	3	R5
6	3	2	1	R6
7	1	2	3	R7
8	2	3	1	R8
9	3	1	2	R9

Step 2. Formulate the array of the means objective function values.

	Factor 1	Factor 2	Factor 3
Level 1	$\frac{1}{3}(R_1 + R_4 + R_7)$	$\frac{1}{3}(R_1 + R_5 + R_9)$	$\frac{1}{3}(R_1 + R_6 + R_8)$
Level 2	$\frac{1}{3}(R_2 + R_5 + R_8)$	$\frac{1}{3}(R_2 + R_6 + R_7)$	$\frac{1}{3}(R_2 + R_4 + R_9)$
Level 3	$\frac{1}{3}(R_3 + R_6 + R_9)$	$\frac{1}{3}(R_3 + R_4 + R_8)$	$\frac{1}{3}(R_3 + R_5 + R_7)$

Step 3. Choose the level of each factor with the best mean. 'Best' means 'maximum' if the objective function is to be maximized; it means 'minimum' if the objective function is to be minimized.

General Guidelines for Applying Simulation

The design of simulation experiments is only one portion of the overall simulation effort. The following general steps should be included in a sound simulation study [27]. Although these steps reveal no new simulation concepts, they are important for successfully applying simulation to practical, industrial settings.

1. *Formulate the problem and plan the study.* State study objectives clearly. Delineate the system design to be studied if possible. Specify the criteria for comparing alternative system designs. Plan the study in terms of the number of people, the cost and the time required for each aspect of the study.
2. *Collect data and define a model.* Data should be collected on the system of interest to specify input parameters and probability distributions (e.g., machine repair time distribution). Accurate data is needed for valid results. Data includes processing times, travel and conveyance times, time to failure of various machines, repair times, etc. Not all data may be readily available. Some potential sources of data (in decreasing order of likely accuracy) include time studies, historical records, vendors' claims, the client's best guess or the modeler's best guess.
3. *Model statistics* of system randomness, such as machine breakdowns.
4. *Ensure validity.* Involve people who are intimately familiar with the operations of the system (machine operators, industrial engineers, etc.) in the model building process.
5. *Construct and verify the computer model.* Decide whether to use a simulation language or a manufacturing simulator. The modeler can make a complex model more accessible to the users by adding user interface features, which allow manufacturing personnel to make certain modifications to the model without programming.
6. *Make pilot runs and check validity.* This can be done to test sensitivity of the model's output to small changes in an input parameter, for instance. For existing systems, output data can be compared with existing data of the actual system.
7. *Design experiments.* Specify the system designs to be simulated, the number of independent simulation runs for each alternative, the length

of each run and the initial conditions for each simulation run (e.g., initial state of each machine and worker).

8. *Perform the runs* specified in Step 7. Note that multiple runs are preferred to a single run in order to get a sense of the distribution of variables under examination. It is typical for each run to begin with a "warm-up period" in order to allow the system to approach a steady state before output data is collected for measurement purposes [28].

9. *Analyze output data.* Estimate measures of performance for a particular system design, then determine the best system design relative to some measure(s) of performance that are of concern.

10. *Document and implement results.* Document the model's assumptions as well as the model code itself. Implement the results of the simulation study.

Virtual Reality

The rapid progress in computer graphics and digital simulation, in the past two decades, introduced Virtual Reality (VR) technology in many engineering applications. Virtual Reality technology is often defined as "the use of real-time digital computers and other special hardware and software to generate the simulation of an alternate world or environment, believable as real or true by the users" [29]. There are four basic attributes that a VR system utilizes in order to enhance the human-computer interface. These are immersion, presence, navigation and interaction [29]. Immersion refers to the VR user's feeling that the virtual world is real. Presence requires a self-representation of the human body within the virtual environment. Navigation is the ability to move within a virtual environment, whilst interaction denotes the ability to select and move objects.

VR, with its aforementioned capabilities, provides designers and engineers with the means of evaluating products and processes at the early stages of the design phase. Planning for manufacturing using VR, involves virtual manufacturing process plan generation, virtual assembly planning, virtual factory planning etc. Advanced modeling techniques including behavior modeling, motion modeling, machines' operation modeling, are required to be employed in order to build a VR manufacturing environment. Moreover, Computer Aided-Design (CAD) and Product Lifecycle Management (PLM) software tools can be incorporated with virtual applications and help manufacturers accelerate the design and build cycles of new products while lowering production costs and plant deficiencies. With digital manufacturing solutions, manufacturers in aerospace and automotive industry have implemented virtual factory systems that allow greater re-use of engineering data, better control of late engineering changes in the design cycle, and more sophisticated simulations of NC machining processes and factory-floor layouts [30, 31]. Figure 5.32 shows a factory planning simulation that enables visualizing plant-floor layouts, prior to part production.

Fig. 5.32. Factory-floor layout with Tecnomatix' eMPower [32]

In manufacturing related work, human operators are very often involved, with the flexibility that a human brings with it, along with all the difficulties in modeling their behavior. In this case, the interaction among humans, machines and products in all the phases of the product life cycle, must be studied. Moreover, ergonomic considerations demand empirical data on human capabilities and should be examined at the early design stages. There are two ways of simulating human interaction within virtual environments. The first one is by immersing a real person into the virtual environment, which interacts directly with the elements of a simulated virtual world. In Figure 5.33, a fixing operation during an immersive and interactive assembly simulation is shown. To perform immersive execution in a VR environment, special peripheral devices, such as data gloves and head mounted displays, are required together with motion tracking systems.

The second way is to replace the human by a computational human model, also called "virtual human" or "mannequin", and use it as a subject in simulated virtual environment tests. Figure 5.34 shows an example of an ergonomic simulation, using a mannequin from a commercially available software package. Such an application allows the study of unsafe situations in a plant.

Fig. 5.33. Immersive assembly [33]

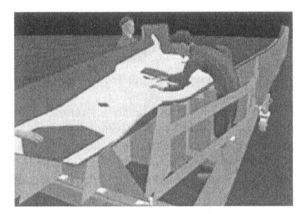

Fig. 5.34. Ergonomic simulation on the plant floor [34]

VR can be supplementary to traditional simulation, which fails to pro-vide a virtual environment computer-generated in which the operator can be immersed [35]. VR technologies present a new paradigm that affects each of the product life-cycle stages, and integrates various life-cycle con-cerns via advanced information technologies. This new paradigm allows interactivity, tele-presence, collaboration and augmented capabilities of rapid creation and analysis of virtual product models during the design phase. Manufacturing planning is made easier, through the integration of manufacturing knowledge-based and assembly knowledge that bases up front into the CAD models [29].

5.3 Applications

The academic literature on manufacturing system design concentrates on the solution of simplified, individual portions of the overall manufacturing system design problems: the *resource requirements problem*, the *resource layout problem*, the *material flow problem*, and the *buffer capacity problem* (Fig. 5.10). The intent of formulating and solving these subproblems, is not to arrive at a detailed engineering plan for constructing a factory, but rather to capture the essential difficulty of the aspect of a manufacturing system design. Contrary to reality, the subproblems are treated as if the solution of one was unaffected by the solution of the others. For most of the part, methods and tools from operations research (OR) and artificial intelligence (AI) are applied to the solution of these subproblems; their limited scope and complexity and well-defined measures of solution 'goodness' make them amenable to the decision making and optimization tools of the OR and AI fields.

In industrial applications, problem size and complexity, and even the difficulty of defining suitable measures of design 'goodness', make the use of decision making and optimization tools difficult. Here, the most frequently used tool is that of simulation. It is applied as an analysis tool to evaluating designs, generated by human designers. The goal here is merely to avoid egregious failure by finding an acceptable design solution; typically, it is not possible to guarantee that the optimum design will be achieved, nor is it possible to determine how close to the optimum a particular design lies. Some efforts have been made, however, to incorporate optimization tools, such as search into simulation applications.

In this section we will be reviewing the application of OR, AI, and simulation methods and tools to both the simplified design subproblems and to more complex problem formulations that more closely reflect industrial need.

5.3.1 The Resource Requirements Problem

The resource requirements problem can be defined as the determination of the number of each type of production resource, in a manufacturing facility, during some planning horizon. A resource can be anything, regarded as an individual production unit on the factory floor, such as a machine, an operator, a machining center with associated material hand-ling equipment, or an automated guided vehicle (AGV).

The problem is difficult because its solution depends on a number of inter-related factors:

- The *facility layout* affects the number of resources needed because it determines which resources are accessible from each point in the manufacturing system. Lack of accessibility increases the number of resources required.
- The *process plans* and the *required volumes* of the parts to be manufactured in the system must be accounted for, since together, they dictate the amount of demand that will arise for each type of resource.
- The *operational policy* affects the solution because it determines how efficiently each resource of a manufacturing system will be loaded over time.
- *Constraints*, such as the equipment acquisition budget, floor space, and number of workers, are interrelated. Thus, decisions regarding one type of resource may violate constraints imposed by other type of resources.

Analytical Approach to the Resource Requirements Problem

The first scientific approaches to the resource requirements problem have employed *descriptive* analytical models. A descriptive model, in the context of the resource requirements problem, is an equation, which expresses the required number of resources as a function of production factors, such as the required production rate, the part scrap rate, and the breakdown frequency and duration of the resources. The resulting number is usually a fractional number, which must be rounded to a neighboring integer value, on the basis of intuitive considerations [36]. As an example of a descriptive analytical model, the following equation has been proposed for determining the number of resources for a single work center. Here, a work center is defined as a group of a particular type of resource, or of the same manual processing operations [37]:

$$n_r = \frac{(st \times n)}{(60 \times h \times sf)}$$

n_r ≡ number of machines

n ≡ total required production in units per day

st ≡ standard time required to process one unit on a machine

sf ≡ scrap factor (the number of good pieces/number of scrapped pieces)

h ≡ standard numbers of hours available per day per machine

This is a single-period model, which applies only to a single work center, single product and single operation facility. As with other analytical models, although it is very easy to use, it solves a very limited kind of problems. It neglects the dynamic nature of production requirements over a planning horizon and the probabilistic nature of breakdowns and scrap parts. More importantly, it does not consider any of the interactions with the layout or the scheduling methods of the facility.

Extensions of the analytical approach have been made to accommodate the single product, multiple operation case [38, 39]. Other extensions have been accounted for the uncertainty of the problem parameters in actual manufacturing facilities by modeling the parameters as random variables. Both single operation [40] and multiple operation [41] cases have been addressed in this context.

Mathematical Programming Approach to the Resource Requirements Problem

A mathematical programming formulation of the resource requirements problem has the advantage of limiting on various design and operating quantities that are shared between resources and can be explicitly modeled as constraints in the mathematical program. Examples of such shared quantities include the budget, floor space, and overtime hours. Each resource acquired "consumes" a certain amount of each quantity; in this way, interactions between the quantities of different types of resources can be captured.

One of the more comprehensive mathematical programming formulations of the resource requirements problem [42] treats the case in which "resources" are machines. It will be described below in detail, as an example of how the resource requirements problem can be treated via mathematical programming.

The problem is to determine the number of machines to have in each of N work centers in each of T time periods. The manufacturing system is assumed to be a flow line, which produces multiple products. Each of the N work centers in the flow line contains only one type of machine, but different work centers contain different machine types. The production characteristics of a work center (e.g., the production rate per machine and the scrap factor) vary from one period to the next, as does demand for the finished products. The objective is to find the number of machines in each work center, at the least cost, in each time period. Machines in a work center can either be purchased at the beginning of a time period, or be left over from the previous time period. It is assumed that the relevant costs are machine investment costs, overtime operating expenses, undertime opportunities costs, and machine disposal costs. The present values of these costs are used for all calculations.

Let x_{it} be the number of machines on hand, in work center i, at the beginning of (and during) period t. The objective function, to be minimized, is the sum of a number of costs:

- Investment Cost $= \sum_{t=1}^{T} (P/F, I, t-1) \sum_{i=1}^{N} F_{it} \left[x_{it} - x_{i,t-1} \right]^{+}$

    where:

    $(P|F,I,t)$ $\equiv$ Present worth interest factor for discrete compounding when the discount rate is $I\%$ per period and the discount interval is t periods

    F_{it} $\equiv$ The first cost of a machine in work center i when it is purchased at the beginning of period t

    $|x_{it}-x_{i,t-1}|^{+}$ $\equiv$ The number of additional type i machines procured at the beginning of period t

    $= x_{it}-x_{i,t-1}$ if $x_{it} \geq x_{i,t-1}$, 0 otherwise

- Machine Disposal Cost $= \sum_{t=1}^{T} (P/F, I, t-1) \sum_{i=1}^{N} S_{it} \left| x_{i,t-1} - x_{it} \right|^{+}$

    where:

    S_{it} $\equiv$ Average cash inflow resulting from the disposal of one machine from work center i at the beginning of period t

    $|x_{i,t-1}-x_{it}|^{+}$ $\equiv$ The number of type i machines, which are disposed of at the beginning of period t

$$= x_{i,t-1} - x_{it} \text{ if } x_{i,t-1} \ge x_{it}, 0 \text{ otherwise}$$

- Operating and Maintenance Costs $= \sum_{t=1}^{T} (P/F, I, t) \sum_{i=1}^{N} m_{it} x_{it}$

where:

m_{it} $\equiv$ The cost of operating and maintaining a machine of type i during period t

The remaining two costs are functions of the deviation of the number of machines on hand, in a period, x_{it}, from the number M_{it} needed to meet production requirements, scheduled for that period. In order to calculate M_{it}, we note that if each machine at work center i has a production rate of r_{ij} units/period for product type j ($j = 1, 2, \ldots, J$) in the absence of disruptions, and a scrap loss factor e_{ij} for product type j, then the number of machines needed to exactly satisfy production requirements of P_{ijt} units ($j = 1, 2, \ldots, J$) is

$$M_{it} = \sum_{j=1}^{J} \frac{P_{ijt}}{(1 - e_{ij}) \, r_{ij}}$$

Since the system is a flow line, the output from work center i is the input to work center $i+1$. Final product requirements can thus, be designated P_{Njt} ($j = 1, 2, \ldots, J$). To ensure that these are met, the output requirements for work center N-1 must be

$$P_{N-1, jt} = \frac{P_{Njt}}{(1 - e_{Nj})}$$

In a similar manner, the output requirement for work center N-2, $P_{N-2,jt}$, can be calculated from $P_{N-1,jt}$, and so forth, so that the number of machines, needed at work center i during the time period t, should be

$$M_{it} = \sum_{j=1}^{J} \frac{P_{Njt}}{\prod_{k=1}^{N-i}\left(1 - e_{i+kj}\right)\left(1 - e_{ij}\right)r_{ij}}$$

When $x_{it} > M_{it}$, excess production capacity is present, resulting in an opportunity cost for money tied up in idle workers and equipment. This cost is referred to as *undertime cost*.

- Undertime Cost $= \sum_{t=1}^{T}\left(P/F,I,t\right)\sum_{i=1}^{N}C_{it}^{+}\left|x_{it} - M_{it}\right|^{+}$

    Where:

    C_{it}^{+} $\equiv$ The cost of an excess machine of type i carried during period t

    $|x_{it}-M_{it}|^{+}$ $=$ $x_{it}-M_{it}$ if $x_{it} \geq M_{it}$, 0 otherwise

When $x_{it} < M_{it}$, there is not enough production capacity to meet the final product requirements (P_{Njt}, for all j), and additional production must be obtained through overtime. (The option of subcontracting is not considered in this model.)

- Overtime Cost $= \sum_{t=1}^{T}\left(P/F,I,t\right)\sum_{i=1}^{N}C_{it}^{-}\left|M_{it} - x_{it}\right|^{+}$

    Where:

    C_{it}^{-} $\equiv$ The overtime labor cost of a shortage of one machine of type i during period t

    $|M_{it}-x_{it}|^{+}$ $\equiv$ $M_{it}-x_{it}$ if $M_{it} \geq x_{it}$, and 0 otherwise

The main constraints of the problem are the following:

- *Budget constraint.* The machine investment cost in period t must not exceed the sum of the budget for period t plus the cash inflow from the disposal of machines at the beginning of period t.

$$\sum_{i=1}^{N} \left(F_{it} \left| x_{it} - x_{i,t-1} \right|^+ - S_{it} \left| x_{i,t-1} - x_{it} \right|^+ \right) \le B_t, \text{ For all t}$$

where:

B_t ≡ The budget for period t

- *Floor space constraint.* The machines planned for period t must fit within the floor space available in that period.

$$\sum_{i=1}^{N} a_i x_{it} \le L_t, \text{ for all t}$$

where:

a_i ≡ Floor area of machine type i

L_t ≡ Total floor area available in period t

- *Overtime constraint.* The maximum overtime hours for each work center in each period must not be exceeded.

$$b_i \left| M_{it} - x_{it} \right|^+ \le U_{it}, \text{ for all } i, t$$

where:

b_i ≡ Overtime hours per machine per period for machine type i

U_{it} ≡ Maximum overtime hours for work center i in period t

- *Solution feasibility constraints.* The number of machines, planned in each period, must be a non-negative integer. In addition, the number of machines disposed at the beginning of each period must not be greater than the number of machines on hand during the previous period.

$$x_{it} \ge 0 \text{ and integer, for all } i, t$$

$$\left| x_{i,t-1} - x_{it} \right|^+ \le x_{i,t-1} \text{ for all } i, t$$

As posed, the mathematical model is non-linear because of the many absolute value operators. However, it can be transformed into a linear model with the aid of a few definitions:

Let:

- $y_{it}^+ = [x_{it} - x_{i,t-1}]^+ = (x_{it} - x_{i,t-1})$ if $x_{it} \geq x_{i,t-1}, 0$ otherwise
- $y_{it}^- = |x_{i,t-1} - x_{it}|^+ = (x_{i,t-1} - x_{it})$ if $x_{i,t-1} \geq x_{it}, 0$ otherwise
- $\mu_{it}^+ = |x_{it} - M_{it}|^+ = (x_{it} - M_{it})$ if $x_{it} \geq M_{it}, 0$ otherwise
- $\mu_{it}^- = |M_{iT} - x_{it}|^+ = (M_{it} - x_{it})$ if $M_{it} x \geq x_{it}, 0$ otherwise

such that

- $y_{it}^+ \cdot y_{it}^- = 0, \; \mu_{it}^+ \cdot \mu_{it}^- = 0$
- $x_{it} = y_{it}^+ - y_{it}^- + x_{i,t-1}$, or

$$x_{it} = x_{io} + \sum_{k=1}^{t} \left(y_{ik}^+ - y_{ik}^- \right)$$

and

- $x_{it} = \mu_{it}^+ - \mu_{it}^- + M_{it}$

The complete model is a mixed-integer, nonlinear programming problem, and can be expressed as follows:

$$\text{Minimize } z = \sum_{t=1}^{T} \sum_{i=1}^{N} \begin{bmatrix} (P/F,I,t-1)\left(F_{it} y_{it}^+ - S_{it} y_{it}^-\right) \\ + (P/F,I,t)\left(C_{it}^+ \mu_{it}^+ - C_{it}^- \mu_{it}^-\right) \\ + (P/F,I,t)(m_{it})\left(\mu_{it}^+ - \mu_{it}^- + M_{it}\right) \end{bmatrix} \quad (5\text{-}16a)$$

Subject to:

$$\sum_{i=1}^{N} \left(F_{it} y_{it}^+ - S_{it} y_{it}^-\right) \leq B_t, \text{ for all } t \quad (5\text{-}16b)$$

$$\sum_{i=1}^{N} a_i \left(\mu_{it}^+ - \mu_{it}^- + M_{it}\right) \leq L_t, \text{ for all } t \quad (5\text{-}16c)$$

$$b_i \mu_{it}^- \leq U_{it}, \text{ for all } t \quad (5\text{-}16d)$$

$$y_{it}^- - \sum_{k=1}^{t-1}\left(y_{ik}^+ - y_{ik}^-\right) \le x_{io}, \text{ for all } i, t > 1 \tag{5-16e}$$

and

$$y_{i1}^- \le x_{i0}$$

$$\sum_{k=1}^{t}\left(y_{ik}^+ - y_{ik}^-\right) - \mu_{it}^+ + \mu_{it}^- = M_{it} - x_{io}, \text{for all } i, t \tag{5-16f}$$

$$y_{it}^+ \cdot y_{it}^- = 0, \mu_{it}^+ \cdot \mu_{it}^- = 0, \text{ for all } i, t \tag{5-16g}$$

$$y_{it}^+, y_{it}^- \ge 0 \text{ and integer; } \mu_{it}^+, \mu_{it}^- \ge 0 \text{ for all } i, t \tag{5-16h}$$

$$\left(\mu_{it}^+ - \mu_{it}^- + M_{it}\right) \text{ integer, for all } i, t \tag{5-16i}$$

Constraints (5-16b) through (5-16e) are the budget, floor space, over-time, and solution feasibility constraints respectively. Constraint (5-16f) is necessary to ensure that the definition of x_{it} be the same when defined by the y's as when defined by the μ's. The last constraint, (5-16i), ensures that x_{it} will be integer.

By studying this model, we may conclude that for a realistic problem, the formulation of a mathematical program may be very difficult: much effort is required to develop analytical expressions for the objective function and the constraints, expressions, which must then be linearized. Not apparent from the model, however, is that the *formulation* of a problem does not guarantee its *solution*. The formulation may be difficult to solve due to its size (i.e., great number of variables and constraints), or it might be virtually impossible to collect all of the required data [43]. On the other hand, simpler formulations may not capture the full difficulty of the problem.

Simulation Approach to the Resource Requirements Problem

In the design of relatively simple systems it is possible to employ simulation in a manner, which is more efficient than that of blind trial and error. This is the case when determining the resource requirements of a flow line, which must achieve a given production rate.

The production rate of a flow line is absolutely limited by the "slowest" resource, the *bottleneck* resource. Resources upstream of the bottleneck may experience blocking because they are constrained by the rate at which the bottleneck accepts input, while resources downstream of the bottleneck may experience starvation because they cannot be fed quickly enough. A bottleneck always exists, because in practice it is not possible to perfectly match the output rates of all of the resources in the flow line. The bottleneck may be alleviated by:

- redistributing the work performed at the bottleneck resource,
- changing the design or the operating settings of the bottleneck resource, or
- adding extra resources to the bottleneck.

Once one of these remedial actions has been taken, however, another bottleneck (albeit less severe than the first) develops elsewhere in the flow line. Note that addition of resources is only one of the options available for dealing with a bottleneck. In industry, it would be the least desirable option, because of the capital expense involved.

One procedure for designing a flow line, then, is to start with a minimal initial configuration (a given number of resources of each type). To be conservative, this configuration can be that, with just one resource of each type. The next step is to determine the bottleneck via simulation. The production rate of the flow line can then be increased by taking one of the above remedial actions on the bottleneck. If the required production rate still has not been reached, then remedial action is taken on the new bottleneck, and so forth (Fig. 5.35) [44].

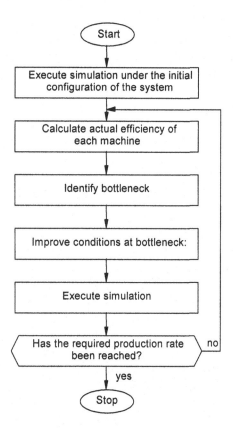

Fig. 5.35. A Procedure for Designing a Flow Line via Simulation

The role of simulation in this procedure is to determine the bottleneck. This can be done by calculating the actual efficiency E_r of each resource, defined as [31]

$$E_r = \frac{J_r}{T_r - S_r - B_r}$$

where $J_r \equiv$ Number of completed products at Resource r, except for those products which are reworked
$T_r \equiv$ Total time at Resource r
$S_r \equiv$ Total starving time at Resource r
$B_r \equiv$ Total blocking time at Resource r

The resource with the lowest actual efficiency E_r is deemed to be the bottleneck resource.

Statistics, such as total starving time at Resource r and total blocking time at Resource r, are difficult to track by any method other than that of simulation, especially when the system is large, the details of its operation are intricate, and the actual system does not exist yet.

This procedure has been applied in industry for the configuration of a flow line with 25 processes for the production of large diameter steel pipes. The desired production rate in this case was 7,000 pipes per month [44].

Queuing Theory Approach to the Resource Requirements Problem

In the design of manufacturing systems, the role of the queuing theory is similar to that of simulation. For simple manufacturing systems, the results normally provided by simulation, can be provided by numerical solution instead of the algebraic equations, which make up a queuing model.

As with all analytical models, queuing models have a limited range of applicability. Only certain types of manufacturing systems are easily modeled. In particular, systems in which the work in process (W.I.P.) remains constant are chiefly amenable to queuing analysis. The constant W.I.P. condition may hold in many Flexible Manufacturing Systems (FMS), in which parts circulate about the system on a fixed number of pallets.

Let us discuss how the queuing theory may be used to model an FMS. We will restrict our attention to the case of an FMS, which produces only a single part type, and which consists of M machine groups of one machine each. This case will be sufficient to provide the flavor of the required analysis. The analysis can be extended without difficulty in covering the general case of multiple part types and multiple machines at each machine group [45].

Let

L $\equiv$ Number of pallets in the FMS

M $\equiv$ Number of machine groups (= the number of machines)

X $\equiv$ Production rate

$V(m)$ $\equiv$ Mean number of visits to machine m

$T(m)$ $\equiv$ Mean processing time at machine m, at each visit

$R(m)$ $\equiv$ Mean response time (waiting + processing) at machine m, at each visit

$L_q(m)$ $\equiv$ Mean queue length at machine m (includes jobs waiting or being processed)

$W_q(m) \equiv$ Mean waiting time at machine m, at each visit

$U(m) \equiv$ Utilization of machine m

In these definitions, the term "mean" stands for an average, over all routes, for one instance of a part type. The objective is to determine the production rate X, the mean queue length $L_q(m)$ at each machine m, and the utilization $U(m)$ of each machine m.

For this analysis, servers are machines and customers are parts. First, we note that for each machine m,

$$R(m) = T(m) + W_q(m)$$

A part arriving at machine m will, on average, finds $L_q(m) * (L-1)/L$ parts waiting for processing ahead of it. This number is the mean queue length at m, corrected by the fact that it is being "observed" by one of the parts in the FMS. Since each part that is ahead in the queue has a processing time of $T(m)$,

$$R(m) = T(m) + \frac{(L-1)}{L} L_q(m) T(m)$$

Multiplied through by $V(m)$,

$$R(m)V(m) = T(m)V(m)\left[1 + \frac{(L-1)}{L}\right]L_q(m) \quad m = 1,...,M \tag{5-17}$$

The left hand side of this equation represents the mean time spent by a part at machine m, over all visits. Therefore, the average time that a part spends on the FMS is

$$W = \sum_{m=1}^{M} R(m) \, V(m)$$

Little's formula (5-10) applied to the FMS as a whole, gives $\bar{\lambda} = L/W$, where λ is the mean rate at which parts arrive into the FMS. However, since the number of parts (or pallets) in the FMS is constant at L, λ must just be equal to X, the production rate:

$$X = \frac{L}{\sum\limits_{m=1}^{M} R(m) V(m)}$$

(5-18)

The mean queue lengths $L_q(m)$ may be found by applying Little's formula (5-11) at each machine m:

$$L_q(m) = R(m) \, V(m) \, X$$

(5-19)

The machine utilizations $U(m)$ may be calculated as follows. Suppose that a total of p parts are produced in ΔT time units, during which machine m is actually busy for $B(m)$ time units. Then

$$B(m) = T(m) \, V(m) \, p$$

But $U(m) = B(m) / \Delta T$, and $X = p / \Delta T$, so that

$$U(m) = T(m) \, V(m) \, X$$

(5-20)

Looking at Equations 5-17, 5-18, and 5-19, we see that by starting with an initial set of values for $L_q(m)$, we can evaluate Equations 5-17, 5-18, and 5-19 in that order so as to obtain a new set of values for $L_q(m)$, $m = 1$, ..., M. The accepted procedure is to make this iteration until convergence has been achieved. The outcome will be the values of the production rate X and average queue lengths $L_q(m)$. After this has been completed, the utilizations $U(m)$ can be calculated from Equation 5-20.

The queuing theory, like simulation, may be applied in a trial and error fashion for the designing of manufacturing systems. The queuing model of a given system will only provide a subset of the performance measures, provided by the simulation model. However, it will generally provide results within a shorter span of time [45] – roughly speaking, in minutes as opposed to hours.

5.3.2 The Resource Layout Problem

The resource layout problem is concerned with the placement of resources on the factory floor so that some set of production requirements are met. The problem has been formulated in several ways, with varying degrees of sophistication.

1. The template shuffling formulation
2. The quadratic assignment problem (QAP) formulation
3. The relationship (REL) chart formulation

In the following sections, these formulations will be briefly discussed.

Template Shuffling Approach to the Resource Layout Problem

The template shuffling formulation is a manual method in which a number of templates (geometric replicas of machines, material handling units, etc.) are arranged by *trial and error* on the prespecified floor area of the facility. It is the most widely used of all the resource layout formulations in industry.

This formulation is not amenable to automatic solution because it is not well structured and layouts must be ranked by visual inspection. However, it is supported by practically all of the commercially available facilities planning software packages. These packages provide features that make the shuffling of templates more convenient than the actual manual manipulation of physical templates. Editing features on screen enables templates to be copied, moved, deleted, and resized easily (Fig. 5.36).

Features, such as layers are typically supported, and the designer can focus on a subset of templates by making the others invisible.

Quadratic Assignment Problem Approach to the Resource Layout Problem

The Quadratic Assignment Problem (QAP) formulation of the resource layout problem is to assign n resources to n spaces of equal area in order to optimize an objective. The most common objectives include the total material handling distance, the total cost for material handling, and the cost for relocating existing resources. Material handling costs are usually assumed to be proportional to the product of the distance and the rate of material flow between each pair of resources.

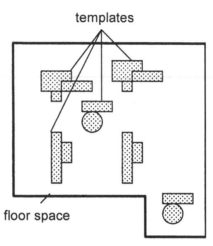

Fig. 5.36. The Template Shuffling Formulation of the Resource Layout Problem

The QAP formulation has several limitations:

- Realistically sized problems are difficult to solve optimally.
- Individual constraints, such as requiring that particular machines be located no further than a certain distance from each other, are not incorporated into the current problem formulation.
- It can lead to irregularly shaped resource areas in the final layout. This comes from having to break up resource areas into smaller ones so that all areas can be of equal size.
- It allows no explicit way of incorporating into the model factors other than those of the resource relocation cost and the material movement cost.
- It assumes a simple linear relationship between the cost of material movement, amid any pair of resources, and the distance between those resources.

The mathematical programming formulation of the QAP is given by:

$$\text{Minimize} \quad \sum_i \sum_j \sum_k \sum_l c_{ijkl} x_{ik} x_{jl} \qquad (5\text{-}21a)$$

Subject to:

$$\sum_j x_{ij} = 1, \quad \text{for } i = 1, ..., n \qquad (5\text{-}21b)$$

$$\sum_i x_{ij} = 1, \quad \text{for } j = 1,...,n \tag{5-21c}$$

$$x_{ij} \in \{0, 1\} \tag{5-21d}$$

The variable x_{ik} represents a (0–1) variable equal to 1 if and only if facility i is assigned to location k and c_{ijkl} is the cost of assigning facilities i and j to locations k and l respectively. The constraints (5-21b) express the fact that each resource i must be assigned to exactly one location and the constraints (5-21c) express that the fact that each location j must have one facility assigned to it. This is an integer program.

Many solution approaches to this program and their variations have been proposed [46-50]. However, the problem is non-polynomial-hard (NP-hard) [51], meaning that the time taken by any algorithm to find the optimum solution must increase exponentially as the problem size (i.e., the number of resources) increases linearly. Consequently, optimal solutions for problems involving more than 20 or so resources 51] cannot be obtained within the realm of practical computational effort.

Due to this computational difficulty, many different heuristics have been devised for finding sub-optimal but 'good' solutions to the QAP. These are either *construction procedures*, which put resources on the factory floor one after another, until all resources have been placed, or *improvement procedures*, which start from an initial complete solution and then attempt to improve the solution by interchanging resources [53].

Relationship Chart Approach to the Resource Layout Problem

The relationship (REL) chart formulation of the resource layout problem is a more qualitative formulation, which overcomes the stringent data requirements of the QAP formulation. A REL chart gives the desirability of having each pair of resources adjacent to each other [54]. This desirability is usually expressed in a letter code (Fig. 5.37):

A	Absolutely essential that two departments be located adjacently
E	Essential that two departments be located adjacently
I	Important that two departments be located adjacently
O	Marginally beneficial that two departments be located adjacently
U	Unimportant that two departments be located adjacently
X	The two departments should not be adjacent

These codes are transformed into numerical ratings when scoring a layout. For example, the ratings used by the Automated Layout Design Program (ALDEP) [55] are: $A = 4^3$, $E = 4^2$, $I = 4^1$, $O = 4^0$, $U = 0$, $X = -4^5$. A layout's score is simply the sum of the ratings for each pair of adjacent resources in the layout. The objective is to find the layout with the highest score.

Existing solution methods for the REL formulation are heuristics; they find only "good" solutions, not optimal solutions. The methods are mostly construction procedures, which add to the facility one resource at a time, until the layout is complete [55, 56]. The quality of the layout is determined by the quality of the heuristic.

The REL chart formulation is based on the premise that maximizing adjacency ratings is good for the overall layout. However, the correlation between adjacency ratings and more global measures of performance, such as production rate or equipment acquisition cost, may be very weak.

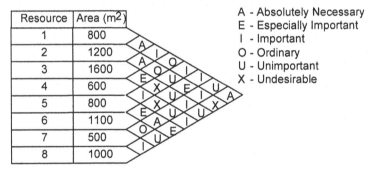

Fig. 5.37. A REL Chart for Eight Resources Containing Adjacency Ratings

Rule-Based Approach to the Resource Layout Problem

The application of rule-based systems to the resource layout problem has been limited. Because of the daunting combinatorial nature of the problem, it is impossible to establish generally applicable rules for its solution. Simplifications must be made.

One approach is to divide resource layouts into a few generic classes. Four classes of layouts that have been proposed are: linear single-row, circular single-row, linear double-row, and multi-row. In this approach, these layouts are to be mated with one of two classes of material handling systems: automated guided vehicles (AGVs) or robots (Fig. 5.38). Furthermore, only certain combinations of layout and material handling are considered: linear single-row/AGV, circular single-row/robot, linear double-row/AGV, and multi-row/AGV [57].

With the problem thus simplified, a rule-based system can be applied in two ways. First, based on floor space restrictions, the system can select one of the four layout/material handling combinations. The existing system always recommends the combinations in the order circular single-row/robot, linear single-row/AGV, linear double-row/AGV, and multi-row/AGV, passing to the next combination in the order only when the current combination cannot fit within the specified factory floor space.

Once this step has been completed, additional rules in the rule base are used to select one of a number of analytical models (similar to Equation 5-21, the Quadratic Assignment Problem model), plus a solution algorithm for the model. The selection of a model is based on the selected layout structure, as well as on the numbers and sizes of the resources. The function of the selected analytical model is to locate the individual resources within the selected layout structure in order for that material handling cost to be minimized.

It is important to note that the rule-based system itself does not generate any new design knowledge; it cannot prescribe designs that the author of the system's rules does not know how to design[58]. Specifically, it can only "design" one of the four layouts in Figure 5.38. The role of the rule-based system is to apply the intuition of a human expert, as set forth in a set of rules, to prescribe the basic structure of the resource layout; then, again by following the expert's judgment, prescribe an analytical model for optimizing the selected layout structure.

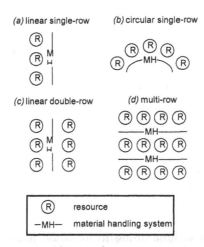

Fig. 5.38. Classes of Resource Layouts

5.3.3 The Material Flow Problem

Material flow decision variables, which must be specified in the design of a material handling system, can be divided into two broad categories: these which specify the *type* of the material handling system, and those which specify the *configuration* of a given type of material handling system.

Configuration decision variables depend on the type of material handling system. For a material handling system based on conveyors, these variables may be the layout of the conveyors on the factory floor and their operating directions and speeds.

A different set of decision variables apply to material handling systems, based on automated guided vehicles (AGVs). AGVs are unmanned vehicles used to transporting loads from one location in the factory to another. They are operated with or without wire guidance and are controlled by a computer. AGVs are often used in flexible manufacturing systems [59]. The configuration decision variables that must be considered in the design and operation of an AGV-based material handling system are [60, 61]:

- The travel aisle layout
- The number and the locations of the pickup and delivery stations
- The pattern of material flow within the travel aisles (unidirectional, bidirectional or combinations)
- The number of vehicles required
- The routes used by vehicles during specific operations
- The dispatching logic used during operation
- The storage capacities of pickup and delivery stations

Mathematical programming has been applied to determining the pattern of material flow in an AGV-based material handling system [62,63]. The objective of the proposed approach is to find the flow path, which will minimize the total loaded travel of AGVs. The required inputs are: a layout of the departments of a manufacturing system, the aisles where AGV travel may occur, the locations of pickup (P) and delivery (D) stations for each department, and the material flow intensities between departments. The output of the program is a solution that indicates which aisles should be used for AGV travel, and what the direction of travel should be on each of these aisles. It is assumed that travel only in a single direction is permitted in each aisle.

The mathematical programming formulation is based on an abstraction of the input information in the form of a graph. The graph consists of nodes, which represent pickup/delivery (P/D) stations and intersections and corners of aisles, and arcs connecting the nodes, which represent possible directions of travel along the aisles. Each aisle is therefore associated with two arcs, one for each possible direction of travel. Each node is identified by a number. An arc from Node i to Node j is identified by an integer variable x_{ij}. If x_{ij} equals 1, then the final material flow pattern will include AGV travel from the location represented by Node i to the location represented by Node j; if x_{ij} equals 0, then no material flow from location i to location j will be present in the final solution.

This abstraction can be demonstrated with the aid of a simple example [62] with two-departments (Fig. 5.39). The department and aisle layout is shown in Figure 5.39(a). The numbers represent distances between adjacent nodes (which are P/D stations, intersections, or corners). P is a pickup station, and D is a delivery station. The material flow intensity between these two stations (from Department 1 to Department 2) is 100. This layout is then converted into the graph of Figure 5.39(b). Deletion of arcs from this graph results in the final material flow pattern (Fig. 5.39(c)).

The objective is to minimize the loaded travel distance of the AGVs. To capture this in an algebraic objective function, we note that an AGV must leave from P on either arc x_{32} or on arc x_{36}. Furthermore, it must arrive at D on either x_{45} or x_{65}. Since travel is unidirectional along each aisle, the four possibilities are: travel on 1) x_{32} and x_{45}, 2) x_{32} and x_{65}, 3) x_{36} and x_{45}, and 4) x_{36} and x_{65}. For the first possibility, the shortest travel path would be 3-1-2-4-5, with a total distance of $4+7+10+3 = 24$ (Fig. 5.39(a)). The shortest travel distances for the other possibilities can be similarly computed to be 52, 36, and 10 respectively. The objective function can then be stated as:

$$\text{Minimize } 100 \, [24x_{32}x_{45} + 52x_{32}x_{65} + 36x_{36}x_{45} + 10x_{36}x_{65}]$$

Since in a legal solution only one of the four products $x_{32}x_{45}$, $x_{32}x_{65}$, $x_{36}x_{45}$, and $x_{36}x_{65}$ can be 1, while the others must be 0, the objective function is minimized when the shortest path (3-6-5) is selected (making $x_{36}x_{65}$ equal to 1 and the objective function value equal to 1,000.)

In order to ensure a legal solution, a number of constraints must be observed. Travel must be unidirectional, meaning that only one of the two arcs, which connect each pair of nodes can be in the final solution. Therefore,

$$x_{14} + x_{41} = 1$$
$$x_{47} + x_{74} = 1$$
$$x_{78} + x_{87} = 1$$

:

Furthermore, it must be possible to reach every node; there must be at least one incoming arc for each node. For Node 4, this translates into

$$x_{41} + x_{45} + x_{47} \geq 1$$

A node which violates this constraint is called a source node. It must also be possible to leave every node. For this reason, for Node 4,

$$x_{14} + x_{54} + x_{74} \geq 1$$

A node which violates this constraint is called a sink node. Similar source and sink constraints apply to other nodes.

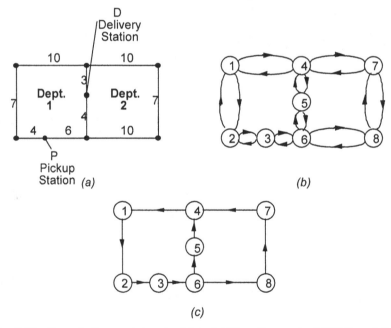

Fig. 5.39. Steps in Determining the Material Flow Pattern in AGV-Based Systems

Finally, we note that source and sink constraints must be observed for *groups* of nodes as well. Consider the group of nodes {1,2,3,4,5,6}. The constraint

$$x_{74} + x_{86} \geq 1$$

prevents this group from becoming a source, while the constraint

$$x_{47} + x_{68} \geq 1$$

prevents this group from becoming a sink.

The objective function in conjunction with the unidirectional constraints and the source and sink constraints together, constitute the mathematical programming formulation of the example problem. The formulation is a nonlinear integer program. Solution of this program yields (Fig. 5.39(c)):

$$x_{12} = 1 \quad x_{41} = 1 \quad x_{23} = 1$$
$$x_{68} = 1 \quad x_{36} = 1 \quad x_{87} = 1$$
$$x_{65} = 1 \quad x_{74} = 1 \quad x_{54} = 1$$

all other variables = 0

Since its introduction, the above approach has been expanded to incorporate the function of optimally locating the P/D stations. The expanded approach [64] consists of two phases. Phase 1 is the original approach. In Phase 2, the locations of the P/D stations are altered by a heuristic to reduce the estimated total distance traveled by the AGVs. Since this relocation potentially spoils the optimality of the flow directions, derived in Phase 1, both phases must be executed iteratively until Phase 2 no longer alters the optimality of Phase 1. In a further expansion of the approach, *unloaded* AGV travel has been incorporated into the objective function as well [65].

5.3.4 The Buffer Capacity Problem

A buffer is a storage space in a manufacturing system for pieces between processing stages. Buffer serves for decoupling the separate processing stages in a manufacturing system. By providing buffer space for inventory between machines, starvation and blockage are reduced, resulting in increased production rate. This, however, comes at the expense of increased inventory.

Buffer allocation is a difficult problem because in general it is not possible to derive an analytical relationship between performance requirements and the proper buffer locations and sizes. Another aspect of the problem is that existing factory floor layouts often impose constraints upon the locations and capacities of buffers that can be implemented.

Dynamic programming has been used [66] to address the allocation of a fixed buffer capacity of Z pieces over the N-1 possible buffer locations in an N-machine automatic transfer line (Fig. 5.40).

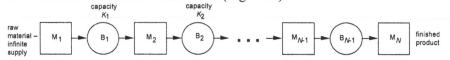

Fig. 5.40. Transfer Line System Studied for the Dynamic Programming Application

In the dynamic programming formulation, the total allocated capacity must be exactly Z, and each individual buffer capacity K_n must not exceed a given local capacity constraint C_n. The objective is to maximize the production rate $f(\mathbf{K}_{N-1})$, where $\mathbf{K}_{N-1}$ is the vector of N-1 individual buffer capacities. If the total capacity, Z, exceeds the sum of the local capacities, $C_1 + C_2 + \dots + C_{N-1}$, the trivial solution is to make each $K_n = C_n$. This case is excluded from further consideration.

The above problem is cast in the form of the following dynamic programming problem:

$$\text{Maximize:} \quad f\left(K_{N-1}\right)$$

Subject to:

$$\sum_{n=1}^{N-1} K_N = Z$$

$$0 \le K_n \le C_n \tag{5-22}$$

In the dynamic programming approach, the multi-variable optimization problem involving the N-1 individual buffer capacities $K_1, \dots, K_{N-1}$ is broken down into N-1 single-variable optimization problems involving the separate K's. These are solved in N-1 consecutive stages. The decisions at each stage n, $1 \le n \le N$-1, require the calculation of the production rate of the portion of the transfer line consisting of M_1–B_1–...–B_n–M_{n+1}. Furthermore, the production rate calculated at the nth stage must be expressible in terms of parameters calculated at the n-1th stage. This recursive quality is a requirement of dynamic programming objective functions. Production rate can be recursively calculated via an approach in which the system is decomposed into N-1 two-stage transfer lines (M_1, B_1, M_2), (RM_1, B_2, M_3), ..., (RM_{N-2}, B_{N-1}, M_N), where RM_i is an equivalent single machine replacing (RM_{i-1}, B_i, M_{i+1}). The production rate of each two-stage transfer line is analytically calculated, under the simplifying assumption of no second-stage blocking. The necessary recursion is achieved by defining the breakdown characteristics of RM_i in terms of the breakdown characteristics of RM_{i-1}, the capacity of the buffer B_i, and the breakdown characteristics of M_{i+1}.

The main requirement of this formulation is the recursive quality of the objective function. The production rate objective can be made to satisfy this condition, but generalization of the approach to other types of performance measures is difficult. For example, as inspection of Equation 5-22 will verify, the cost related to providing buffers is determined solely by the total storage space allocated and not by the actual inventory levels in the storage space. In addition, the decomposition approach to production rate calculation can be implemented primarily for systems with serial material flow.

5.3.5 Complex Design Problems

Most of the design approaches mentioned so far have only been applied to academic research. The real manufacturing system design is still very much the domain of human experts, who put systems together based on personal experience and conviction. At best, system designs are merely verified by simulation before being implemented. Creating the initial design or modifying the design, should initial simulation results prove unfavorable, is still an unstructured, creative process.

The immediate obstacles to the infusion of greater scientific content into the manufacturing system design process are the limitations of the problem formulations in engineering science. Any approach, which addresses only the resource requirements problem or only the buffer capacity problem, is missing the bigger design problem, in which consideration of the fact that buffer capacities will affect the number of resources required and vice versa, must play a role.

In this context, the engineering science approaches, which view a manufacturing system design as a vector of decision variables, show particular promise. Using these approaches, it is possible to formulate a problem in which some of the decision variables to be determined are buffer capacities, some are resource requirements and others are material handling parameters, etc. In literature, these approaches are often referred to as *parametric design* approaches.

Simulation Plus Search Approach to Manufacturing System Design

One parametric design approach seeks to supplement the descriptive capabilities of simulation with *prescriptive* techniques, capable of generating new manufacturing system designs. The prescriptive technique is usually a type of search algorithm. This approach views the manufacturing system design as a combinatorial optimization problem in which a performance measure, which depends on the values of the decision variables, must be optimized.

The approach begins with an initial vector of decision variables, which represent the current design. The performance measures of the current design are evaluated via simulation. Next, the search algorithm is used to modify the current design. If the performance of the modified design is better than that of the current, then the modified one becomes the current design. In either case, the search algorithm is used to modify the current design again, and the process is repeated (Fig. 5.41).

This approach has been applied to the problem of assigning buffer capacities for an automatic assembly system (AAS) consisting of N machines and $N+1$ buffers [67]. Material flow in the category of systems considered, consists of a main assembly loop and a repair loop for the repair of imperfectly-assembled parts (Fig. 5.42). A fixed number P of pallets carry parts about the system. The pallets ride on transfer chains or conveyors.

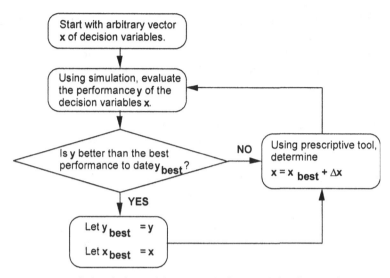

Fig. 5.41. Use of Simulation with a Prescriptive Tool for Generating New Designs

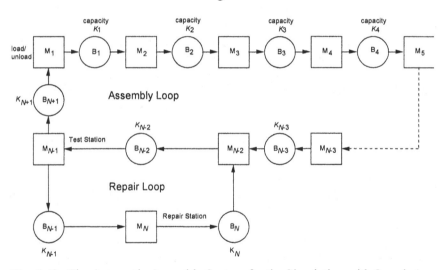

Fig. 5.42. The Automatic Assembly System for the Simulation with Search Application

The major factors which affect the performance of an AAS are: the cycle times of the machines, the rate of occurrence of parts jams for each machine, the distribution of jam clearing times, the pallet transfer times, the sampling policy for the inspection of the finished assemblies, and the buffer sizes between machines. In this application, only the buffer factor is addressed; the other factors are assumed to be given.

The problem is viewed as a multi-variable combinatorial optimization problem in which the buffer capacities K_1, ..., K_{N+1} are the state variables. The objective is to find the state variable values, which maximize an objective function, defined to be the production rate of the AAS. The size of the state space is defined by the consideration that any buffer in the real system must have a capacity in the range $[1, P]$. Based on empirical evidence, the buffer capacities may be further restricted to the range $[1, P/3]$. The resulting state space size is $(P/3)^{N+1}$.

The combinatorial optimization problem defined above is approached via simulation in conjunction with simulated annealing. Simulated annealing is a search procedure [68] based on an analogy between the process of annealing in solids, as described by the models of statistical mechanics, and the process of combinatorial optimization. In general terms, when a solid is annealed, (that is, heated and then gradually cooled), the positions of a vast number of molecules are gradually evolved into a configuration of very low internal strain energy in the solid. If we make the analogy between the molecular positions and internal strain energy of a solid and the state variables and objective function of a combinatorial optimization problem, then the statistical mechanical models, which describe solid annealing can provide a "recipe" for the minimization of an objective function of many state variables. In the combinatorial optimization problem, as in the solid, the annealing performance is dictated by the time history of a temperature parameter T; this is the so-called "cooling schedule." The application of this procedure to the determination of buffer capacities is summarized in Figure 5.43.

Since simulation is used to dealing with the complexity of the relationship between the performance measure (production rate) and the decision variables, this approach requires few restrictive assumptions about the nature of the performance measures and decision variables of the system to be designed. While only buffer capacities are optimized in this particular application, complicated decision variables, such as inspection policy, are incorporated into the simulation model and hence, the solutions achieved are applicable to the assumed inspection policies.

One possible limitation of the simulated annealing/simulation approach is that the required number of numerical simulations may be high (in the thousands). The computational burden of the procedure may therefore be prohibitive.

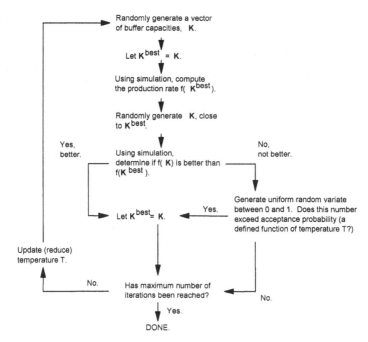

Fig. 5.43. Use of Simulated Annealing for Determining Buffer Capacities

Neural Network Approach to Manufacturing System Design

The design of manufacturing systems may be considered a generalization application of neural networks, and requires a *training phase* and a *use phase*. In the *training phase*, simulations are run to provide sample correlations between decision variable (DV) and performance measure (PM) values. A neural network is then trained by exposure to these correlations, which are expressed in the form of training pairs $\{I_1\text{-}O_1, I_2\text{-}O_2, ..., I_n\text{-}O_n\}$, as in Figure 5.27. I_k is a vector of performance measure values, while O_k is the corresponding vector of decision variable values. On the basis of this training, the neural network generalizes the relationship between the I_k's and the O_k's.

In the *use phase*, the approach assumes that for a certain period during the operation of a manufacturing system, there is a general goal that can be expressed in terms of several performance measures. An example might be "Average cost per part of $35, mean tardiness 20 minutes." These goal performance measures (say, I_{n+1}) are input to the trained neural network, which will then output the decision variable values (O_{n+1}) that will come closest to realizing the performance goal. This approach is summarized in Figure 5.44.

This approach has been applied to the determination of the number of resources (machines) in each of the N work centers of a job shop [51]. A test job shop consisting of $N = 3$ work centers was used. Each feasible job shop configuration was thus described by three positive integers which specified the number of resources in each work center. All resources in a given work center were assumed to be identical. The work load of the job shop consisted of five job types, each containing one to three tasks. Each task could be processed by any resource from its suitable work center. Eight hours of job arrivals were considered.

In the *training phase*, a particular job shop configuration, work load, and operational policy for the assignment of production tasks to resources, was input to a simulator, which produced a corresponding set of four performance measures: mean tardiness (the average time by which the completion time of a job exceeded its due date), mean flowtime (the average time that a job spent in the job shop), mean utilization (the percentage of time, on average, that each machine spent on actually processing a task), and time of completion (of the last job).

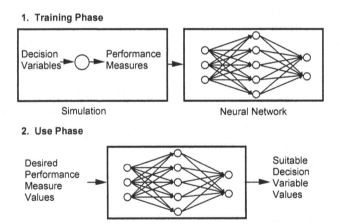

Fig. 5.44. The Application of Neural Networks to Manufacturing System Design

The simulation was performed five times, each time with a different job shop configuration. The outcome of the training phase was five input-output pairs $\{I_1\text{-}O_1, I_2\text{-}O_2, \ldots, I_5\text{-}O_5\}$ for which each input I_k was a set of values for the four performance measures, and each output O_k was the corresponding job shop configuration (Tbl. 5.6). These were used to train a 4-15-3 neural network (this terminology is explained in Figure 5.28). This network structure was suitable because the number of input nodes in the network (4) matched the number of performance measures, and the number of output nodes (3) matched the number of work centers in the test job shop. In the *use phase*, five sets of desired performance measures were submitted to the trained neural network. For each set, the network output a job shop configuration (Tbl. 5.7). This completed the use phase.

	Training Inputs (I_k): Performance Measures				Target Outputs (O_k): Job Shop Configurations		
k	Mean Tard [min]	Mean Flow [min]	Mean Util [%]	Time of Comp [min]	# Res WC-1	# Res WC-2	# Res WC-3
1	30.25	116.05	77.98	642.00	2	3	3
2	135.25	226.60	89.50	895.00	1	2	2
3	1.95	76.60	66.93	544.00	3	4	4
4	96.80	185.60	56.26	791.00	4	2	3
5	69.00	151.05	56.57	885.00	1	4	3

Table 5.6. Input-Output Pairs Used for Neural Network Training

Network Inputs: Desired Performance Measures				Network Outputs: Job Shop Configurations		
Mean Tard [min]	Mean Flow [min]	Mean Util [%]	Time of Comp [min]	# Res WC-1	# Res WC-2	# Res WC-3
0	50	70	600	1	4	4
50	100	75	500	4	2	4
25	75	90	550	2	3	3
0	50	85	600	1	4	4
0	50	95	650	1	4	3

Table 5.7. Job Shop Configurations Prescribed by the Neural Network

In order to gain more insight into the capabilities of the neural network, the configurations output by the network were compared to those of regression, a standard mapping technique. A linear regression was performed on the five input-output pairs in Table 5.6, meaning that each resource quantity n_{WC-i} was expressed as a function of the form

$$n_{WC-i} = a_0 + b_i \text{ (mean tardiness)} + c_i \text{ (mean flowtime)}$$

$+ d_i$ (mean utilization)$+ e_i$ (time of completion),

with the coefficients a_0, b_i, c_i, d_i and e_i being evaluated from the data in Table 5.6. The desired performance measures of Table 5.7 were then substituted into these functions, yielding the configurations shown in Table 5.8. Several infeasible configurations with negative numbers of resources were prescribed. This demonstrates the advantage of neural networks in modeling relationships that are highly non-linear.

Regression Inputs: Desired Performance Measures				Regression Outputs: Job Shop Configurations		
Mean Tard [min]	Mean Flow [min]	Mean Util [%]	Time of Comp [min]	# Res WC-1	# Res WC-2	# Res WC-3
0	50	70	600	1	7	5
50	100	75	500	4	5	5
25	75	90	550	1	6	5
0	50	85	600	-1	7	5
0	50	95	650	-2	8	4

Table 5.8. Job Shop Configurations Prescribed by Regression

Another aspect of the approach tested, was the degree of agreement between the desired performance measures and the "actual" performance achieved by the job shops, prescribed by the neural network. "Actual" performance was determined by simulating each configuration output by the neural network under the same work load and operational policy used in the training phase simulations. The degree of agreement, for a particular performance measure (*PM*), was quantified using a fractional difference index:

$$FD_{PM} = \frac{1}{5} \sum_{i=1}^{5} \left| \frac{desPM_i - achPM_i}{desPM_{avg}} \right| \qquad (5\text{-}23)$$

where $desPM_i \equiv$ Value of *PM* in the ith set of desired performance measures

$achPM_i \equiv$ Value of *PM* achieved by the job shop prescribed by the network in response to the ith set of desired performance measures

$$\text{des}PM_{avg} \equiv \text{Average of the desired values of } PM \text{ over the five sets of desired performance measures shown in Table 5.7}$$

The less the fractional difference index is, the better the agreement between desired and achieved performance measures. The FDs obtained are shown in Table 5.9.

FD Mean Tard	FD Mean Flow	FD Mean Util	FD Comp Time	Mean FD
3.43	1.31	0.31	0.41	**1.36**

Table 5.9. Degree of Agreement Between Desired and Achieved Performance Measures, as Measured by FD (Eq. 5-23), for a Neural Network Trained with Five Input-Output Pairs

When the same neural network, trained with seven input-output pairs instead of five, was given the same desired performance measures (Tbl. 5.7), improved FDs were achieved (Tbl. 5.10).

FD Mean Tard	FD Mean Flow	FD Mean Util	FD Comp Time	Mean FD
2.58	1.10	0.25	0.33	1.06

Table 5.10. Degree of Agreement Between Desired and Achieved Performance Measures, as Measured by FD (Eq. 5-23), for a Neural Network Trained with Seven Input-Output Pairs

These results (Tbls. 5.9 and 5.10) illustrate the ability of the neural network to *learn* as it is trained with increasing numbers of examples. However, there is not enough data to determine limits on the degree to which network-prescribed job shops can satisfy the given desired performance measures.

Several features of this approach are worth noting.

1. The design goals are expressed as desired values of whatever performance measures may be relevant. This is advantageous in several respects:

   - multiple objectives can be accommodated;
   - the design goals can be changed without changing the approach;
   - the design goals, being *performance measures*, are commonly used and accepted expressions of the desired performance of a manufacturing system.

   The problem *formulation* is thus, very flexible. With this approach, the formulation can be adapted to the problem; with other (e.g. analytical) approaches, the problem must be adapted to the formulation.

2. Through the use of a simulator, this approach takes the effect of specific operational policies into account. Other approaches typically assume (sometimes implicitly) a First-Come-First-Served policy for simplicity.

3. The design procedure is decoupled from the simulation model. The simulation model can be increased in sophistication, without a corresponding increase in the sophistication of the neural network design procedure.

Several issues concerning this approach remain to be resolved:

1. Neural network training data for the approach is generated from a limited number of simulation runs in which the decision variables (resource quantities) are varied in some fashion, resulting in a limited number x of {decision variable}–{performance measure} mappings. In most cases, the ratio x/n, where n is the total number of feasible mappings or equivalently the total number of feasible decision variable permutations, will be very small indeed. How large must x/n be? Certainly, if n can be reduced in some fashion, then the likelihood of success of the proposed approach would be increased. Some kind of easy-to-apply screening criterion/criteria for eliminating poor decision variable permutations would be very desirable. Then the neural network could be trained with purely "good" mappings – mappings which result in good performance measure values.

2. Another concern is that the results of the proposed approach are derived from simulations involving the processing of a single work load by various job shop configurations. How much these results generalize in other work loads must be determined.

The use of neural networks shows promise as a toll for the design of manufacturing systems. Its promise lies in its potential to guide the highly iterative process of the manufacturing system design and hence, reduces the number of simulations required in the process. Most importantly, a neural network is imbued with an ability to be guided not by the codification of a complicated set of heuristic rules, but rather by a simple and automatic training process.

While the parametric design approaches described above are flexible enough to accommodate sufficiently realistic formulations of the manufacturing system design problem, it remains to be seen whether any of these formulations can be effectively solved. Whether the solution approach employs simulation in conjunction with search [67], neural networks [69], or other methods, it seems likely that considerable computational resources will be required to address industrial problems realistically.

Further Reading

The design of manufacturing systems requires the *synthesis* of many ideas from economics, operations research, artificial intelligence, and other disciplines. Although each discipline is well established in its own right, the assemblage of disciplines can be viewed in a new light when brought together so as to treat the problem of manufacturing system design. This is the intended contribution of this chapter.

Consequently, the descriptions of individual methods and tools have been drawn extensively from existing works, and it is to these that we will refer the interested reader. In particular, the sections on operations research methods and tools have been primarily adapted from *Theory and Problems of Operations Research*, by Bronson [13]. This work provides an efficient handbook-style survey of operations research concepts. Some material was also drawn from *Operations Research: Applications and Algorithms*, by W.L. Winston [12]. The treatment here is more extended, with an emphasis on motivations and applications of the various methods and tools. For those particularly interested in the queuing theory, an in-depth treatment is provided in *Fundamentals of Queuing Theory* [14], by Gross and Harris.

The sections on search and rule-based systems have been adapted from P.H. Winston's *Artificial Intelligence* [15], a lucid and very readable account of the techniques of artificial intelligence. The abstraction of fundamental concepts from very technical tools is a strong suit of this book. MATLAB [70] Neural Network toolbox provides a very good environment for modelling and training neural networks and the toolbox documentation can be used for introductory purposes in neural networks. Further insight into neural network learning algorithms is available in *Neural Network Learning and Expert Systems* [17].

Up to date information regarding virtual reality technology can be accessed in *Understanding Virtual Reality: Interface, Application, and Design* [71] while one may refer in *Virtual reality for industrial applications* [72] and *Virtual and Augmented Applications in Manufacturing* [73] for recent VR applications in industrial practice.

Finally a good mixture of both conceptual and practical issues in simulation is covered in *Simulation Modeling and Analysis*, by Law and Kelton [23]. An emphasis is placed in this work on the application of simulation to manufacturing systems.

References

1. Primrose, P.L. and R. Leonard, "The Use of a Conceptual Model to Evaluate Financially Flexible Manufacturing System Projects," *Flexible Manufacturing Systems Current Issues and Models.* Choobineh, F. and R. Suri, Editors. Industrial Engineering and Management Press, Atlanta, Georgia, (1986), pp. 282-288.

2. Lefle, F., "The payback method of investment appraisal: A review and synthesis", *International Journal of Production Economics* (Vol. 44, 1996), pp. 207-224.

3. Chan F.T.S, H.M., Chan, L.K. Mak and H.N.K. Tang, "An Integrated Approach to Investment Appraisal for Advanced Manufacturing Technology", *Human Factors and Ergonomics in Manufacturing* (Vol. 9, No.1, 1999), pp. 69-86.

4. Kengpol, A. and C. O'Brien, "The development of a decision support tool for the selection of advanced technology to achieve rapid product development", *International Journal of Production Economics* (Vol. 69, No.2, 2001), pp. 177-191.

5. Kurtoglu A., "Flexibility analysis of two assembly lines", *Robotics and Computer-Integrated Manufacturing* (Vol. 20, No. 3, 2004), pp. 247-253.

6. Ramasesh, R.V.and D.M. Jayakumar, "Inclusion of flexibility benefits in discounted cash flow analyses for investment evaluation: A simulation/optimization model", *European Journal of Operational Research* (Vol. 102, No.1, 1997), pp. 124-141.

7. Elkins, A.D., N. Huang and J.M. Alden, "Agile manufacturing systems in the automotive industry", *International Journal of Production Economics*, In Press, Corrected Proof, Available online, (December 2003).

8. Holton, A.G., *Value-at-Risk: Theory and Practice*, Academic Press, 2003.

9. Muther, R., *Systematic Planning of Industrial Facilities*, Management and Industrial Publications, Kansas City, Missouri, 1979.

10. Muther, R. and E.J. Phillips, "Facility Planners Cite Clear Objectives And Proper Input Data As Main Success Factors," *Industrial Engineering* (March 1983), pp. 44-48.

11. Tompkins, J.A. and J.D. Spain, "Utilization Of Spine Concept Maximizes Modularity In Facilities Design," *Industrial Engineering* (March 1983), pp. 34-42.

12. Winston, W.L., *Operations Research: Applications and Algorithms*, Duxbury Press, Boston, 1987.

13. Bronson, R., *Schaum's Outline of Theory and Problems of Operations Research*, McGraw-Hill, New York, 1982.

14. Gross, D. and C.M. Harris, *Fundamentals of Queueing Theory*, John Wiley & Sons, New York, 1985.

15. Winston, P.H., *Artificial Intelligence*, 3rd edition, Addison-Wesley, 1993

16. Kohonen, T., *Self–Organization and Associative Memory*, Springer Verlag, Berlin, 1984.

17. Galant S., *Neural Network Learning and Expert Systems*, MIT Press, 1993.

18. Hecht-Nielsen, R., *Neurocomputing*, Addison-Wesley, Reading, Massachusetts, 1990.

19. Dutta, S., and S. Shekhar, "Bond Rating: A Non-Conservative Application of Neural Networks", *IEEE International Conference on Neural Networks* (Vol. 2, 1988), pp. 443-450.

20. Soulie, F.F., P. Gallinari, Y. Le Cun, and S. Thiria, "Evaluation of Neural Network Architectures on Test Learning Tasks", *IEEE International Conference on Neural Networks* (Vol. 2, 1987), pp. 653-660.

21. Rumelhart, D.E., G.E. Hinton and R.J. Williams, "Learning Internal Representation by Error Propagation", in Rumelhart, D.E. and J.L. McClelland, *Parallel Distributed Processing: Explorations in the Microstructure of Cognition* MIT Press, Cambridge, Massachusetts, (Vol. 1, 1986), pp. 318–362.

22. Jägstam, M. and P. Klingstam, "A handbook for integrating discrete event simulation as an aid in conceptual design of manufacturing systems", *Proceedings of the 2002 Winter Simulation Conference,* eds. E. Yücesan, C.-H. Chen, J. L. Snowdon, and J. M. Charnes, (2002), pp. 1940-1944.

23. Law, A.M. and W.D. Kelton, *Simulation Modeling and Analysis*, McGraw-Hill, Science, 3rd Edition, 1999.

24. Law, A.M. and M.G. Mccomas, "Simulation of manufacturing systems," *Proceedings of the 1998 Winter Simulation Conference*, eds. D.J. Medeiros, E.F. Watson, J.S. Carson and M.S. Manivannan, (1998), pp. 49-52

25. http://www.w3.org/

26. Phadke, M.S., *Quality Engineering Using Robust Design*, Prentice Hall, Englewood Cliffs, New Jersey, 1989.

27. Law, A., "Introduction to Simulation: A Powerful Tool for Analyzing Complex Manufacturing Systems", *Industrial Engineering* (Vol. 5, 1986), pp. 46-59.

28. Law, A.M. and M.G. Mccomas, "Pitfalls to Avoid in the Simulation of Manufacturing Systems", *Industrial Engineering* (Vol. 5, No. 5, 1989), p. 28.

29. Lu, S.C-Y., M. Shpitalni and R. Gadh, "Virtual and Augment Reality Technologies for Product Realization", *Keynote Paper, Annals of the CIRP*, (Vol. 48, No. 2, 1999), pp. 471-494.

30. Waurzyniak, P., "Visualizing the Virtual Factory. Collaborative 3-D visualization tools help drive manufacturers closer to the true e-factory", *Manufacturing Engineering* (Vol. 132 No. 4, 2004).

31. Smith, P. R. and A.J. Heim, "Virtual facility layout design: the value of an interactive three-dimensional representation", *International Journal of Production Research*, (Vol. 37, No. 17, 1999), pp. 3941-3957.

32. Technomatix Inc., http://www.tehnomatix.com

33. Chryssolouris, G., D. Mavrikios, D. Fragos and V. Karabatsou, "A Virtual Reality based experimentation environment for the verification of human related factors in assembly processes," *International Journal of Robotics and Computer Integrated Manufacturing* (Vol. 16, No.4, 2000), pp. 267-276.

34. Delmia Corp, http://www.delmia.com

35. Wenbin, Z., F. Xiumin, Y. Juanqi and Z. Pengsheng, "An integrated simulation method to support Virtual Factory Engineering", *International Journal of CAD/CAM*, (Vol.2, No.1, 2002), pp. 39-44.

36. Miller, D.M., and R.P. Davis, "The Machine Requirements Problem", *International Journal of Production Research*, (Vol. 15, 1977), pg. 219.

37. Shubin, J.A. and H. Madeheim, *Plant Layout*, Prentice-Hall, New York, 1951.

38. Apple, J.M., *Plant Layout and Material Handling*, The Ronald Press Company, New York, 1950.

39. Francis, R.L. and J.A. White, *Facility Layout and Location: An Analytical Approach*, Prentice-Hall, New York, 1974.

40. Morris, W.T., "Facilities Planning", *Journal of Industrial Engineering* (Vol. 9, 1958).

41. Reed, R., Jr., *Plant Layout: Factors, Principles, and Techniques*, R. D. Irwin, Homewood, Illinois, 1961.

42. Miller, D.M., and R.P. Davis, "A Dynamic Resource Allocation Model for a Machine Requirements Problem", *AIIE Transactions* (Vol. 10, No. 3, 1978), pp. 237-243.

43. Kusiak, A., "The Production Equipment Requirements Problem", *International Journal of Production Research*, (Vol. 25, No. 3, 1987), pp. 319-325.

44. Ueno, N., S. Sotojima, and J. Takeda, "Simulation-Based Approach to Design a Multi-Stage Flow-Shop in Steel Works", *Proceedings of the 24th Annual Simulation Symposium*, IEEE Computer Society Press, Los Angeles, California, (1991), pp. 332-337.

45. Suri, R. and R.R. Hildebrant, "Modelling Flexible Manufacturing Systems Using Mean Value Analysis", *Journal of Manufacturing Systems* (Vol. 3, No. 1, 1984), pp. 27-38.

46. Lawler, E.L., "The Quadratic Assignment Problem", *Management Science* (Vol. 9, 1963), pg. 586.

47. Gavett, J.W. and N.V. Plyter, "The Optimal Assignment of Facilities to Locations by Branch and Bound", *Operations Research* (Vol. 14, 1966), pg. 210.

48. Graves, G.W. and A.B. Whinston, "An Algorithm for the Quadratic Assignment Problem", *Management Science*, (Vol. 17, 1970), pg. 453.

49. Kaufmann, L. and F. Broeckx, "An Algorithm for the Quadratic Assignment Problem Using Benders' Decomposition", *European Journal of Operations Research*, (Vol. 2, 1978), pg. 204.

50. Bazaraa, M.S. and H.D. Sherali, "Bender's Partitioning Scheme Applied to a New Formulation of the Quadratic Assignment Problem", *Naval Research Logistics Quarterly*, (Vol. 27, 1980), pg. 29.

51. Hahn, P., T. Grant and N. Hall, "A branch-and-bound algorithm for the quadratic assignment problem based on the Hungarian method", *European Journal of Operational Research*, (Vol. 108, No. 3, 1998), pp. 629-640.

52. Ahuja, K. R., B.J Orlin. and A. Tiwari, "A greedy genetic algorithm for the quadratic assignment problem", *Computers & Operations Research*, (Vol. 27, No. 10, 2000), pp. 917-934.

53. Evans, G.W., M.R. Wilhelm, and W. Karwowski, "A Layout Design Heuristic Employing the Theory of Fuzzy Sets", *International Journal of Production Research*, (Vol. 25, No. 10, 1987), pp. 1431-1450.

54. Tompkins, J., J. White, Y. Bozer and J. Tanchoco. *Facilities Planning*, 3rd edition, John Wiley and Sons, 2002.

55. Seehof, J.M. and W.O. Evans, "Automated Layout Design Program", *The Journal of Industrial Engineering*, (Vol. 18, No. 12, 1967), pp. 690-695.

56. Lee, R.C. and J.M. Moore, "CORELAP – Computerized Relationship Layout Planning", *Journal of Industrial Engineering*, (Vol. 18, 1967), pg. 194.

57. Heragu, S. and A. Kusiak, "Knowledge Based System for Machine Layout (KBML)", *1988 International Industrial Engineering Conference Proceedings*, pp. 159-164.

58. Burdorf, A., B. Kampczyk, M. Lederhose and H. Schmidt-Traub, "CAPD-computer-aided plant design", *Computers and Chemical Engineering*, (Vol. 28, 2004), pp. 73-81

59. Maughan, F.G., and H.J. Lewis, "AGV controlled FMS", *International Journal of Production Research*, (Vol. 38, No. 17, 2000), pp. 4445-4453.

60. Mantel, R.J. and H.R.A. Landeweerd, "Design and operational control of an AGV system", *International Journal of Production Economics* (Vol. 41, 1995), pp. 257-266.

61. Wilhelm, M.R. and G.W. Evans, "The State-of-the-Art in AGV System Analysis and Planning", *Proceeding of the AGVS '87*, Pittsburgh, Pennsylvania (October 1987).

62. Gaskins, R.J. and J.M.A. Tanchoco, "Flow Path Design for Automated Guided Vehicle Systems", *International Journal of Production Research*, (Vol. 25, No. 5, 1987), pp. 667-676.

63. Chen, M., "A mathematical programming model for AGVs planning and control in manufacturing systems", *Computers & Industrial Engineering*, (Vol. 30, No. 4, 1996), pp. 647-658.

64. Usher, J.S., G.W. Evans, and M.R. Wilhelm, "AGV Flow Path Design and Load Transfer Point Location", *1988 International Industrial Engineering Conference Proceedings*, pp. 174-179.

65. Rabeneck, C.W., J.S. Usher, and G.W. Evans, "An Analytical Models for AGVS Design", *1989 International Industrial Engineering Conference & Societies' Manufacturing and Productivity Symposium Proceedings*, pp. 191-195.

66. Jafari, M.A. and J.G. Shanthikumar, "Determination of Optimal Buffer Storage Capacities and Optimal Allocation in Multistage Automatic Transfer Lines", *IIE Transactions* (Vol. 21, No. 2, 1989), pp. 130-135.

67. Bulgak, A.A. and J.L. Sanders, "Integrating a Modified Simulated Annealing Algorithm with the Simulation of a Manufacturing System to Optimize Buffer Sizes in Automatic Assembly Systems", *Proceedings of the 1988 Winter Simulation Conference*, pp. 684-690.

68. Kirkpatrick, S., C.D. Gelatt Jr., and M.P. Vecchi, "Optimization by Simulated Annealing", *Science*, (Vol. 220, 1983), pp. 671-680.

69. Chryssolouris, G. and M. Lee, "Use of Neural Networks for the Design of Manufacturing Systems", *Manufacturing Review*, (Vol. 3, No. 3, Sept. 1990), pp. 187-194.

70. MATLAB web site, http://www.mathworks.com

71. Sherman, W.R., and Craig A.B., *Understanding Virtual Reality: Interface, Application, and Design*, Morgan Kaufmann, 2002.

72. Dai, F., *Virtual Reality for industrial applications*, Berlin: Springer, 1998.

73. Ong, S.K. and A.Y.C Nee, *Virtual and Augmented Applications in Manufacturing*, Springer, 2004.

Review Questions

1. Define manufacturing system design.

2. How are the decision variables of a job shop different from the decision variables of a flow line?

3. How are the decision variables of continuous systems different from the decision variables of discrete-part systems?

4. Why is manufacturing system design difficult?

5. Why is discounted cash flow preferred over return on investment and payback as a method for the appraisal of a manufacturing system?

7. What are the difficulties of applying discounted cash flow techniques to the appraisal of a manufacturing system?

6. If a new manufacturing system requires an initial cash outlay of $30,000,000, and is expected to generate a steady net cash inflow of $k per year for 20 years, how large does k have to be in order for the investment in the manufacturing system to be justifiable from a financial point of view?

8. What are the advantages and disadvantages of organizing a manufacturing system as a job shop?

9. Consider the five methods of structuring the processing area of a manufacturing system. For each method, name five products that might be processed in that way.

10. Why do stationary part assembly systems tend to have a higher floor area than moving part assembly systems? Why do moving part assembly systems tend to be more expensive than stationary part assembly systems?

11. Describe the "trial and error" approach to the design of manufacturing systems.

12. How do some research formulations of manufacturing system design

problems differ from the actual design problems encountered in industry?

13. What sub-problems of the manufacturing system design problem are typically addressed by researchers?

14. When referring to a mathematical program, what is a *point*? What is the *feasible region*? What is the *optimal solution*?

15. All steel manufactured by a steel company must meet the following requirements: 3.2-3.5% carbon; 1.8-2.5% silicon; 0.9-1.2% nickel; tensile strength of at least 310 MPa. The company manufactures steel by combining two alloys. The cost and properties of each alloy are given in the table below. Assume that the tensile strength of a mixture of the two alloys can be roughly determined by averaging the tensile strength of the alloys that are mixed together. For example, a 1,000 kg mixture that is 40% Alloy 1 and 60% Alloy 2 has a tensile strength of 0.4 (290 MPa) + 0.6 (345 MPa). Formulate a mathematical program to find the most economical way of manufacturing 1,000 kg of steel. What kind of mathematical program is this?

	Alloy 1	Alloy 2
Cost per 1,000 kg	$190	$200
Percent silicon	2%	2.5%
Percent nickel	1%	1.5%
Percent carbon	3%	4%
Tensile strength	290 MPa	345 MPa

16. A steel company manufactures two types of steel at three different steel mills. During a given month, each steel mill has 200 hours of blast furnace time available. Because of differences in the furnaces at each mill, the time and cost to produce 1,000 kg of steel differs for each mill. The time and cost for each mill are shown in the table below. Each month the company must manufacture at least 500,000 kg of Steel 1 and 600 tons of Steel 2. Formulate a mathematical program to minimize the cost of manufacturing the desired steel. What kind of mathematical program is this?

	Steel 1		Steel 2	
	Cost	Time (min.)	Cost	Time (min.)
Mill 1	$10	20	$11	22
Mill 2	$12	24	$9	18
Mill 3	$14	28	$10	30

17. Is it easier to solve a linear program with few constraints or a linear program with a greater number of constraints? Why?

18. In goal programming, what is the significance of the coefficients of the objective function?

19. Production employees at a manufacturing firm work ten hours per day, four days per week. Each day of the week at least the following number of employees must be working: Monday–Friday, 7 employees; Saturday and Sunday, 3 employees. The company has set the following goals, listed in order of priority.

    Goal 1.   Meet employee requirements with 11 workers.

    Goal 2.   The average number of weekend days off per employee should be at least 1.5 days.

    Goal 3.   The average number of consecutive days off an employee gets during the week should exceed 2.8 days.

    Formulate a goal programming model that could be used to determine how to schedule the company's employees.

20. A company must complete three jobs. The amounts of processing time (in minutes) required to complete the jobs are shown in the table below. A job cannot be processed on machine j unless for all $i < j$ the job has completed its processing on machine i. The flow time for a job is the difference between the job's completion time and the time at which the job begins its first stage of processing. Formulate a mathematical program which can be used to minimize the flow time of the three jobs. What kind of mathematical program is this?

	Machine			
	1	2	3	4
Job 1	20	–	25	30
Job 2	15	20	–	18
Job 3	–	35	28	–

21. What is a multistage decision process? Briefly describe how dynamic programming may be used to optimize a multistage decision process.

22. At the beginning of each period a company must determine how many units to produce. A setup cost of $5 is incurred during each period in which production takes place. The production of each unit also incurs a $2 variable cost. All demand must be made on time, and there is a $1 per-unit holding cost on each period's ending inventory. During each period it is equally likely that demand will equal 0 or 1 unit. Assume that each period's ending inventory cannot exceed 2 units. Use dynamic programming to minimize the expected cost incurred during three periods.

23. What are the major components of a queueing system?

24. A CNC machine tool can process a part in 10 minutes. Part arrivals occur an average of 15 minutes apart, and are exponentially distributed. (a) On average, how many parts are waiting to be processed? (b) If the machine could be speeded up, what processing time would reduce the average wait time to 5 minutes?

25. Draw the search tree corresponding to the network shown below. The goal is to find a path from the start node, S, to the goal node, G. Determine the sequence of node visits when using depth-first search.

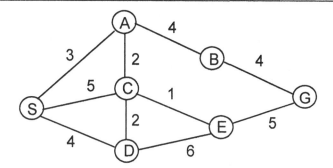

26. Repeat Question 25 for breadth-first search.

27. Describe how branch and bound search would proceed in order to find the optimal path from node S to node G for the network in Question 25.

28. Use A^* search to find the optimal path from node S to node G for the network in Question 25. For the underestimate of remaining distance, u(distance remaining), use the maximum of {actual remaining distance–2, 0}.

29. What are the major components of a rule-based system?

30. Under what circumstances is it advantageous to use a forward chaining inference engine? Under what circumstances is it advantageous to use a backward chaining inference engine?

31. Describe two potential generalization applications of neural networks.

32. What are the inputs and outputs of a computer simulator?

33. What are the advantages and disadvantages of computer simulation models versus analytical models of manufacturing systems?

34. What aspects of a manufacturing system require the generation of random variates when modeled using computer simulation?

6. The Operation of Manufacturing Systems

6.1 Introduction

The operation of a manufacturing system is the complex task of planning the material and information flows in the system. Proper material flow is what enables a manufacturing system to produce products on time and in sufficient quantity. It is a direct consequence of the system's information flows: *command* information from human planners or from planning software, prescribes the material flow in the system, while *sensory* information about the status of the system's resources is used to decide on the appropriate commands. The fundamental activity in the operation of a manufacturing system is thus determining the commands, which prescribe the material flow in the system.

Command information flows can be organized into a bi-level hierarchy. High-level commands dictate the flow of materials *into* the manufacturing system. Therefore, they determine the system's workload – namely, the number of commands for each type of the part to be manufactured in each time period. The time periods involved are typically long, in the order of days or weeks. For this reason, the decision-making activity of generating high-level commands is called *long-term planning*. Planning is concerned with determining the *aggregate timing of production* (when to produce and what quantities to produce).

Low-level commands dictate the flow of materials *within* and *out of* a manufacturing system. They determine which production resources are to be assigned to each operation on each part, and when each operation is to take place. The role of lower-level commands is to resolve contention for the resources of a manufacturing system. This contention occurs among the parts released into the system as a consequence of long-term planning decisions. Since low-level commands control individual operations, they must be generated much more frequently than high-level commands. The time between commands is typically in the order of seconds or minutes. For this reason, the decision-making activity of generating lower-level commands is called *short-term dispatching*. Dispatching is concerned with the *detailed assignment of operations to production resources*.

Long-term planning and short-term dispatching are collectively labeled, in broad terms, as "production scheduling". The specific terms, "(long-term) planning" and "(short-term) dispatching", will be used for greater precision whenever possible.

6.2 Academic Versus Industrial Perspectives

Academic research in production scheduling has been based primarily on optimization techniques (more specifically, mathematical programming). Many recent research efforts have employed other approximation techniques, including artificial intelligence methods and rule-based systems. Unfortunately, the use of academic results in industry has been rather minimal [1]; many academic approaches make assumptions that do not reflect reality. Actual manufacturing environments are extremely variable, and usually cannot be rigidly classified into one of the classical scheduling models often assumed in academic research. Perhaps, in no other field is the dichotomy between academic research and actual practice more pronounced.

The *academic* literature contains an abundance of theoretical work on a number of classical scheduling problems. The entities to be scheduled are typically referred to as *jobs*. Usually, each job corresponds to an individual part. The individual production operations, which make up a job, are referred to as *tasks*. Classical scheduling problems can be categorized on the basis of the following dimensions [2]:

- Requirements generation
- Processing complexity
- Scheduling criteria
- Scheduling environment

Requirements are generated either directly by customers' orders or indirectly by inventory replenishment decisions. This distinction is often made in terms of an *open shop* versus a *closed shop*. In an open shop, all jobs are by customer request and no inventory is stocked; production scheduling becomes a *sequencing problem* in which tasks corresponding to the requested jobs are sequenced at each facility. In closed shop, all customer requests are serviced from the inventory and jobs are generally the result of inventory replenishment decisions. For a closed shop, production scheduling involves lot-sizing decisions associated with the inventory replenishment process in addition to sequencing. Depending on how the requirements are generated, the specification of the requirements may be termed as *deterministic* or *stochastic*. For example, in an open shop, the processing time for each task of a job may be known, or may be a random variable with a specified probability distribution. Similarly, in a closed shop, the customer demand process, which drives the inventory replenishment decisions, may be assumed to be stochastic or deterministic.

Processing complexity is concerned with the number of tasks associated with each job, and is often classified as follows:

- One-stage, one processor (facility)
- One-stage, parallel processors (facilities)
- Multistage, flow shop
- Multistage, job shop

In the *one-stage, one processor* problem, which is also known as the one-machine problem, all jobs consist of one task, which must be performed on the one production facility. The *one-stage, parallel processors* problem is similar to the one-machine problem except that the jobs, which still contain a single task, may be performed on any of the parallel processors. For the *multistage* problem, each job requires processing at a set of distinct facilities, where there is a strict precedence ordering of the tasks for a particular job. The *flow shop* problem assumes that all jobs are to be processed on the same set of facilities with an identical precedence ordering of the tasks. In the *job shop* problem, there are no restrictions on the task sequence for a job and alternative routings for a job may be allowed. The $n/m/A/B$ notation is often used to represent processing complexity [3], where:

n	is the number of jobs. Each job is assumed to consist of m tasks.
m	is the number of resources (machines).
A	describes the flow pattern or discipline within the system. When $m = 1$, A is left blank. A may be:

	F	for the flow shop case, i.e., the machine order for all jobs is the same.
	P	for the permutation flow shop case. Here not only is the machine order the same for all jobs, but the job order is the same for each machine. Thus a schedule is completely specified by a single permutation of the numbers 1, 2, 3, ..., n, giving the order in which the jobs are processed on each and every machine.

G   for the general job shop case where there are no restrictions on the form of the technological constraints (constraints on the order in which the machines may process the jobs or the order in which the jobs may visit the machines).

B   describes the performance measure by which the schedule is to be evaluated.

As an example, *n/2/F/(total production time)* represents the *n* job, 2 machine, flow-shop problem where the aim is to minimize the total production time of the *n* jobs.

Scheduling criteria are measures by which schedules are to be evaluated, and may be classified broadly into *schedule costs* and *schedule performance measures*. Schedule *costs* include:

- Fixed costs associated with production setups or changeovers
- Variable production and overtime costs
- Inventory holding costs
- Shortage costs for not meeting deadlines or for stocking out
- Expediting costs for implementing "rush" jobs
- Costs for generating the schedule and monitoring the progress of the schedule

Common measures for schedule *performance* are:

- The utilization level of the production resources
- The percentage of late jobs
- The average or maximum tardiness for a set of jobs
- The average or maximum flow time for a set of jobs

Tardiness is the positive part of the difference between a job's actual completion time and its desired completion time. The flow time for a job is the difference between the completion time of the job and the time at which the job was released to the manufacturing system. In most production environments, the schedule evaluation is based on a mixture of both cost and performance criteria.

The *scheduling environment* deals with assumptions about the certainty of information regarding future jobs, and can be classified as being *static* or *dynamic*. In a static environment, the problem contains a finite set of fully specified jobs; no additional jobs will be added to this set, and none of the jobs in this set will be altered. In a dynamic environment, the problem contains not only an initial set of known jobs, but also additional jobs, which arrive over future time periods. The scheduling problem in most production environments is stochastic and dynamic; however, most models for scheduling problems are deterministic and static [2].

A number of scheduling methods that either give some guarantee as to the insensitivity of the schedule to future disruptions or explicitly reflect the uncertain nature of the available information, originate from the *control theory* [4]. These methods attempt to command material flow in a way that is robust with respect to disruptions, such as machine failures, operator absences, material shortages, surges in demand [5,6], and with respect to the absence or inaccuracy of status information [7]. Control theory views scheduling as a dynamic activity (Fig. 6.1) where the production scheduling problem is really one of understanding how to re-schedule [4]. The system under control operates on a sequence of inputs $u(t)$, yielding a sequence of outputs $y(t)$. The outputs at any time are a function of the state of the system $x(t)$. A general statement of the control problem is to determine a means of guiding the system state (and thus the system output) through time, according to some trajectory that satisfies the constraints of the system (e.g., maximum number of working hours per week) and that simultaneously satisfies some set of performance criteria.

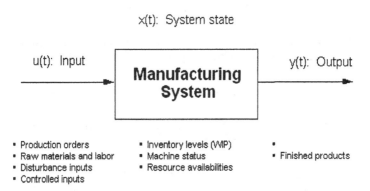

Fig. 6.1. A Manufacturing System from a Control Theory Perspective

The inputs to a manufacturing system include production orders, raw material and labor, disturbance inputs and controlled inputs. Production orders specify the quantities of various jobs to be processed and the dates on which these jobs are due. Disturbance inputs include machine failures and labor outages, over which the scheduler has little influence. Controlled inputs include scheduling, maintenance and overtime decisions, which the scheduler can regulate within bounds. The state of a manufacturing system defines the levels for all completed and partially completed jobs, the status of all machines (whether active, idle or under repair), the availability of labor and the inventories for all materials. Outputs of the manufacturing system may be defined as any portion of the state – for example, the inventory levels of all jobs ready for shipment on a specified date (Fig. 6.1 [4]).

Artificial Intelligence (AI) techniques, in particular rule-based systems, have been an early focus of academic research on production scheduling. Rule-based systems seek to capture generic scheduling rules that are applicable to a wide range of situations. In general, these rules must be tailored to meet specific system requirements [10]. Knowledge in a rule-based system can be classified into static knowledge (or data) and dynamic knowledge (or solution methods). Static knowledge includes all information about the manufacturing system itself (e.g., the number and types of machines) and the production objectives (e.g., the part types to be produced, along with their processing sequences, quantities, due dates etc.). Dynamic knowledge describes available expertise as to the way of deriving feasible schedules, and consists of the following:

- *Theoretical expertise*, which refers to operations research techniques that deal with the management of time and resources;
- *Empirical expertise*, which consists of heuristic dispatch rules;
- *Practical dedicated expertise*, which is provided by experienced shop floor supervisors, who are aware of the constraints that have to be taken into account when preparing the schedule.

In a rule-base, these types of expertise are captured in the form of rules. They are invoked by higher-level rules, which select a particular type of expertise on the basis of the status of the manufacturing system and the production scheduling problem to be solved.

In *industry*, it is customary to divide the operation of a manufacturing organization into three major levels: strategic planning, operations planning and detailed planning and execution [11]. At the strategic level, few decisions are made, but each decision takes a long time, and its impact is felt throughout the organization. At the detailed planning and execution level, many decisions are made, each requiring a much shorter time. Although the impact of each decision is local in time and in place, the great number of decisions, taken together, can have a significant impact on the performance of the organization. The characteristics of decisions at the operational planning level lie between those of the other two levels.

Strategic planning is performed by the top management of the organization. Several basic questions are asked and answered at this level, including: "What is our business?" and "How will we compete?" At this level the decision is made as to which manufacturing system structure (flow line, cellular system, job shop, project shop or continuous system) is most consistent with the organization's overall strategy. Furthermore, a timetable is established for the major activities of the organization, such as the entry into a new market or the acquisition of additional manufacturing capacity. Finally, performance measures are established for the organization, such as return on investment, market share, earnings, growth levels, and so forth.

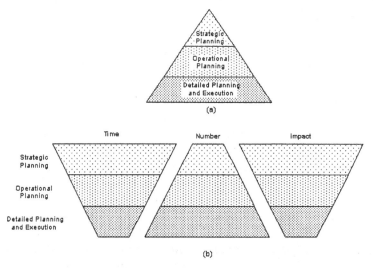

Fig. 6.2. (a) The Levels of Manufacturing Operation and (b) The Characteristics of Decisions Made at Each Level

The resulting overall corporate strategy is used to guiding the development of functional plans, primarily the marketing plan, the financial plan, and the manufacturing plan. The marketing plan specifies how the organization will compete in the marketplace. Included in this plan are forecasts of dollar sales, unit sales, and descriptions of distribution channels. Similarly, the financial plan specifies the role that finance will play in the overall corporate strategy. Plans for cash flow and capital expenditures are also included. Finally, the manufacturing plan specifies what manufacturing resources will be needed, how the resources will be deployed and how much is to be produced. The output component of the manufacturing plan is stated in aggregate terms, such as annual or quarterly units of production, for each major product line [12]. This plan also sets guidelines and policies within which manufacturing is expected to perform.

Ideally, any conflicts in strategies and policies among functions will be solved at the strategic level. Once acceptable strategic functional plans have been developed, the next step is to develop a more detailed operations plan.

In operational planning, the objectives of the strategic manufacturing plan are converted into more detailed and specific plans. One of the most important of these plans is the Master Production Schedule (MPS). The MPS is expressed in specific product configurations, quantities and dates and is frequently the key link between top management and the manufacturing facility. It is also a key link between manufacturing and marketing and serves to balancing the demands of the marketplace against the constraints of the manufacturing facility.

The master production schedule is one of the inputs to material requirements planning (MRP). The other inputs to MRP include, for each type of final product, the existing inventory level, bill of materials, and the production time or order lead time for each individual component. A bill of materials lists the subassemblies of a final product and the individual components which comprise each subassembly. The bill of materials and its role within MRP will be described in more detail, in the following section on long-term planning. The output of MRP is a plan, which describes the quantity of each individual component or subassembly to order or to begin producing in each time "bucket". Since the generation of the material plan requires a vast amount of data processing when many part types with possibly many layers of subassemblies are involved, it is usually handled by computer programs called MRP systems. MRP systems calculate material requirements on the basis of infinite capacity, meaning that the production and assembly times assumed for individual components and subassemblies do not incorporate waiting times due to resource contention.

Neither of the planning activities introduced thus far, master production scheduling nor material requirements planning, does it incorporate any consideration of whether the manufacturing system has enough capacity to achieve the demanded levels of production. This is the role of capacity planning. Capacity planning provides projections of the capacity needs implied by the material plan, so that timely actions can be taken to balance the capacity requirements with available capacity. In general, four variables may be adjusted to balancing required and available capacity:

- Inventory levels
- The number of shifts worked
- The quantity of resources used for production
- The amount of orders accepted for production

The first three options adjust the effective available capacity of the manufacturing system to fitting the required capacity. The last option results in the modification of the master production schedule, which in turn, results in a more feasible material plan. It adjusts the "required" capacity to meeting the available capacity.

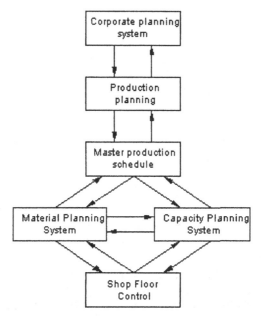

Fig. 6.3. Major Components of the Manufacturing Planning System

The third and final level of planning is the detailed planning and execution level. The shop floor control system is the main operational system at this level. The shop floor control system is responsible for detailed planning, such as determining the priority sequence of orders in a queue, and for monitoring the execution of the detailed plans, such as capacity utilization at the machine center (i.e. input/output control). The shop floor control system is also responsible for collecting data from the shop floor and directing the flow of information back up to the operations planning level. After interpretation and aggregation, this information will pass from the operational level to the strategic level (Fig. 6.3 [13]).

6.2.1 Master Production Scheduling

Master production scheduling (MPS) is a form of aggregate planning. Typically it considers only finished products, not the individual components, which make up those products. The objective of MPS is to develop an aggregate production schedule that will meet seasonal requirements for finished products, and at the same time, will minimize the incremental costs incurred. It is important to identify and measure these costs so that alternative master production schedules may be evaluated on the basis of total cost. Some costs that may be relevant are:

- Payroll costs
- Costs of overtime, extra shifts, and subcontracting
- Costs of hiring and laying off workers
- Costs of excess inventory and backlog
- Costs of production rate changes

The costs that are included in the total cost model should vary with changes in the decision variables. In a MPS, these decision variables typically denote the size of the work force to be employed, the number of overtime hours to be scheduled, and the quantity of each finished product to be produced during each time "bucket" of some planning horizon. However, the behavior of cost with respect to changes in these decision variables, is not necessarily easy to quantify. Often, approximations are made by assuming the costs to be linear or quadratic functions of the appropriate decision variables. This permits the easier use of optimization techniques, such as that of linear programming.

The simplest structure for the MPS is represented by the single-stage system (Fig. 6.4 [16]), in which the planning horizon is only one period long. The state of the system at the end of the previous period, is defined by W_O, the aggregate work force size, P_O, the production rate, and I_O, the inventory level. The ending state conditions for one period, become the initial conditions for the upcoming period. A forecast of the finished product requirements for the upcoming period results through some process in decisions that set the size of the work force and production rate for the upcoming period; the projected ending inventory is then, $I_1 = I_O + P_1 - F_1$, where F_1 is forecasted sales.

The decision making objective of a multistage planning structure (Fig. 6.5 [16]) is the same as that of the single stage, that is, to decide on a suitable work force size and production rate for the upcoming period. However, the decision for the upcoming period is affected by the future period forecasts, and the decision process must consider the cost effects of the sequence of decisions. The connecting links among the several stages are the W, P and I values at the end of one period and the beginning of the next. The feedback loop from the decision process may involve some iterative or trial-and-error procedures in order to obtain a solution.

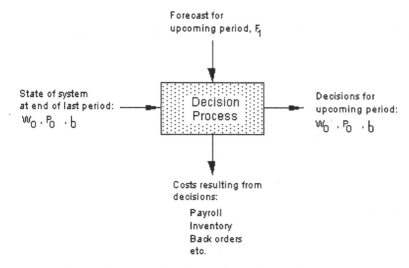

Fig. 6.4. Single-Stage Planning Decision for a Horizon of One Period

Many approaches for making MPS decisions have been proposed in the literature. These make use of a variety of tools from operations research and artificial intelligence, including mathematical programming, dynamic programming, and search [14, 16].

To perceive the effect of different MPS decisions, consider a numerical example. Assume that the normal plant capacity is 350 units per day and an additional capacity up to a maximum of 410 units per day, may be obtained through overtime, at an additional cost of $10 per unit. Column 6 of Table 6.1 shows buffer inventories, which are the minimum stocks required. Their purpose is to foresee that market requirements could be greater than those expected. The buffer inventories for each month, if added to the cumulative production requirements in column 5, result in the cumulative maximum requirements shown in column 7. Column 8 provides the basis for weighting the buffer inventory by production days and for computing the average buffer inventory of 3045.9 units in the table footnote.

The following three alternative master production schedules, Plans 1, 2 and 3, can be considered:

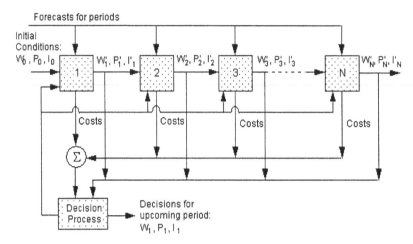

Forecasts for periods

Fig. 6.5. Multistage Planning Decision for a Planning Horizon of N Periods

Plan 1 - Level Production. The simplest production plan is to establish an average output level that meets annual requirements.

(1)	(2)	(3)	(4)	(5)	(6)	(7)	(8)	(9)
						Cumulative		Production
						Maximum		Requirements
		Cumulative	Expected	Cumulative	Required	Production		per Production
	Production	Production	Production	Production	Buffer	Requirements		Day
Month	Days	Days	Requirements	Requirements	Inventories	col.5 + col.6	col.2 x col.6	col.4 / col.2
January	22	22	5000	5000	2800	7800	61600	227.3
February	20	42	4000	9000	2500	11500	50000	200
March	23	65	4000	13000	2500	15500	57500	173.9
April	19	84	5000	18000	2800	20800	53200	263.2
May	22	106	7000	25000	3200	28200	70400	318.2
June	22	128	9000	34000	3500	37500	77000	409.1
July	20	148	11000	45000	4100	49100	82000	550
August	23	171	9000	54000	3500	57500	80500	391 3
September	11	182	6500	60500	3000	63500	33000	590.9
October	22	204	6000	66500	3000	69500	66000	272.7
November	22	226	5000	71500	2800	74300	61600	227.3
December	18	244	5000	76500	2800	79300	50400	277.8
							743200	

[a]Average Buffer Inventory = 743200/244 = 3045.9 units

Table 6.1. Forecast of Production Requirements and Buffer Inventories: Cumulative Requirements, Average Buffer Inventories[a], and Cumulative Maximum Production Requirements

The total annual requirements are the last figure in the cumulated requirements schedule in column 5 of Table 6.1, which is 76,500 units.
Since there are 244 working days, an average daily output of 76,500/244 =
314 units should cover requirements. The level production strategy is
simple: accumulate seasonal inventory during the slack requirements
months for use during peak requirements months. The level production
plan is shown in relation to the production requirements per day in Figure
6.6 [16] as Plan 1.

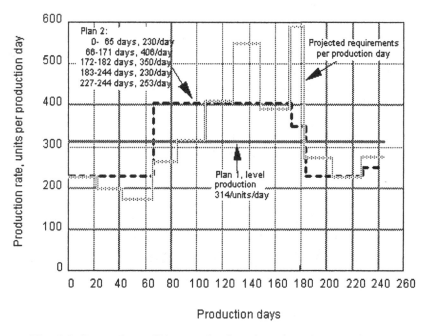

Fig. 6.6. Comparison of Two Production Plans that Meet Requirements

The inventory requirements for Plan 1 are calculated in Table 6.2 [16].
The production in each month is computed in column 3 and is cumulated
in column 4 to produce a schedule of units available each month, starting
with a beginning inventory of 2800 units, the required buffer inventory in
January. Then by comparing the units available in column 4 with the cumulative maximum requirements schedule in column 5, one can generate
the schedule of seasonal inventories in column 6. The significance of the
negative seasonal inventories is that the plan calls for dipping into buffer
stocks. In August, 1006 units will be used out of the planned buffer of
3500 units, but in September, the planned buffer of 552 units will be exceeded. In other words, the plan would actually require either back-
ordering or the expected loss of the sale of 552 units in September. The

plan recovers in subsequent months and meets aggregate requirements, but it incurs total shortages for the year of 7738 (1006+3552+2664+536) units.

The buffer inventory should not be used since the buffer was designed to absorb unexpected increases in sales and if they were used, they would lose their buffering function. The negative seasonal inventories can be taken into account by increasing the beginning inventory by the most negative seasonal inventory balance, -3352 units in September. This new beginning inventory level has the effect of increasing the entire schedule of cumulative units available in column 4 of Table 6.2 by 3552 units. Then, average seasonal inventories will also be increased by 3352 units.

The average seasonal inventory for Plan 1 is calculated in Table 6.2 as 3147.6 units, weighted by production days, assuming that buffer stocks are used and shortages are as indicated in column 6. If the plan was revised so that the buffer inventories would not be used, the average seasonal inventory would be 3147.6+3552 = 6699.6 units. Assuming that inventory holding costs are $50 per unit per year and that shortage costs are $25 per unit short, the relative costs of the variants of Plan 1 can be computed. If at the beginning the inventories are only 2800 units, the annual inventory costs are 50x147.6 = $157,380, and the shortage costs are 25x7738 = $193,450. The total incremental costs are then $350,830. By comparison, if the buffer inventory was not used, the average seasonal inventories would be 6699.6 units at a cost of $50x699.6 = $334,980.

	(1)	(2)	(3)	(4) Cumulative units Available. Cumulative Production + Beg. Inventory (2800)	(5) Cumulative Maximum Requirements. col.7 of Table 6.2	(6) Seasonal Inventory* col.4 - col.5
Month	Production Days	Production Rate Units/day	Production per Month col.1 x col.2			
January	22	314	6908	9708	7800	1908
February	20	314	6280	15988	11500	4488
March	23	314	7222	23210	15500	7710
April	19	314	5966	29176	20800	8376
May	22	314	6908	36084	28200	7884
June	22	314	6908	42992	37500	5492
July	20	314	6280	49272	49100	172
August	23	314	7222	56494	57500	-1006
September	11	314	3454	59948	63500	-3552
October	22	314	6908	66856	69500	-2644
November	22	314	6908	73764	74300	-536
December	18	314	5652	79416	79300	116

*Average seasonal Inventory (positive values in column 7/days) = 768,010/244 = 3147.6 units

Table 6.2. Calculation of Seasonal Inventory Requirements for Plan 1

Given these costs for holding inventories and incurring shortages, it is obviously more economical to plan on larger inventories whilst in other situations, the reverse might be true. If the cost of shortages was only $20 per unit, it would be slightly more economical to take the risk. Alternatively, if the costs of holding inventories were $62 per unit per year and the shortage costs were $25 per unit, then the balance of costs would again favor taking the risk of incurring shortages. Of course, there are other factors that enter into the decision of whether or not to risk incurring shortages, such as the potential of losing market share permanently.

Plan 1 has several advantages. First, it does not require the hiring or laying off of personnel. It provides stable employment for workers and would be favored by the organized labor. Also, scheduling is simple: 314 units per day. From an incremental production cost viewpoint, however, it fails to consider whether or not there is an economic advantage in trading off the large seasonal inventory and shortage costs for overtime costs and/or costs incurred from hiring or laying off personnel to meet seasonal variations in requirements.

Plan 2-Using Hiring, Layoff and Overtime. Note from Figure 6.6 that normal plant capacity allows an output of 350 units per day and that an additional 60 units per day can be obtained through overtime work. Units produced on overtime, cost an additional $10 per unit in this example.

Up to the normal capacity of 350 units per day, we can increase or decrease output by hiring or laying off labor. A worker hired or laid off affects the net output by 1 unit per day. The cost of changing output levels in this way, is $200 per worker hired or laid off, owing to hiring, training, severance, and other associated costs.

Plan 2 offers the additional options of changing basic output rates and using overtime for peak requirements. Plan 2 is shown in Figure 6.6 and involves two basic employment levels: labor to produce at normal output rates of 230 and 350 units per day. Additional variations are achieved through the use of overtime when it is needed. The plan has the following schedule:

0 to 65 days   – produce at 230 units per day.
66 to 171 days   – produce at 406 units per day (hire 120 workers to increase basic rate without overtime from 230 to 350 units per day; produce 56 units per day at over time rates).
172 to 182 days   – produce at 350 units per day (no overtime).
183 to 226 days   – produce at 230 units per day (layoff 120 workers

to reduce basic rate from 350 to 230 units per day again).

227 to 244 days   – produce at 253 units per day (23 units per day at overtime rates).

To counterbalance the inventory reduction, we must hire 120 workers in April and lay off an equal number in October. Also, we have produced a significant number of units at overtime rates, from April to September and in December. The costs of Plan 2 are summarized in Table 6.3 [16]. The total incremental costs of Plan 2 are $229,320, which is 65 percent of Plan 1 and 68 percent of Plan 1 without shortages.

Plan 2 is a more economical plan, but it requires substantial fluctuations in the size of the work force. Perhaps some of the variations can be absorbed by more overtime work, and in some kinds of industries, subcontracting can be used to meet the most severe peak requirements.

Month	(1) Production Days	(2) Production Rate Units/day	(3) Units of Production Rate Change	(4) Units Produced at Overtime Production Rates Greater Than 350 or 230 x col. 1
January	22	230	0	0
February	20	230	0	0
March	23	230	0	0
April	19	406	120	1064
May	22	406	0	1232
June	22	406	0	1232
July	20	406	0	1120
August	23	406	0	1288
September	11	350	0	0
October	22	230	120	0
November	22	230	0	0
December	18	253	0	414
			240	6350

Production rate change costs = 240x200 = (A change in the basic rate of one unit requires the hiring or layoff of one worker at $200 each).	$48,000
Overtime costs at $10 extra per unit = 10x6,350 =	$63,500
Seasonal Inventory cost (2356.4 units at $50 per unit per year) = 50x2356,4 =	$117,820
Total incremental cost =	$229,320

Table 6.3 Calculation of Incremental Costs for Plan 2

Plan 3 - Adding Subcontracting as a Source. A third alternative that involves smaller work force fluctuations by using overtime, seasonal inventories, and subcontracting to absorb the balance of requirements fluctuations can be considered and has the following schedule:

0 to 84 days	– produce at 250 units per day.
85 to 128 days	– produce at 350 units per day (hire 100 workers to increase basic rate without overtime from 250 to 350 units per day).
129 to 148 days	– produce at 410 units per day (60 units per day produced on overtime, plus 1700 units subcontracted).
149 to 171 days	– produce at 370 units per day (20 units per day produced on overtime).
172 to 182 days	– produce at 410 units per day (60 units per day produced on overtime, plus 1380 units subcontracted).
183 to 204 days	– produce at 273 units per day (23 units per day produced on overtime; layoff 100 workers to reduce employment level from basic rate of 350 to 250 units per day).
205 to 244 days	– produce at 250 units per day.

Plan 3 reduces seasonal inventories even further to an average of only 1301 units. Employment fluctuation is more modest, involving the hiring and laying off of only 100 workers. Only 2826 units were produced at overtime rates, but a total of 3080 units are subcontracted at an additional cost of $15 per unit.

Table 6.4 [16] summarizes the costs for all three plans. For the particular example, Plan 3 is the most economical. The buffer inventories are nearly the same for all the plans, so their costs are not included as incremental. Even though Plan 3 involves less employment fluctuation than Plan 2, it may still be felt to be so severe. Other plans involving less fluctuation could be developed and their incremental costs are determined in the same way.

Although Figure 6.6 [16] shows the effects of the production rate changes quite clearly, it is actually easier to work with the cumulative curves shown in Figure 6.7 [16]. The procedure is to plot first, the cumulative production requirements. The cumulative maximum requirements' curve is then simply the former curve with the required buffer inventories added for each period. The cumulative graph of maximum requirements can then be used as a basis for generating alternative program proposals.

Any production program that is feasible, must fall entirely above the cumulative maximum requirements line. The vertical distances between the proposed curves and the cumulative maximum requirements' curve represent the seasonal inventory accumulation for each plan.

Costs	Plan 1		Plan 2 ($)	Plan 3 ($)
	With Shortages ($)	Without Shortages ($)		
Shortages[a]	193,450	-	-	-
Seasonal Inventory[b]	157,380	334,980	117,820	65,070
Labor turnover[c]	-	-	48,000	40,000
Overtime[d]	-	-	63,500	28,260
Subcontracting[e]	-	-	-	46,200
Totals	350,830	334,980	229,320	179,530

[a]Shortages cost $25 per unit.

[b]Inventory carrying costs are $50 per unit per year.

[c]An increase or decrease in the basic production rate of one unit requires the hiring or layoff of one employee at a hiring and training, or severance, cost of $200 each.

[d]Units produced at overtime rates cost an additional $10 per unit.

[e]Unit subcontracted cost an additional $15 per unit.

Table 6.4. Comparison of costs of Alternate Production Plans

The graphical methods discussed above are simple and have the advantage of allowing alternative programs to be visualized over a broad planning horizon. The difficulty with graphical methods, is the static nature of the model. In addition, the process does not itself generate good programs; it simply compares proposals that have been made. Mathematical and or search models attempt to find optimal combinations of sources of short-term capacity, such as seasonal inventories, overtime capacity and subcontracting. As an example, we will briefly discuss the Linear Decision Rule and a linear programming approach.

The *Linear Decision Rule* (LDR) is based on the development of a quadratic cost function for the company in question which incorporates costs due to regular payroll, hiring and layoff, overtime, inventory holding, back-ordering, and machine set-up. The problem is to minimize the sum of the monthly combined cost function over the planning horizon time of N periods. Two linear decision rules are used to compute the aggregate size of the work force and the production rate for the upcoming period.

Both rules require as inputs the forecast for each period of the planning horizon in aggregate terms, the size of work force, and the inventory level in the last period.

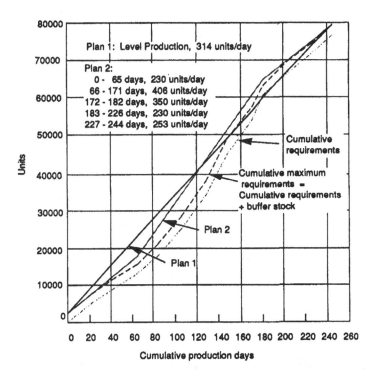

Fig. 6.7. Cumulative Graphs of Requirements and Alternate Introduction Programs

The Linear Decision Rule Model is:

$$\text{Minimize } C = \sum_{t=1}^{N} C_t \tag{6-1}$$

where

$$
\begin{aligned}
C_t = [\,(\,c_1 W_t\,) &\qquad \text{(regular payroll costs)} \\
+ c_2 (\,W_t - W_{t-1}\,)^2 &\qquad \text{(hiring and layoff costs)} \\
+ c_3 (P_t - c_4 W_t)^2 + c_5 P_t - c_6 W_t &\quad \text{(overtime costs)} \\
+ c_7 (I_t - c_8 - c_9 S_t)^2\,] &\qquad \text{(inventory} \qquad \text{connected}
\end{aligned}
\tag{6-2}
$$

costs)

subject to the constraints

$$I_{t-1} + P_t - F_t = I_t \qquad t = 1, 2, ..., N \qquad (6\text{-}3)$$

where

P_t =   The number of units of product that should be produced during the forthcoming month t.

W_{t-1} =   The number of employees in the work force at the beginning of the month (the end of the previous month).

I_{t-1} =   The number of units of inventory, minus the number of units on back-order, at the beginning of the month.

W_t =   The number of employees that will be required for the current month. The number of employees that should be hired is therefore $W_t - W_{t-1}$.

S_t =   A forecast of number of units of product that will be ordered for shipment during the current month t.

The total cost for N periods is given by Equation 6-1 and the monthly cost, C_t, is given by Equation 6-2. Equation 6-3 states the relationship between the initial inventory, production, and sales during the month and the final inventory. The decision rules for the work force level and the production rate are obtained by differentiating Equation 6-2 with respect to each decision variable.

Most *linear programming* approaches to MPS are variations of the model shown in Figure 6.8 [16] (Eqs. 6-4 to 6-8). The objective function (Eq. 6-4) minimizes, over the entire planning horizon of T periods, the sum of regular production costs, regular and overtime labor costs, inventory and backlogging costs, and work force change costs.

Constraint 6-5 merely states that the final inventory, I_t, equals the initial inventory plus production during the period, minus demand during the period. Constraint 6-6 states a further restriction on the final inventory, I_t, to take into account the balance between inventory on hand and back-orders. Constraint 6-7 states that the work force level in period t, W_t, measured in regular time hours, is equal to the level of the previous period, W_{t-1}, plus any additions or decreases due to hiring or layoff. Constraint 6-8 states a required balance between the sum of overtime hours scheduled less undertime or unused capacity in period t and the hours used to produce the scheduled number of units, X_t, less the work force size in hours available.

Minimize

$$Z = \sum_{t=1}^{T} \left(c_t X_t + L_t W_t + L'_t O_t + h_t I_t^+ + \pi_t I_t^- + \theta_t w_t^+ + \theta'_t w_t^- \right) \quad (6\text{-}4)$$

Subject to, for $t = 1, 2, \ldots, T$

$$I_t = I_{t-1} + X_t - S_t \quad (6\text{-}5)$$

$$I_t = I_t^+ - I_t^- \quad (6\text{-}6)$$

$$W_t = W_{t-1} + w_t^+ - w_t^- \quad (6\text{-}7)$$

$$O_t - U_t = m X_t - W_t \quad (6\text{-}8)$$

$X_t, I_t^+, I_t^-, W_t, w_t^+, w_t^-, O_t, U_t$ all ≥ 0

where:

S_t	=	Demand in period t (in units of product).
W_t	=	Work force level in period t, measured in regular time hours
w_t^+	=	Increase in work force level from period t-1 to t (in hours)
w_t^-	=	Decrease in work force level from period t-1 to t (in hours)
O_t	=	Overtime scheduled in period t (in hours)
U_t	=	Undertime (unused regular time capacity) scheduled in period t (in hours)
X_t	=	Production scheduled for period t (in hours)
I_t^+	=	On-hand inventory at the end of period t
I_t^-	=	Back order position at the end o period t
m	=	Number of hours required to produce one unit of product
L_t	=	Cost of an hour's worth of labor on regular time in period t
L'_t	=	Cost of an hour's worth of labor on overtime in period t
θ_t	=	Cost to increase the work force level by one hour in period t
θ'_t	=	Cost to increase the work force level by one hour in period t
h_t	=	Inventory carrying cost, per unit held from period t to t+1
π_t	=	Back order cost, per unit carried from period t to t-1
c_t	=	Unit variable production cost in period t (excluding labor)

Fig. 6.8. Generalized Linear Optimization Model for Aggregate Planning

This equation ensures that all production occur on regular time or overtime. The undertime variable, U_t, is a slack variable that takes into account the fact that the work force may not always be fully utilized. Finally, all variables are restricted to be nonnegative.

A computer search procedure may be used to evaluate systematically a cost or profit criterion function at trial points. When using this procedure, it is hoped that an optimum value may be eventually found, but there is no guarantee. In direct search methods, the cost criterion function is evaluated at a point, the result is compared with previous trial results, and a move is determined on the basis of a set of heuristics ("rules of thumb"). The new point is then evaluated, and the procedure is repeated until it is determined that a better solution, resulting in an improved value of the objective function, cannot be found or until the predetermined computer time limit is exceeded.

The costs to be minimized are expressed as a function of production rates and work force levels in each period of the planning horizon. Therefore, each period included in the planning horizon requires that two dimensions be added to the criterion function, one for the production rate and one for the work force size.

In rolling horizon master production schedules, the model is solved and implemented for the immediate decision period and then updated and resolved one or more periods later. In this case, MPS performance is often influenced by design options, such as re-planning frequency and forecast window interval as well as lot-sizing and inventory policies [15].

6.2.2 Material Requirements Planning

Material Requirements Planning (MRP) obtains future requirements for finished products from a master production schedule and uses this and other information to generate the requirements for all the sub-assemblies, the components and the raw materials that will be making up the finished product. Although MRP is a generic methodology, it is almost always implemented via computer software called MRP systems.

MRP is specifically concerned with the manufacture of multi-component assemblies and relies on the fact that the demands for all sub-assemblies, components and raw materials are known and are dependent upon the demand for the finished products. The sub-assemblies, components, and raw materials are said to have "dependent demands." There are also some items, which have "independent demands," that is, the demand for them does not depend on the demand for any other item.

In order to carry out its calculations, the MRP system requires three types of information:

1. The *master production schedule* for the finished products,
2. The *bill of material* which contains details as to the kind of raw materials, components and sub-assemblies that go to make up the finished products,
3. Information concerning the *status of all materials* held in the inventory.

Having performed its calculations, MRP then produces reports, the majority of which are concerned with *what* and *when* to *manufacture* or *order* (Fig. 6.9).

Bills of material (BOM) play a key role in MRP. They are sets of files which contain the "recipe" for each finished product. Each "recipe" consists of information regarding which materials, components and sub-assemblies go together to make up a finished product, held on what is often known as a product structure file. Furthermore, all the standard information about each item, such as part number, description, unit of measure, lead time for manufacturing or procurement etc., is held on what is often known as a part master file.

For each finished product a bill of material is originally created from the design and production engineering information. This information will initially be in the form of drawings (Fig. 6.10 [17]) and assembly charts (Fig. 6.11 [17]), which, together with information on the relevant lead times, form the basis of the inputs to the BOM (Fig. 6.12).

The information on the BOM needs to be accurate, since inaccuracies can lead to incorrect items or incorrect quantities of items being ordered. However, the problem of file accuracy is complicated by the fact that in many operating environments there are continual changes to the BOM in the form of product modifications. These modifications may originate from many sources, for example, safety legislation, production process changes and improvements for marketing purposes.

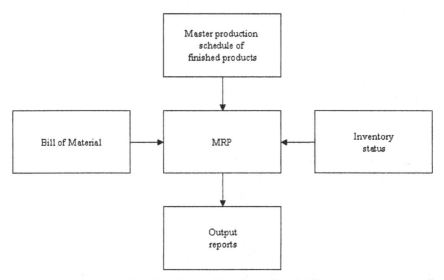

Fig. 6.9. General MRP System Outline

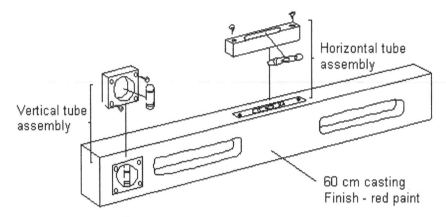

Fig. 6.10. Design Drawing (Simplified) of Part No. A1234 - A 60 cm Spirit Level

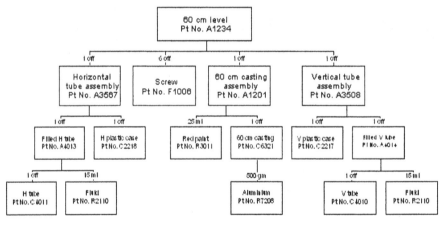

Fig. 6.11. Assembly Chart (Simplified) for Part No. A1234 - A 60 cm Spirit Level

Figure 6.12 is a bill of materials for the 60 cm spirit level (Fig. 6.10). It is called an *indented bill* because the dependence of parts and components is indicated by indenting items in the list. All the numbers preceding the materials indicate the quantity of that item required for one unit of the item on which it is dependent. For example, each horizontal tube assembly (Part #A3567) requires 15 ml of fluid (Part #R2110).

The *inventory status file* keeps a record of all transactions and balances of all stock throughout the organization. The transactions are mainly receipts and issues, starting with the "on order" condition and eventually finishing with the issue of the completed product out of the factory.

	Part #
1 60 cm spirit level	A1234
1 Horizontal tube assembly	A3567
1 Filled Horizontal tube	A4013
1 Horizontal tube	C4011
15 ml Fluid	R2110
1 Horizontal plastic case	C2218
6 Screw	F1006
1 60 cm casting assembly	A1201
25 ml Red paint	R3011
1 60 cm casting	C6321
500 gm Aluminium	R7208
1 Vertical tube assembly	A3508
1 Vertical plastic case	C2217
1 Filled vertical tube	A4014
1 Vertical tube	C4010
15 ml Fluid	R2110

Fig. 6.12. Bill of Materials for 60 cm Spirit Level

Other transactions may record occurrences, such as inspection rejects or stock adjustments as a result of physical stock checks. Many organizations use the concept of the "stock condition" to describe the stage to which the material has reached, for example: 500 kg of aluminium part #R7208 has been inspected and is in the raw material store ready for issue into work in progress.

The MRP system carries out calculations on a level by level basis which converts the master schedule of finished products into suggested orders for all the sub-assemblies, components and raw materials. These calculations or "requirements generation runs" are likely to be carried out on the computer every one or two weeks so that the situation is kept constantly up to date.

At each level of assembly breakdown, the MRP system undertakes three steps in its calculations before continuing to the next lower level. These steps are as follows:

1. It generates *gross requirements* for the item by "exploding" the "planned start" quantities of the next higher level assembly, by reference to the bill of material structure file. For example, a "planned start" of 200 spirit levels (Part No. A 1234) in week 15 would be ex-

ploded to give gross requirements of 1,200 screws (Part No. F1006) in week 15. (Refer to Figure 6.11 for an assembly chart).

2. The gross requirements are amended by the amount of inventory of that item that is expected to be available in each week i.e. on hand plus scheduled receipts. This information is obtained from the inventory status file and the amended requirements are called the *net requirements*, e.g. in week 15 a total of 800 screws are expected to be available, so the gross requirement of 1,200 is amended to give a net requirement of 400 screws in week 15.

3. The net requirements are then offset by the relevant lead time for initiating the manufacture or purchase of the item, e.g. if the lead time for the screws (Part No. F1006) is four weeks, the *net* requirement of 400 screws in week 15 are offset as follows:

Week No.	11	12	13	14	15	16
Net requirements					400	
Planned starts	400					

To summarize, in its simplest form, the MRP system would calculate the requirement of screws for each period of the planning horizon as in Figure 6.13 [17].

This calculation assumes that the only use of the screw (Part No. F1006) is in the assembly of the 60 cm spirit level. If this were not the case and if its usage were common to other products assembled by the organization, for example 40 cm and 80 cm spirit levels, then the gross requirements for the screw would have been the aggregated requirements generated from the planned starts of all the assemblies using that screw.

Safety stocks are held in any manufacturing system to cater for uncertainty. In an MRP system the major cause of uncertainty, that of the future usage of the item, has been mainly eliminated since items should be produced to meet a plan – the master schedule. However, safety stocks are still needed because of uncertainties in supply both in terms of the variation of actual lead times and the variation of quantities supplied, caused by inspection rejects and material shortages. There will also be changes of demand, caused by short-term emergency changes to the master schedule and unexpected demands for items, such as spares.

Statistical techniques of establishing safety stocks have been developed for application to MRP; these fall into three main categories: fixed quantity safety stock, safety times, and percentage increases in requirements.

Pt No. A1234	60 cm spirit level assembly						
Week No.	10	11	12	13	14	15	16
(Master schedule) Planned starts	400	300	200	200	300	200	400

Pt No. F1006	Screw		LT = 4				
Week No.	10	11	12	13	14	15	16
Gross requirements	2,400	1,800	1,200	1,200	1,800	1,200	2,400
Scheduled receipts		6,000					
Projected stock on hand 3,200	800	5,000	3,800	2,600	800	0	0
Net requirements						400	2,400
Planned starts		400	2,400				

Fig. 6.13. MRP Calculation Example

Fixed quantity safety stocks are introduced by triggering a net requirement, whenever the projected stock on hand reaches a safety stock level, as opposed to zero. Figure 6.14 [17] shows examples of MRP calculations with and without a fixed quantity safety stock.

The calculation of the size of the fixed quantity stock should be related to the cause of the unexpected usage during the lead time. For example, if the unplanned demand is primarily due to unforecasted spares demand for the item, then a historical analysis of this variation may lead towards the setting of a satisfactory safety stock level.

However, since in most cases, variations in usage and supply could be the result of many factors and since there are not any "scientific" methods of setting safety stocks in MRP systems, "rule of thumb" approximations have been applied. For example, it may be satisfactory to initially set the safety stock level at one week's average requirement or one week's "maximum" requirement. However, it is essential that the usage of the safety stock be monitored and the level then be adjusted accordingly, i.e. too frequent use of the safety stock would suggest the need for a higher safety level whereas; infrequent use would suggest a lower one.

The *safety time* approach for setting safety margins is essentially "planning to make items available earlier than they are required." The introduction of safety time is straightforward in that the net requirements are offset by the lead time plus, the safety time to produce planned starts. Figure 6.15 [17] shows examples of MRP calculations with and without a safety time.

Pt No. C2218
H. Plastic case   LT = 1   Safety quantity = 0

Week No	13	14	15	16	17	18
Gross requirements	200	300	200	400	300	500
Scheduled receipts	--	--	--	--	--	--
Projected stock on hand 600	400	100	0	0	0	0
Net requirements	--	--	100	400	300	500
Planned starts	--	100	400	300	500	--

a) Without safety stock

Pt No. C2218
H. Plastic case   LT = 1   Safety quantity = 150

Week No	13	14	15	16	17	18
Gross requirements	200	300	200	400	300	500
Scheduled receipts	--	--	--	--	--	--
Projected stock on hand 600	400	150	150	150	150	150
Net requirements	--	50	200	400	300	500
Planned starts	50	200	400	300	500	--

b) With safety stock

Fig. 6.14. The Effect of a Fixed Quantity Safety Stock

The choice of the length of the safety time could, perhaps, be related to the variability of the manufacturing or procurement lead time of the item being considered. However, since other factors may influence the use of the safety stock generated by the use of safety time, an arbitrary setting of the safety time and subsequent adjustment, based on the monitoring of the usage of the safety stock, is necessary.

The setting of safety margins by the *percentage increases in requirements* method is particularly suitable for dealing with the variations in supply, caused by scrap or process yield losses and are often implemented as "scrap factors" or "shrinkage factors." This type of safety margin is introduced by increasing the net requirements by a factor to produce planned starts. An example is shown in Figure 6.16 [17].

MRP is an effective ordering method for dependent demand situations, where the placement of an order or the start of a production batch for a part, is determined by the timing and usage of the part in a subsequent stage of production.

Pt No. C2218
H. Plastic case   LT = 1   Safety Time = 0

Week No	13	14	15	16	17	18
Gross requirements	200	300	200	400	300	500
Scheduled receipts	--	--	--	--	--	--
Projected stock on hand 600	400	100	0	0	0	0
Net requirements	--	--	100	400	300	500
Planned starts	--	100	400	300	500	--

a) Without safety time

Pt No. C2218
H. Plastic case   LT = 1   Safety Time = 1

Week No	13	14	15	16	17	18
Gross requirements	200	300	200	400	300	500
Scheduled receipts	--	--	--	--	--	--
Projected stock on hand 600	400	200	400	300	500	0
Net requirements	--	--	100	400	300	500
Planned starts	100	400	300	500	--	--

b) With safety time

Fig. 6.15. The Effect of Safety Time

Pt No. C2218
H. Plastic case   LT = 1   Scrap factor = 0.05

Week No	13	14	15	16	17	18
Gross requirements	200	300	200	400	300	500
Scheduled receipts	--	--	--	--	--	--
Projected stock on hand 600	400	100	0	0	0	0
Net requirements	--	--	100	400	300	500
Planned starts	--	105	420	315	525	--

Fig. 6.16. The Effect of a Scrap Factor

Since MRP decisions for a production stage (what to manufacture, how many, and when) are coordinated with other decisions for the stages, it is natural to extend MRP to include capacity planning, shop floor control, and purchasing. In this case the term MRP denotes Manufacturing Resource Planning as opposed to Material Requirements Planning. Although this extended MRP, referred to as closed loop MRP or MRP II, ties the financial and marketing functions to the operations function (Fig. 6.17 [13]), including additional features, such as costing and simulation possibilities, the basic mechanism of the MRP and MRP II systems, the time-phased material or resource preparation process, remains the same [18]. MRP II therefore, provides a vehicle for coordinating the efforts of manufacturing, finance, marketing, engineering, and personnel departments toward a common business plan.

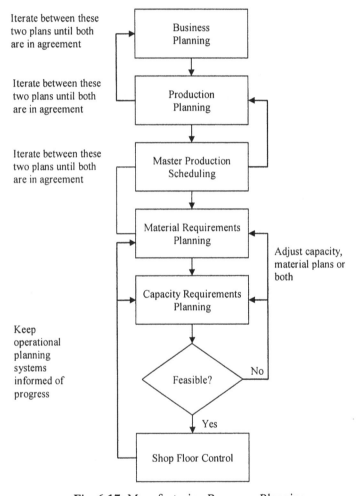

Fig. 6.17. Manufacturing Resource Planning

6.2.3 Capacity Planning

The requirements plans that have been discussed have shown ways of exploiting the knowledge of demand dependence and product structure in order to develop production order schedules. These schedules take into account the necessary timing of production orders, but they assume that the capacity is available when needed. However, capacity constraints are a reality and must be taken into account.

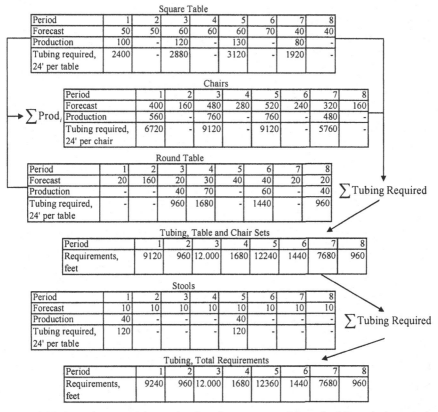

Square Table

Period	1	2	3	4	5	6	7	8
Forecast	50	50	60	60	60	70	40	40
Production	100	-	120	-	130	-	80	-
Tubing required, 24' per table	2400	-	2880	-	3120	-	1920	-

Chairs

Period	1	2	3	4	5	6	7	8
Forecast	400	160	480	280	520	240	320	160
Production	560	-	760	-	760	-	480	-
Tubing required, 24' per chair	6720	-	9120	-	9120	-	5760	-

$\sum$ Prod.

Round Table

Period	1	2	3	4	5	6	7	8
Forecast	20	160	20	30	40	40	20	20
Production	-	-	40	70	-	60	-	40
Tubing required, 24' per table	-	-	960	1680	-	1440	-	960

$\sum$ Tubing Required

Tubing, Table and Chair Sets

Period	1	2	3	4	5	6	7	8
Requirements, feet	9120	960	12.000	1680	12240	1440	7680	960

Stools

Period	1	2	3	4	5	6	7	8
Forecast	10	10	10	10	10	10	10	10
Production	40	-	-	-	40	-	-	-
Tubing required, 24' per table	120	-	-	-	120	-	-	-

$\sum$ Tubing Required

Tubing, Total Requirements

Period	1	2	3	4	5	6	7	8
Requirements, feet	9240	960	12.000	1680	12360	1440	7680	960

Fig. 6.18. Requirements determination for square and round tables, chairs, stools, and finally tubing. Requirements for chairs are dependent on the production schedules for tables. Tubing requirements are dependent on the production schedules for all four products.

As an example, consider the processing of metal legs for tables beginning with the information contained in Figure 6.18 [16]. There are two tables that use the same legs, and Table 6.5 [16] summarizes the leg requirements from Figure 6.18. The bottom line of Table 6.5 gives the production requirements for legs if we accumulate four weeks' future requirements as production orders. Therefore, it is necessary to receive lots of 1320 legs in period 1 and 1240 legs in period 5. Because there is a production lead time of three weeks, these production orders must be released three weeks ahead of the schedule shown in Table 6.5.

Period, weeks	1	2	3	4	5	6	7	8
Leg requirements, square	400	--	480	--	520	--	320	--
Leg requirements, round	--	160	--	280	--	240	--	160
Total leg requirements	400	160	480	280	520	240	320	160
Production requirements	1320	--	--	--	1240	--	--	--

Table 6.5. Requirements for Table Legs from Production Schedules for Square and Round Tables Shown in Figure 6.18 (Production Lead Time = 3 Weeks)

From the operation process chart of Figure 6.19 [16], let us consider only the load requirements for the fabrication operations of (1) cut to length, (2) weld support brackets, and (3) drill brackets. The expected time requirements for all three operations are shown in Table 6.6 [16] for each of the two lot sizes that must be considered.

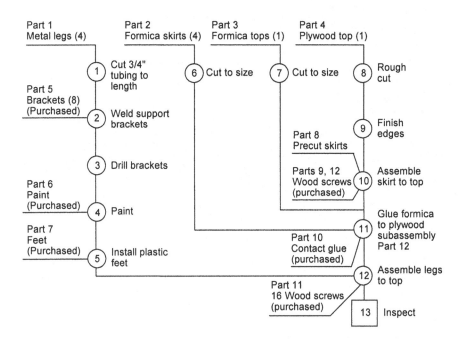

Fig. 6.19. Operation Process Chart

If the production orders are released three weeks before the legs are needed for assembly, assume that the cutting operation is planned for the first week, the welding operation for the second week, and the drilling and other minor operations for the third week. Then, the machine hour load for each of the three processes can be projected, as shown in Figure 6.20 [16], so that all operations are completed and the two orders are available in the first and fifth weeks.

	Process	Setup Time, Minutes	Run Time, Minutes/unit	Total Time, Hours, in Lots of 1320	1240
1.	Cut to length	5	0.25	5.58	5.25
2.	Weld brackets	10	1.00	22.16	20.83
3.	Drill brackets	20	0.75	16.83	15.50
4.	Paint	10	0.25	5.66	5.33
5.	Install feet	5	0.10	2.28	2.15

Table 6.6. Process Time Requirements for Legs in Lots of 1320 and 1240

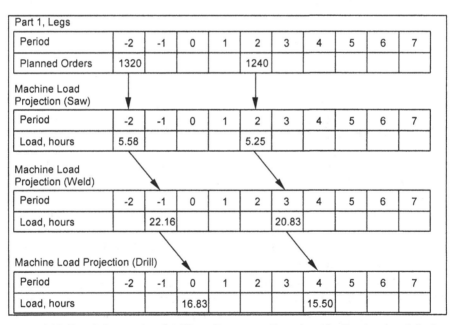

Fig. 6.20. Load Generation for Three Processes, Based on the Production Schedule for Legs. The Time Requirements are from Table 6.6.

Following the same rationale that determined the load effects for the three processes shown in Figure 6.20, a computer program can pick up the loads, for all orders for all parts and products in the files of the requirements program and estimate the projected load for each work center. For example, the projected weekly load on the drill press work center is shown in Figure 6.21 [16]. The accumulated load by weeks is shown as "released load," in hours. The capacity for the eight-week horizon is shown as 80 hours, the equivalent of two available machines. The available hours in Figure 6.21 then indicate whether or not capacity problems are projected. In periods 4, 5 and 6, there are projected overloads. Given this information, we may wish to anticipate the problem by changing the timing of some orders, meet the overload through the use of overtime, or possibly subcontract some items. In the example shown in Figure 6.21, substantial slack is projected for periods 3 and 7, so it might be possible to smooth the load by releasing some orders earlier.

Projected weekly machine load report Work Center 21, Drill Presses							Date: 02/01/80	
Period	1	2	3	4	5	6	7	8
Released load, hours	65	71	49	90	81	95	48	62
Capacity, hours	80	80	80	80	80	80	80	80
Available hours	15	9	31	-10	-1	-15	32	18

Fig. 6.21. Sample Projected Load Report for One Work Center

6.2.4 Shop Floor Control

The industrial planning systems reviewed so far have been responsible for long-term planning activities. In contrast, shop floor control systems implement short-term dispatching activities. Shop floor control systems are responsible for allocating resources in the forms of labor, tooling, and equipment to the various orders on the shop floor. They use information provided by the planning systems in order to identify those orders for which action must be taken [13undertaken [12].

Shop floor control complements other planning systems, such as material requirements planning and capacity requirements planning. These planning systems provide the resources, required by the shop floor control system and set the objectives to be achieved by this system. Shop floor control is then responsible for using these resources to achieve its objectives in an effective and efficient fashion. Shop floor control is supposed to close the loop between the planning and execution phases of the manufacturing system by feeding back information from the shop floor to the planning systems.

The Shop Order

A central focal point for any shop floor control system is the shop order, which is released by the planning system to the shop floor. All of the activities undertaken by the shop floor control system are directed at ensuring the timely and efficient completion of the shop order.

The shop order is an authorization given by the planning system for the shop to produce a predetermined quantity of a particular item (as identified by its part number) to arrive in inventory at a prespecified time (i.e., the order due date). The shop order allows the shop floor control system to allocate the physical resources (i.e., inventory, machines, labor, and tooling) needed against the order.

There are two flows accompanying the shop order that the shop floor control system must manage. The first flow is the product flow and the attendant physical allocation of resources. The second flow is the flow of information. As the shop order progresses through the various stages of processing, it generates information, which is then used in monitoring its progress. This flow of information is used by both the planning system and the shop floor control system. It is this information flow that enables the shop floor control system to close the loop, initiated by the planning system.

Initially, the shop order is a statement of intent. That is, it describes the finished form of the item released to the shop floor. This form is described in terms of such attributes as quantity, part number, and due date. As the shop order moves through the various stages of processing on the shop floor, it undergoes a process of physical transformation. At each stage, the addition of components, labor, tooling, and machine capacity changes the order, bringing it closer to the required finished form. At each stage, a decision must be made about the future handling of the order. This decision is based on a comparison of the actual progress of the order with its planned progress.

The Major Resources

The shop floor control system manages the flow of shop orders through the shop floor and allocates various quantities of four resources to the shop order. These resources are:

- *Manpower.* This resource includes all of the personnel that the shop floor can draw on in executing the plans released to it. The manpower resource can take various forms, such as overtime, workers transferred in from other locations, part-time help, and multiple shift operations. Manpower includes both direct and indirect labor.

- *Tooling.* This is all of the equipment and special fixtures that are used during the setup and operation of a machine or assembly operation.
- *Machine Capacity.* This is the total amount of productive capacity offered by the equipment available.
- *Material.* This is the total stock of components that can be used in completing shop orders.

The shop floor control system does not determine the level of each resource that can be drawn. That task is a primary responsibility of the planning system. The planning system determines the total amount of material available, and it also sets the number of manpower hours that are available in any period. In this sense, the planning system constrains the shop floor control system by placing an upper limit on the availability of these resources.

The shop floor control system is responsible for working within these constraints. It makes the detailed allocation of resources to the various shop orders, and it controls and monitors the use of these resources when they are assigned.

The Major Activities of Shop Floor Control

The activities governed by shop floor control can be divided into five groups. These are:

- Order review/release
- Detailed assignment
- Data collection/monitoring
- Feedback/corrective action
- Order disposition

These five groups of activities (Fig. 6.22 [13]) encompass the entire process of transforming a planned order into a completed order, which is then available to support the further activities of the planning system.

Order review/release: This includes those activities, which must take place before an order can be released to the shop floor. These activities are necessary for controlling the flow of information and the orders passing from the planning system to the execution system and to ensure that the orders released have a reasonable chance of being competed by the time and in the quantity required. The first of the Order review/release activities is order documentation.

An order, once it matures, can be regarded as an authorization for producing a specified quantity of a specified item by a specified time. The *order documentation* activities provide information not given by the planning system but needed by the shop floor. Typically the following information is added to the order at this stage:

- *Order identification*: The order is given a number or code, which can be used to help in tracking the order on the floor and to retrieve necessary information about the order (e.g., the processing time or next operation). The order identification (which may be distinct from the part number) links the shop floor with the planning system.
- *Routings:* The order is described in terms of the various operations through which it must pass. The routing also helps identify the order's resource requirements.
- *Time standards*: The order is described in terms of the resources (machine and labor) required at each stage in its transformation. Such information is important for activities, such as order sequencing (dispatching), monitoring, and capacity management.
- *Material Requirements*: The order is described in terms of the raw material and components needed and the stage in the process at which these components are needed.
- *Tooling Requirements*: Orders, as they progress through their various stages, may require special tools. The tooling needed by the order and the stage at which this tooling is required must be identified and the information provided to the shop order. This information forms the basis for issuing the tool order.
- *Other*: Other information provided at this stage may include report forms, operation due dates, and any special handling requirements.

This information is taken from a set of records, which can either be centrally located (in the form of a central database) or dispersed throughout the firm in various departments. In the latter case, for example, information on routings and time standards might be obtained from the engineering files.

Another part of the order review/release activity is checking the *inventory status* of those components and the raw materials required by the shop order in order to ensure that they will be available in a sufficient quantity at the necessary time.

Inventory availability, by itself, is a necessary but not sufficient condition for the successful completion of a shop order. Another requirement is the availability of capacity.

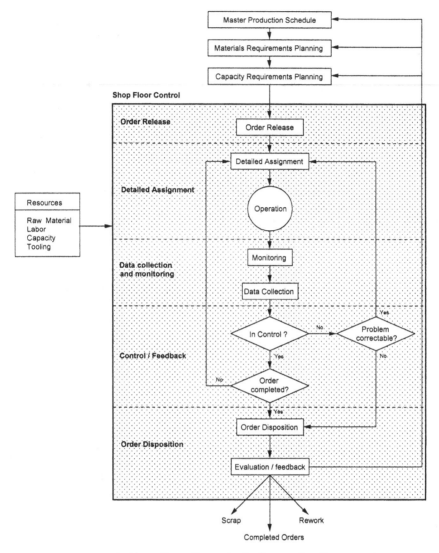

Fig. 6.22. Shop Floor Control: An Integrating Framework

In the capacity evaluation activity, the capacity required by the shop order is compared with that available in the system. At this stage, a judgment must be made as to whether or not the capacity is adequate. If it is not, then the release of the order may be delayed until such capacity becomes available. Shop overload, with the resulting increasing queues and lengthening lead times, is avoided by evaluating the capacity.

The final activity assigned to order review/release is *load leveling*. The orders that are recommended for release by the planning system are not released immediately in most shop floor control systems. Instead, they are accumulated (backlogged) for a short period of time, in order to level the load on the shop floor by controlling the rate at which orders are released to the floor. This activity [19,20] seeks to ensure that capacity utilization is high by smoothing out the peaks and the valleys of load on work centers.

The various activities of order review/release are essentially short-term planning activities. They do not entail that the resources required by the various matured orders actually be committed. Instead, the availability of these resources is evaluated to determine if it is sufficient to justify the release of the order. The actual commitment of shop floor resources to a given order is the major focus of the second set of shop floor control activities collectively termed detailed assignment.

Detailed assignment. The shop floor control system has to allocate four major resources available to it:

- Material
- Labor
- Tooling
- Machine capability

These resources must be used to satisfy not only the demand of competing orders but also other activities required to continue the provision of these resources (e.g., scheduled preventive maintenance). The activities included in detailed assignments are responsible for formally matching (assigning) the supplies of shop floor resources with the competing demands being placed on these resources. The assignment decisions made, matching supply with demand must be detailed enough to address the following concerns:

1. Type of resources: The assignment of resources must identify the specific type of resource to be assigned.
2. Quantity of resources: The amount of the resource to be used must be identified in terms meaningful for the assignments. For example, the amount of labor to be used may be described in terms of standard labor hours.
3. Timing of assignment: The time, at which the resource(s) are to be assigned, must be identified, including the expected time the resources are to be assigned and the expected time they are to be released.
4. Placement of resources: If the resource to be assigned is available in more than one location, then the assignment procedure should identify the location the resource is to be allocated to, in a given order.

5. Priority of processing: The final aspect of the assignment process is determining the sequence in which competing orders are to be permitted access to the limited resources.

Three major activities make up detailed assignment. These are (1) order sequencing/dispatching, (2) scheduled maintenance, and (3) other assignments.

Once a shop order is released to the shop floor, it must compete with other shop orders for having access to the resources of the shop floor. These orders are often differentiated in terms of such attributes, as order due dates, amount of processing time required at the various operations, amount of slack remaining until the order is needed, and the number of operations left until the order is made. The process of assigning resources to shop orders can be called order sequencing/dispatching.

The major concern of order sequencing/dispatching is the determination of order priority. A crude method of determining order priority is through the use of a dispatching rule, such as earliest due date (EDD), shortest processing time (SPT), etc. This kind of approach orders the tasks waiting to be assigned in a queue, based on the application of a particular dispatch rule and a resource is selected out of those available, and the first task from the ordered queue is assigned to this resource. This approach is characterized by the partition of the assignment problem in two parts, namely the selection of a resource from the many available and the selection of a task to be assigned to the resource. This means that the resources and tasks are not considered simultaneously, but *sequentially,* at a decision point. In general, a rule or a combination of rules may be appropriate for optimizing a particular aspect of the system, but often a number of aspects, which may change dynamically, must be considered. The application of such rules requires a trial-and-error approach, whereby the system is simulated using one rule, the result examined, another rule is applied, etc., until a suitable result is achieved. This may be time consuming and therefore impractical for a dynamically changing environment.

Shop floor resources can be assigned not only to shop orders but also to other activities, such as preventive maintenance, scheduled downtime or indirect labor activities. Such activities may take place in order to level current capacity utilization (in the case of scheduled downtime) or to use capacity that is currently available but not required by the orders on the shop floor (in the case of the transfer of workers to indirect labor activities).

Data collection/monitoring. Information links the planning system with the execution system. The information provided by the shop floor control system is the major means by which the planning system can track the physical flow. The collection, recording, and analysis of this information are the responsibility of the data collection/monitoring activity in shop floor control.

The first task of this activity involves the collection of information pertaining to the actual progress of an order as it moves through the shop. Information collected includes:

* Current location of the shop order
* Current state of completion
* Actual resources used at current operation
* Actual resources used at preceding operations
* Any unplanned delays encountered

Control/feedback. Corrective action by management is required any time that the actual progress of a shop order exceeds some predefined margin of difference from its planned progress. Progress can be monitored along several dimensions: stage of completion, costs, scrap produced, or nearness to due date to name a few.

Capacity control refers to any corrective actions that attempt to correct the problems by means of very short-term adjustments in the level of resources available on the shop floor. Examples of such short-term shop adjustments are:

* Changes in work rate.
* Use of overtime or part-time labor.
* Use of safety capacity.
* Alternate routings.
* Lot splitting.
* Subcontracting of excess work.

Order disposition. The fifth and final set of activities included in shop floor control is that of order disposition. The order disposition activities are those required to transfer out of the shop floor control system responsibility for orders released to the shop floor. Two of the conditions that require such a transfer are (1) order completion and (2) scrap. In the first case, the order has been completed and is no longer part of the shop floor control system and the inventory information is appropriately updated. Responsibility for the completed order passes from shop floor control to inventory control.

In the second case, the items are transferred from the shop floor control system because they are no longer usable by the system. The relevant databases are appropriately revised to reflect the change in status of certain items (from good pieces to scrap).

As part of the order disposition, the quantity received from the shop floor is recorded and the performance of the shop floor system is evaluated. Evaluation of the performance involves recording the actual performance of the shop in completing the order. Performance measures collected may include:

- The number of labor hours required.
- The breakdown of labor hours between regular time and overtime.
- The materials required by the order.
- The number of hours of setup time required.
- The amount of tooling required.
- Completion date of the order.
- Amount of rework or scrap generated by the order.
- The number of machine hours required.

This information is made available for use by other departments in the firm and forms the basis for various cost-based reports. It also enables the evaluation of the shop floor control system by comparing actual performance with planned performance. Finally, the information collected during order disposition is used to identify and solve longer term problems involved in the shop floor (e.g., the lack of capacity) as well as to modify the cost and time standards used by the planning system.

Interactions of Major Activities of Shop Floor Control

The five major activities of the shop floor control framework are neither independent of each other, nor do they take place all at once. Instead, these activities are interrelated and ongoing. For example, consider the detailed assignment, data collection/monitoring, and control/feedback activities. At every stage of the order's processing, the shop floor personnel must decide whether the order is completed, ready to go on to the next stage, or it is in need of special or additional treatment. The determination of the order's status is a function of the data collection/monitoring and control/feedback activities. If the order is ready to go on to the next stage, the activities of detailed assignment will again be invoked. Furthermore, the activity on the shop floor, as related by the data collection/monitoring and control/feedback activities, influences the order review/release and the order disposition activities.

Characteristics of Shop Floor Control Systems

There are four important characteristics of shop floor control systems: level of detail, decision-making latitude, decision time horizon, and level of uncertainty. These characteristics can be tailored for different types of manufacturing systems (Fig. 6.23 [13]).

Level of detail refers to the amount of detailed information required by the shop floor control system. In a batch manufacturing or project environment, the amount of detailed information, considered by the shop floor control system, is enormous. The due date of the order, the routing, the bill of materials, and all engineering change orders are just some of the detailed information. Most orders in a job shop are built to customer specifications, which require all the information about the order being available to the shop floor control system. Very little filtering of the information can take place at higher levels of the manufacturing planning system.

In a repetitive environment, the level of detail in the shop floor control system is considerably reduced. The higher level planning systems specify common routings, standard bills of material, and minimum engineering change orders. The shop floor control system in this case emphasizes execution and does very little planning.

Perhaps the best way to view the level of detail is as a continuum, with low level of detail (in a repetitive environment) at one end, high level of detail (in a project environment) at the other, and intermediate level of detail (in a batch environment) in between the two extremes.

Type of Manufacturing Process	System Characteristics			
	Level of Detail	Decision Making Latitude	Time Horizon	Level of Uncertainty
Repetitive or Continuous	Low	Low	Long	Low
Batch	↕	↕	↕	↕
Project	High	High	Short	High

Fig. 6.23. Characteristics of Shop Floor Control Systems

Decision-making latitude refers to the number of alternatives available to a decision maker. In some shop floor control systems the amount of latitude is very large, while in other systems it is small. The amount of decision-making latitude is related to the manufacturing environment.

In repetitive/continuous manufacturing, the higher level planning system makes the major decisions, leaving the shop floor control system with little latitude. In a project environment, the shop floor control system has a large amount of decision-making latitude, due to the uncertainty about the jobs and the high level of detailed information. In the batch environment, the decision making latitude falls in between these two extremes.

The *time horizon of decisions* includes both the length of time available to the decision maker to make a decision and the impact of the decisions made by the decision maker. The range of possible time horizons varies from short to long. With a short time horizon, the time available to make a decision is short, and the potential impact of any single decision is small. With a long time horizon, there is a large amount of time available to make a decision, and the potential impact of a single decision is large.

In a project environment, the system must make frequent, short-term decisions. The time available is short, and the impact of any one decision is small. In this environment the time horizon is short.

In a repetitive/continuous manufacturing environment, the decisions are of much longer range. Typical decisions derive from asking: what should the standard product be? and what is the best cycle time for each workstation? For these types of decisions there is a large amount of decision time available. However, the impact of each of these decisions on the shop floor, and on the firm as a whole, is large. These types of decisions are typically not made on the shop floor.

In a batch environment, the time horizon of decision is at an intermediate level. There are both short-term decisions, such as "At this moment what is the most appropriate machine for this task?" and long-term decisions, such as "What kinds of machines should be purchased if we continued receiving these types of orders?" Once again, the batch environment shop floor control system falls in between the other two environments.

Even though individual decisions may have only short-term impacts, it is the aggregation of these short-term decisions that determines the effectiveness of the shop floor. If the short-term decisions are consistently wrong, the performance of the shop floor will be poor, and the performance of the total firm may suffer. All the decisions on the shop floor, whether of short or long time horizons, need to be good ones.

The *amount of uncertainty* is concerned with both the degree to which information is known and the tools available for use by the decision maker. If the planning systems above the shop floor control system handle most of the uncertainty, then there will be low uncertainty in the shop floor control system. On the other hand, if the upstream planning systems do not or cannot remove uncertainty, the shop floor control system will have to cope both with high uncertainty and its effects.

As was true with the other three characteristics, the project environment and the repetitive/continuous manufacturing environment are at opposite ends of the spectrum, with the batch environment falling in the middle. In a project environment, with its constantly changing job and varying processing requirements, the shop floor control system must be capable of handling a large amount of uncertainty. In a repetitive/continuous manufacturing environment, with its standard jobs and standard routings, the shop floor control system needs to manage a smaller amount of uncertainty. As the environment progresses from project to repetitive/continuous manufacturing, the shop floor control system must be adapted in order to fit changing needs and requirements [13].

The shop floor control system described above has been presented in its "ideal" functionality, but in reality, only portions" of such a system exist in the industry. Shop Floor Control systems in practice are both partially computerized and manually operated. The subject of computer integrated manufacturing, (CIM), amongst other things, deals with computerizing the shop floor control System and integrating it with the rest of the planning functions.

6.2.5 Just In Time Manufacturing

In just-in-time (JIT) manufacturing, a work station does not perform a manufacturing step until it is called upon by the next station downstream. In this manner, the production system is reactionary to (or "pulled" by) the demand for finished products. This approach differs from typical planning approaches, where finished products, subassemblies and parts are made in anticipation of future demand. A card (or kanban) is a marker (Fig. 6.24) used to control the pace of job activities through sequential processing.

The mechanics of just-in-time can be illustrated by a production system with two work centers, as shown in Figure 6.25. The following sequence is executed:

1. An order is filled by taking a finished product from a cart in output area B. A "move" card is placed in the cart.
2. The empty cart with the "move" card is transported from output area B to output area A. The "move" card is replaced with a "produce" card.
3. The "produce" card signals work center A to take materials from input area A to produce one part. The part is placed on an empty cart in output area A and a "move" card is attached.

4. The filled cart is transported from output area A to input area B. The "move" card is replaced by a "produce" card.
5. The "produce" card signals work center B to take the part from the cart and produce a finished product from it. The finished product is placed on a cart in output area B.

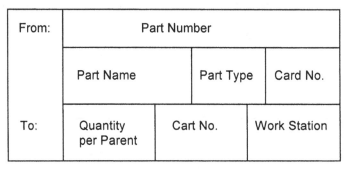

From:	Part Number		
	Part Name	Part Type	Card No.
To:	Quantity per Parent	Cart No.	Work Station

Fig. 6.24. Kanban Configuration

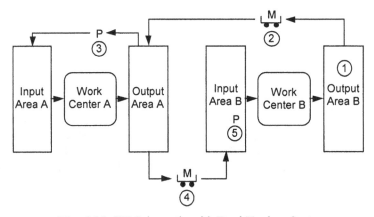

Fig. 6.25. JIT Schematic with Dual Kanban System

In this idealized case, two carts can move simultaneously in the system to keep both workcenters occupied. Each workcenter will not process a part until a "produce" card is received. The production rate of finished products is determined by the workcenter with the lowest production rate.

Just in time manufacturing has the following goals:

- *Lot size reduction* - Ideally, the system should operate with a lot size of 1.

- *Zero inventory* - Minimize the work-in-process inventory (parts in output area A and input area B in the example).
- *Zero defects* - Rejection of a part at a work station due to poor quality, will slow down the entire system, since there is no inventory.

JIT assumes that ordering and setup costs are negligible compared with production costs, leading to an optimal lot size for components equal to that required to assemble one final product. This assumption is only valid if the manufacturing tasks are repetitive and well-defined, materials are highly standardized, and flexible and delivery schedules can be established with materials suppliers. JIT is a reactive system (production reacts to demand), so only short-term production scheduling can be achieved (on a daily basis). Therefore, the demand schedule must not fluctuate significantly or the production system may encounter either shortages or excesses in capacity. JIT does not tolerate defective parts moving forward between stages. A great deal of effort must be taken to detect defective parts, improve rework procedures, and identify and correct causes for defects. This involves both equipment and training of workers.

Order review/release is extremely important to the successful implementation of JIT. Stockless production requires the careful loading of the shop floor. The orders released should not exceed the capacity available on the floor. Queues and lead times are kept short by careful loadings,

In just-in-time manufacturing, the order review/release activities are part of production planning. Once orders are released, all of the necessary components are pulled by the parent assemblies. Under just-in-time manufacturing, detailed assignment of orders can be made quite effectively on the shop floor. There is one complication introduced by just-in-time manufacturing. Because of its increased importance, preventive maintenance must be included in the shop floor control system as a job to be scheduled on a regular basis. Preventive maintenance will consume resources in the short run, but will increase capacity in the long run by improving the overall operation of equipment.

Just-in-time manufacturing may force a change in the methods used in tracking and monitoring the progress of jobs on the shop floor. One of the changes is the result of using small lot sizes. As lot sizes decrease, the number of orders increases, creating more orders for the shop floor control system to track. Ultimately, formal tracking of individual orders may be abandoned altogether, as in the case of many repetitive manufacturing systems.

Monitoring will become an ongoing process with each worker responsible for ensuring that all parts passing through his/her station be of acceptable quality. This monitoring will include not only orders but also equipment and tooling. Monitoring is extremely important since it is needed to ensure that the entire system operate at an acceptable level of quality and efficiency.

6.2.6 Developments in ERP Systems and e-Commerce

Enterprise resource planning (ERP) has been associated with a quite broad spectrum of definitions and applications over the last decades. Earlier concepts, such as MRP and MRP II, were designed to assist planners by combining various forms of process information related to specific business concepts such as manufacturing [21].

From the systems focusing on the inventory control of manufacturing systems in the 1960's, and on the introduction of the MRP systems in the 1970's, when companies could no longer afford maintaining large quantities of inventory, new capacity planning, sales planning and forecasting tools in conjunction with scheduling techniques [22] were developed and incorporated into the MRP systems, leading to the materialization of systems known as closed-loop MRP. In the 1980's companies began to take advantage of the increased and affordable information technology power and the manufacturing resources planning (MRP II) systems of that time incorporated the financial accounting and management systems as well. The MRP II concept was further expanded to incorporate all resource planning and business processes of the entire enterprise, including areas, such as human resources, project management, product design, materials and capacity planning [22]. That was the time when the ERP concept was devised to integrate smaller, otherwise isolated, systems so that real-time resource accountability across all business units and facilities of a corporation could be maintained [22]. The elimination of incorrect information and data redundancy, the standardization of business unit interfaces, the confrontation of global access and security issues [22], the exact modeling of business processes, have all become part of the objectives list to be fulfilled by an ERP system. The fundamental benefits of ERP systems do not, therefore, emanate from their inherent "planning" capabilities but rather from their abilities to process business transactions efficiently [22]. In other words, ERP systems offer a unified enterprise view of the business that encompasses all functions and departments in an enterprise database, where all business transactions are entered, recorded, processed, monitored and reported [22].

ERP systems are highly complicated information technology systems. Large implementation costs, high failure risks, tremendous demands on corporate time and resources [22], complex and often painful business process adjustments are included among the main concerns, pertaining to an ERP implementation.

One of the leading ERP vendors worldwide, SAP AG, has been offering the SAP R/3 ERP system, which is from the beginning of 2000's a part of the mySAP.com solution, a business software ecosystem, featuring web-enabled marketplaces and workplaces, allowing for manufacturers, distributors and customers to perform business transactions in electronic portals, using HTTP and XML communication protocols [23].

The mySAP.com business suite includes the following components [23]:

- The ERP backbone solutions, including Financials (FI), Logistics (LO), Human Resources Management (HR), Industry Solutions (IS),
- The Business Intelligence (BI) solutions, such as Business Information Warehouse (BW), Knowledge Warehouse Management, Strategic Enterprise Management (SEM), and Corporate Finance (CFM),
- Supply Chain Management solutions, such as the Advanced Planner and Optimizer (APO), Logistics Execution System (LES), Business to Business Procurement (BBP), and Environment, Health and Safety (EHS),
- Customer Relationship Management (CRM) solutions, such as Internet Sales Scenarios, Customer Interaction Center and so on.

It is very important for an organization to take into consideration a set of critical factors for increasing the chances for successful ERP implementations [22]:

- Clear understanding of strategic goals,
- Commitment by top management,
- Excellent project management,
- Organizational change management,
- An experienced and capable implementation team,
- Data accuracy and data migration from legacy systems,
- Extensive education and training of end users,
- Focused performance measures,
- Multi-site issues.

ERP implementations usually prove to be huge and complex projects, often resulting in cost and schedule overruns. Statistics show that [24]:

- Only 10% of ERP implementations are considered fully successful in terms of functionality, estimated costs and time frames,
- Average cost overruns reach a 178%,
- Average schedule overruns reach a 230%,
- Average implemented functionality reaches a 41% of what originally desired.

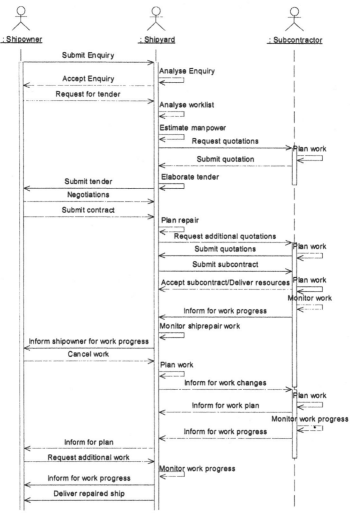

Fig. 6.26. Sequence of supply chain interactions for a maritime business scenario

ERP systems often provide Supply Chain Management (SCM) solutions or provide interfaces for interacting in an integrated way with other external information technology systems. SCM solutions deal with the current trend of manufacturing companies to maximize their communication and collaboration capacity by integrating their operations with those of their business partners [25]. The Internet growth and the associated software technologies provide the means for the realization of this trend.

In the figure above, the ship-repair business process is shown, involving the transactions among the ship-owner, the shipyard, the material suppliers and the shipyard's subcontractors [25]. The corresponding software implementation is based on the 3-tier paradigm (database layer, business process layer, client layer), representing a cost effective integrated environment, which allows enterprises to participate in the supply chain network.

In particular, the business functions, taking place in the process of planning and controlling the preparation and execution of a ship-repair contract, in a number of critical ship-repair nodes, are involved.

6.3 Methods and Tools

The purpose of this section is to present the reader with some methods and tools for the efficient operation of Manufacturing Systems. In addition to the tools already discussed in Chapter 5, Network Analysis, Decision Making, Dispatching Heuristics, Evolution Programs (Metaheuristics), Gantt Charts and Performance Measures will be introduced in this Section. This section intends to provide the reader with an overview of such methodologies; for detailed treatment of these subject areas, the reader is referred to other specialized literature.

6.3.1 Network Analysis

Network analysis can be used to describe the complicated precedence relationships between the activities of a large project. The resulting description can be used to determine a timetable for the activities and to predict a completion date for the project. In manufacturing, the application of network analysis is mainly restricted to "one-time" projects, such as the construction of a new manufacturing facility; continuous, day-to-day operations are usually addressed by the techniques described in Section 6.2. Examples of network planning methods (or project planning methods) are the Critical Path Methods (CPM) and Performance Evaluation and Review Technique (PERT). PERT and CPM are based substantially on the same concepts. PERT is typically based on probabilistic estimates of activity times that result in a probabilistic path, through a network of activities and a probabilistic project completion time. CPM assumes typically deterministic activity times. The development of a project network may be divided into (1) activity, (2) arrow diagramming, and (3) node numbering.

Activity Analysis

The smallest unit of productive effort to be planned, scheduled, and controlled is called an *activity* or *task*. For large projects, it is possible to overlook the need for some activities because of the great complexity. Therefore, although professional planning personnel are commonly used, the generation of the activity list is often partially made in meetings and round-table discussions that include managerial and operating personnel. Data records from previous projects are also used in combination with advanced software applications, allowing for the calculation of cost and time required for the completion of each activity. Table 6.7 [16] is an activity list for the introduction of a new product.

Network Diagramming

A network diagram describes the precedence relationships among the activities; it must be based on a complete, verified, and approved activity list. The important information required for these network diagrams is generated by the following three questions:

1. Which activities must be completed before each activity can be started?
2. Which activities can be carried out in parallel?
3. Which activities can immediately succeed other ones?

The common practice is simply to work backwards through the activity list, generating the immediate predecessors for each activity listed. The network diagram (Fig. 6.27) may then be constructed to represent the logical precedence requirements in Table 6.7. Care must be taken in correctly representing the actual precedence requirements in the network diagram. A "dummy" activity may be used to provide an unambiguous beginning and ending event or node for each activity. For example, a functionally correct relationship might be represented by Figure 6.28*a* [16] in which two activities have the same beginning and ending nodes. If Figure 6.28*a* was used, however, it would not be possible to identify each activity by its predecessor and successor events because both activities *m* and *n* would begin and end with the same node numbers. This is particularly important in larger networks that employ software programs for network diagram generation. Each activity is represented by a pair of node numbers. The problem is solved through the insertion of a dummy activity, as shown in Figure 6.28*b*. The functional relationship is identical because the dummy activity requires zero time, but now *m* and *n* are identified by different pairs of node numbers.

Activity Code	Description	Immediate Predecessor Activity	Time, Weeks
A	Organize sales office	-	6
B	Hire salespeople	A	4
C	Train salespeople	B	7
D	Select advertising agency	A	2
E	Plan advertising campaign	D	4
F	Conduct advertising campaign	D	10
G	Design package	-	2
H	Set up packaging facilities	G	10
I	Package initial stocks	H,J	6
J	Order stock from manufacturer	-	13
K	Select distributors	A	9
L	Sell to distributors	C,K	3
M	Ship stock to distributors	I,L	5

Table 6.7. Precedence Chart Showing Activities, Their Required Sequence, and Time Requirements for the New Product Introduction Project

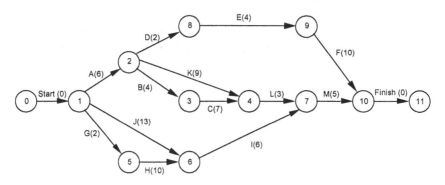

Fig. 6.27. Arcs Network Diagram for the New Product Introduction Project

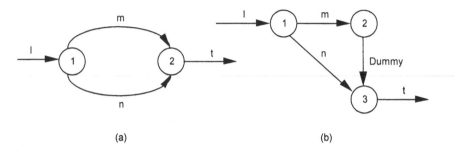

(a)   (b)

Fig. 6.28. (a) Activities m and n may be carried out in parallel, but this results in identical beginning and ending events. (b) The use of a dummy activity makes it possible to separate ending event numbers.

Node Numbering

The node numbering shown in Figure 6.27 has been done in a particular way. Each arc, or arrow, represents an activity. If we identify each activity by its tail (i) and head (j) numbers, the nodes are numbered so that for each activity, i is always less than j, $i < j$. The numbers for every arrow are progressive, and no backtracking through the network is allowed. This convention in node numbering is effective in software programs to develop the logical network relationships and to prevent the occurrence of cycling or closed loops.

A closed loop would occur if an activity was represented as going back in time. This is shown in Figure 6.29 [16], which is simply the structure of Figure 6.28b with the activity n reversed in direction. Cycling in a network can result through a simple error, or when developing the activity plans, one tries to show the repetition of one activity before beginning the next. A repetition of an activity must be represented with additional separate activities, defined by their own unique node numbers.

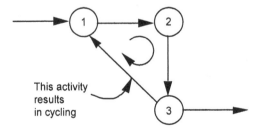

Fig. 6.29. Example of a Closed Loop or Cycling in a Network Diagram

A closed loop would produce an endless cycle in computer programs without built-in routines for the detection and identification of the cycle.

Critical Path Scheduling

With a properly constructed network diagram, it is possible to develop scheduling data for each activity and for the project as a whole. The data of interest are the minimum time required for completion of the project, the critical activities that cannot be delayed or prolonged, and the earliest and latest start and finish times for the activities. If we take zero, as the starting time for the project, then for each activity there is an earliest starting time, ES, relative to that of the project. This is the earliest possible time that the activity can begin, assuming that all the predecessors have also started at their ES. Then, for that activity, its earliest finish time, EF, is simply ES + activity time.

Assume that our target time for completing the project is "as soon as possible." This target is called the latest finish time, LF, of the project and of the finish activity. For the finish activity, LF will be equal to its earliest finish time, EF. The latest start time, LS is the latest time at which an activity can start if the target is to be maintained. Thus, LS of an activity is LF - activity time.

Existing software programs may be used to compute these data automatically, requiring as inputs the activities, their performance time requirements, and the precedence relationships established. The computer output might be similar to that of Figure 6.30 [16], which shows the schedule statistics for all the activities.

The total slack for an activity is simply the difference between the computed late start and the early start time, LS - ES, or between late finish and early finish times, LF - EF. The significance of total slack, TS, is that it specifies the maximum time that an activity can be delayed without delaying the project completion time. Note that all critical activities have zero slack in their schedules whilst all other activities have greater than zero slack.

The free slack, FS, shown in Figure 6.29 indicates the time that an activity can be delayed without delaying the ES of any other activity. FS is computed as the difference between the EF for an activity and the earliest of the ES times of all immediate successors. For example, activity F has FS = 3 weeks. If the earliest finish time is delayed by up to three weeks, neither the ES time for any other activity will it be affected, nor will the project completion time be affected. Note also that activity K can be delayed two weeks without affecting activity L, its successor. To compute the FS manually, one should examine the network diagram in order to take into account the precedence relationships.

On the other hand, total slack is shared with other activities. For example, activities D, E, and F all have TS = 3. If activity D is delayed and thus uses up the slack, then E and F no longer have any slack available. These relationships are most easily seen by examining the network diagram, where the precedence relationships are shown graphically.

THE CRITICAL PATH IS

START -> A.-> B -> C -> L -> M -> FINISH

THE LENGTH OF THE CRITICAL PATH IS 25

NODE	DURATION	EARLY START	EARLY FINISH	LATE START	LATE FINISH	TOTAL SLACK	FREE SLACK
START	0.00	0.00	0.00	0.00	0.00	0.00	0.00
A	6.00	0.00	6.00	0.00	6.00	0.00	0.00
B	4.00	6.00	10.00	6.00	10.00	0.00	0.00
C	7.00	10.00	17.00	10.00	17.00	0.00	0.00
D	2.00	6.00	8.00	9.00	11.00	3.00	0.00
E	4.00	8.00	12.00	11.00	15.00	3.00	3.00
F	10.00	12.00	22.00	15.00	25.00	3.00	3.00
G	2.00	0.00	2.00	2.00	2.00	4.00	0.00
H	10.00	2.00	12.00	4.00	14.00	2.00	1.00
I	6.00	13.00	19.00	14.00	20.00	1.00	1.00
J	13.00	0.00	13.00	1.00	14.00	1.00	0.00
K	9.00	6.00	15.00	8.00	17.00	2.00	2.00
L	3.00	17.00	20.00	17.00	20.00	0.00	0.00
M	5.00	20.00	25.00	20.00	25.00	0.00	0.00
FINISH	0.00	25.00	25.00	25.00	25.00	0.00	0.00

Fig. 6.30. Sample Computer Output of Schedule Statistic and the Critical Path for the New Product Introduction Project

Actually, there are five different paths from start to finish through the network. The longest, most limiting path requires 25 weeks for the activity sequence START-A-B-C-L-M-FINISH, called the critical path.

Manual Computation of Schedule Statistics

Manual computation is appropriate for smaller networks, and it helps convey the significance of the schedule statistics. To compute ES and EF manually from the network, we proceed through the network as follows, referring to Figure 6.31 [16].

1. Place the value of the project start time in both the ES and EF position near the start activity arrow. (See the legend for Figure 6.31.) We will assume relative values, as we did in the computer output of Figure 6.30, so the number 0 is placed at the ES and EF positions for the start of the activity. (Note that it is not necessary in PERT to include the start activity with a zero activity duration. It has been included here to make this example parallel in its activity list with the comparable "activities on nodes" example of Figure 6.33 [16]. The start and finish activities are often necessary in node networks.)

2. Consider any new unmarked activity, all predecessors of which have been marked in their ES and EF positions, and mark in the ES position of the new activity with the largest number, marked in the EF position for any of its immediate predecessors. This number is the ES time of the new activity. For activity A in Figure 6.31, the ES time is 0 because that is the EF time of the preceding activity.

3. Add to this ES number the activity time, and mark the resulting EF time in its proper position. For activity A, EF = ES + 6 = 6

4. Continue through the entire network until the "finish activity has been marked. As we showed in Figure 6.30, the critical path time is 25 weeks, so ES = EF = 25 for the finish activity.

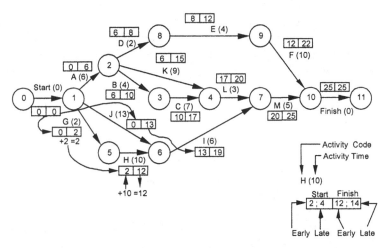

Fig. 6.31. Flow of Calculations for Easy Start, ES, and Early Finish, EF, Times

To compute the LS and LF, we work backwards through the network, beginning with the finish activity. We have already stated that the target time for completing the project is "as soon as possible" or 25 weeks. Therefore, LF = 25 for the finish activity without delaying the total project beyond its target date. Similarly, the LS time for the finish activity is

LF minus the activity time. Since the finish activity requires 0 time units, its LS = LF. To compute LS and LF for each activity, we proceed as follows, referring to Figure 6.32 [16].

1. Mark the values of LS and LF in their respective positions near the finish activity.
2. Consider any new unmarked activity, all successors of which have been marked, and mark in the LF position, for the new activity, the smallest LS time marked for any of its immediate successors. In other words, the LF for an activity equals the earliest LS of the immediate successors of that activity.
3. Subtract from this LF number the activity time. This becomes the LS for the activity.
4. Continue backwards through the network until all the LS and LF times have been entered in their proper positions on the network diagram. Figure 6.32 shows the flow of calculations, beginning with the finish activity and going backwards through several activities.

As discussed previously, the slack for an activity represents the maximum amount of time that it can be delayed beyond its ES, without delaying the project completion time. Since critical activities are those in the sequence with the longest time path, it follows that these activities will have the minimum possible slack. If the project target date coincides with that of the LF for the finish activity, all critical activities will have zero slack. If, however, the target project completion date is later than the EF of the finish activity, all critical activities will have slack equal to this time-phasing difference. The manual computation of slack is simply LS - ES or, alternatively, LF - EF.

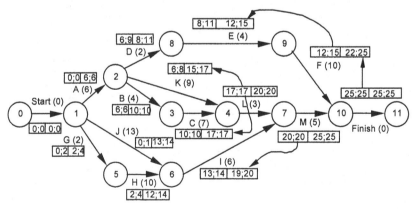

Fig. 6.32. Flow of Calculations for Late Start, LS, and Late Finish, LF, Times

As noted previously, free slack is the difference between the EF for an activity and the earliest ES time of all its immediate successors. Thus, free slack is not affected by any time-phasing difference.

In summary, the project and activity schedule data are computed as follows:

ES = Earliest start of an activity
  = Minimum of EF of all its immediate predecessors
EF = Earliest finish of an activity
  = ES + Activity time
LF = Latest finish of an activity
  = LF - Activity time
TS = Total slack = LS - ES = LF - EF
FS = Free slack
  = Minimum ES of all its immediate successors - EF

Activities on Nodes – Network Diagram Differences

Thus far, we have been using the "activities on arcs" network diagramming procedures. The "activities on nodes" procedure, results in a slightly simpler network system by representing the activities as they occur at the nodes, with the arrows showing only the sequences of activities required. The advantage of this methodology is that it is not necessary to use dummy activities in order to represent the proper sequencing.

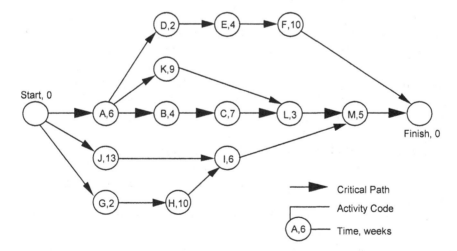

Fig. 6.33. Project Graph of Activities on Nodes for the New Product Introduction Project

Figure 6.33 shows the network for the new product introduction project, which may be compared with the comparable "activities on arcs" network shown in Figure 6.27.

The analysis for developing the early and late starts as well as the finish times and slack times, is identical with the forward and backward pass procedure, previously outlined. The net results of both systems are the schedule statistics that are computed. Since these are the data of interest and because the entire procedure is normally computerized for both methodologies, the choice between the two may depend on other criteria, such as the availability and adaptability of existing computer programs, or the choice may be simply a matter of personal preference.

Probabilistic Network Methods

The network methods that have been discussed so far may be termed "deterministic" because estimated activity times are assumed to be fixed constants. No recognition is given to the fact that activity times could be uncertain.

Probabilistic network methods assume the more realistic situation in which uncertain activity times are represented by probability distributions. With such a basic model of the network of activities, it is possible to develop additional data to help in assessing planning decisions. The nature of the planning decisions might involve the allocation or reallocation of personnel or other resources to the various activities in order for a more satisfactory plan to be derived. Thus, a "crash" schedule involving extra resources might be justified to ensure the on-time completion of certain activities. The extra resources needed are drawn from non-critical activities or from activities for which the probability of criticality is small.

Since the activity time is uncertain, a probability distribution for the activity time is required. This probability distribution is assessed by the engineers, the project manager, or consultants. Once the probability distribution for the activity time is specified, the expected activity time and the variance in activity time can be computed.

Optimistic time, a, is the shortest possible time for the activity to be completed, if all goes well. It is based on the assumption that there is no more than one chance in a hundred for activity to be completed in less than the optimistic time.

Pessimistic time, b, is the longest time for an activity under adverse conditions. It is based on the assumption that there is no more than one chance in a hundred for the activity to be completed in a time greater than b.

Most likely time, m, is the single most likely model value of the activity time distribution. The three time estimates are shown in relation to an activity completion time distribution in Figure 6.33 [16].

By assuming that activity time follows a beta distribution, the expected time of an activity, t_e, is computed as:

$$t_e = \frac{a + 4m + b}{6} \tag{6-9}$$

The *variance*, σ^2, of an activity time is computed using the formula

$$\sigma^2 = \left(\frac{b - a}{6}\right)^2 \tag{6-10}$$

The mean completion time is the mean of a normal distribution, which is the simple sum of the t_e values along a critical path. The variance of the mean project completion time is the simple sum of the variances of the individual activities along a critical path.

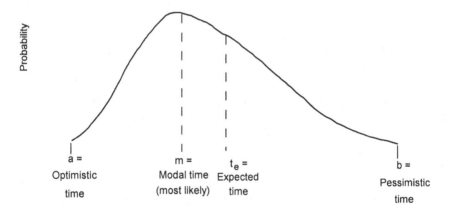

Time duration of activities

Fig. 6.34. Time Values in Relation to a Distribution of Activity Time

We can now determine the critical path by substituting t_e for the activity time. The mean and the variance of the project completion time are obtained by summing t_e and σ^2, respectively, for the activities on the critical path.

We can then use the normal tables to determine the probabilities for the occurrence of given project completion time estimates. For example, the probability that a project will be completed in less than the mean time is only .50. The probability that a project will be completed in less than the mean time plus one standard deviation is about .84; in less than the mean time plus two standard deviations, .98; and so on.

Deployment of Resources

The estimates of the project completion time assume that the resources needed to carry out the activities according to the schedule are available. It is therefore important to understand that the estimates of project completion times are lower bounds that may not be attainable with limited resources. The objective in an activity time/cost trade-off analysis is to compute the trade-off between the project completion time and the additional cost of resources that are needed to achieve it.

In some applications, it may be desirable to minimize the fluctuation in the use of a resource from one period to the next. If we schedule all activities at their earliest times, it is quite possible that a larger amount of resources will be needed in the early periods and relatively lower amounts will be needed in the later periods. The specified project completion time could be achieved by shifting some of the activities so that the profile of resource requirements over time is level or has less fluctuation. This kind of load leveling has the objective of reducing idle labor costs, hiring and separation costs, or the cost of any resource that may be affected by fluctuations in the demand for its use, such as equipment rental.

In many project situations, resources are available only in fixed amounts. This type of problem is often known as resource-constrained project scheduling problem – RCPCP. Shipyards, for instance, deal with such scheduling problems [27]. For example, pieces of machinery or the number of design engineers may be limited. The objective in such situations is to minimize the project completion time (makespan minimization) without exceeding the given limits on available resources. Since the activities that compete for the same limited resources cannot be scheduled simultaneously, the effect of resource constraints will often be to delay the project completion date relative to the completion date that could be achieved without such restrictions. Other objectives include delays minimization in case due dates have been defined, total cost minimization and product quality optimization [28].

An optimal solution for the project completion time with resource constraints requires a mathematical program. In the past, several heuristics had been developed and compared [17, [29]. In some approaches, if all eligible activities cannot be scheduled due to a resource limit, a rule is used to decide which activities should be scheduled and which should be postponed. An example of such a rule is the *minimum activity slack*. Using this rule, priority in scheduling is given to those activities with the least slack (slack = latest start time - earliest start time in the critical path analysis). Activity slack is updated in each period to reflect the change in the available slack, resulting from the postponement of individual activities [30].

6.3.2 Decision Making

Single Attribute Decision Making

In analyzing a decision-making problem, it is necessary that we are able to specify exactly what actions or alternatives are available. We will label the actions as a_1, a_2, a_3, ... In addition, we assume that each action yields a *payoff* (or some type of consequence), which depends on the value of a random variable called the *state of nature*. States of nature will be labeled as θ_1, θ_2, θ_3, ... For example, suppose you are considering buying 100 shares of one of four common stocks (actions a_1, a_2, a_3, or a_4), each of which now costs $10. Action a_1 yields a profit of $500 (a $5 profit on 100 shares), a_2 yields a profit of $100, a_3 yields a profit of - $200 (a loss), and a_4 yields no profit at all. This situation represents *decision making under certainty* since there is no uncertainty about what state of nature (that is, what set of prices) will incur a year from now and the decision would be to buy the first stock (action a_1).

Instead of one trying to figure out what the prices will be in a year, one could probably only guess what the prices might be based either on one's impression of the economy, on the knowledge of various industries, or perhaps on a hot tip from a friend. There may be many different states of nature, each yielding a different set of payoffs for the various actions that could be taken. Thus, the problem is really one of *decision making under uncertainty*.

To extend the stock problem to include uncertainty, suppose you decide that there are three possible states of nature (θ_1, θ_2, and θ_3). For instance, you might decide that the stock prices one year are directly related to the stability of the economy during that year. In this case, θ_1 might correspond to a mild recession, θ_2 to a stable economy, and θ_3 to a mild inflation. Suppose for θ_1 the prices a year from now will be those given previously ($15, $11, $8, and $10); in θ_2 the prices will be $5, $12, $12, $13; while in θ_3 they will be $17, $11, $15, and $15. The payoffs these prices reflect can be expressed in a *payoff table*, as shown in Table 6.8 [31].

		States of Nature		
		θ_1	θ_2	θ_3
	a_1	$500	-$500	$700
Actions	a_2	100	200	100
	a_3	-200	200	500
	a_4	0	300	500

Table 6.8. Payoff Table for Stock Example

There is no one action for this payoff table which is obviously the "best" one. Action a_1 yields the largest payoff if θ_1 or θ_3 occurs, while a_4 is optimal if θ_2 occurs. Note that a_3 can never be the optimal action because a_4 always results in a payoff at least as large as, or larger than, a_3, no matter what state of nature occurs. For example, for θ_1, a_4 yields $0 and a_3 yields only - $200; for θ_2, a_4 results in a payoff of $300 as compared to only $200 for a_3; for θ_3, both actions yield the same payoff, $500. An action, which is no better than any other one, no matter what state of nature occurs, it is said to be a dominated action, such as a_3 in this case.

It might also appear that a_2 can never be optimal for this payoff table. This action is optimal, however, under what is called the *maximin* criterion, which says that the decision maker should focus only on the *worst* possible payoff that could happen for each action. In the stock example, the maximin criterion leads to action a_2, since this stock's lowest payoff, $100, is the largest among the minimum values.

Closely related to the maximin approach is another criterion called the *minimax regret criterion*. Under this approach, the best action for the decision maker is to select that alternative which *minimizes* his *maximum* (that is, minimax) *regret*. We can illustrate how regret is calculated by referring to our stock example. Let's suppose θ_1 occurs. If the decision maker had selected a_1, then he would have zero regret, since $500 is the best payoff he could receive under θ_1. If, however, he had selected a_2, then he would have had a regret of $400, since this is the amount of money he "lost" by not selecting a_1 - that is, $500 - $100 = $400. Similarly, a_3 has a regret of $700 and a_4 has a regret of $500. Regret values are calculated for each column of a payoff table. The regret values for all actions and states can be expressed in what is called a *regret table* (or *opportunity loss* table), as shown in Table 6.9 [31]. The action minimizing the maximum regret is seen to be action a4.

		States of Nature			Maximum regret	Minimax regret
		θ_1	θ_2	θ_3		
	a1	$0	$800	$0	800	
Actions	a2	400	100	600	600	
	a3	700	100	200	700	
	a4	500	0	200	500	◆

Table 6.9. Regret Table for Stock Example (Determined from Table 6.8)

Expected Monetary Value Criterion

In order to be able to use the *expected monetary value criterion*, it is necessary to know or be able to determine the probability of each state of nature. If there is considerable "objective" evidence (e.g., historical data) or a theoretical basis for assigning probabilities, then this task may be a fairly easy one. The difficulty in many real-world problems is that there may be little or no historical data or theoretical basis to decide on probabilities in these circumstances. Bayesian statisticians usually suggest that in assessing the subjective probability of an event, one should ask oneself, at what odds would one be exactly indifferent between the two sides of an even bet. In a decision-making context, one could use this approach to assess the probability of each state of nature, being careful, of course, to see that the sum of these probabilities equals one.

The expected value of a random variable is a "weighted" mean (or arithmetic average) of a probability distribution, where the weights are the probability values (or relative frequency) of the random variable. Since an expectation calculated in decision theory usually involves monetary values, such an expectation is generally referred to as an *expected monetary value* (EMV). To illustrate the calculation of EMV's, let's suppose that in the stock example the probability of our three states of nature are $P(\theta_1) = .30$, $P(\theta_2) = .60$, and $P(\theta_3) = .10$. The EMV of actions a1, a2, and a3 is given by multiplying each payoff by its probability of occurrence and then summing these values:

EMV (a1) = $500(.30) - $500(.60) + $700(.10) = -$80,

EMV (a2) = $100(.30) - $200(.60) + $100(.10) = $160,

EMV (a3) = -$200(.30) - $200(.60) + $500(.10) = $110,

EMV (a4) = $0   (.30) - $300(.60) + $500(.10) = $230,

Suppose we let π_{ij} represent the profit to the decision maker, who, after selecting a_1, finds that the jth state of nature occurs (j = 1, 2, ..., m). Using the notation, we can write the EMV of action a_i as follows:

$$EMV(a_i) = \sum_{j=1}^{m} \pi_{ij} P(\theta_j)$$

Under the EMV criterion the decision maker selects that alternative which yields the highest expected monetary value. For this example, action a_4 results in the highest EMV, with an average payoff of $230. If the three states of nature had been assigned probabilities of .40, .20, and .40 then action a_1 is the optimal EMV. However, no matter what values $P(\theta_1)$, $P(\theta_2)$, and $P(\theta_3)$ take on, a_3 can never yield the largest EMV because it is a dominated action.

The EMV criterion suffers from a major weakness. It considers only the expected or mean profit and does not take into account the variance in the payoffs. If the variance is fairly constant across the relevant alternatives, then this weakness will probably not cause any problems. However, when the variance is large, the EMV criterion might indicate an action which, for some people, won't be the most preferred one. For instance, suppose having to choose between two stocks (a_1 and a_2), and there are only two possible states of nature (θ_1 and θ_2) each having the same probability $P(\theta_1) = .50 = P(\theta_2)$. Let the payoff table be given by Table 6.10 [31]. Even though the EMV of a_2 is 10 times as large as a_1, most people would select a_1, if forced to pick between the two stocks, because they cannot afford to risk losing $10,000. Some people might prefer a_2 over a_1, which merely illustrates the fact that the value of a dollar to one person is not necessarily the same as that of a dollar to another person; neither does the value of a dollar necessarily remain the same to one person over a period of time.

In addition to constructing a payoff table, it may be helpful to visualize the problem in the form of a "decision tree." For the example illustrated in Table 6.10, a decision tree as shown in Figure 6.35 [31] may be drawn. Note that in this diagram, the first set of the tree's "branches" represents the decision maker's possible actions and the second set represents the various states of nature. In tree diagrams it is common to use a square to denote the nodes, where the decision maker must select among various actions as well as use a circle to denote the nodes, where the branch taken is determined by chance.

The encircled numbers represent the EMV resulting from the various decisions. Non-optimal branches on a decision tree are usually marked with a symbol, as shown on the a_1 branch.

		States of nature		
		θ_1	θ_2	EMV
Actions	a_1	-$50	$100	25
	a_2	10,500	-10,000	250
Probability		.50	.50	

Table 6.10. Example of Decision Affected by Risk

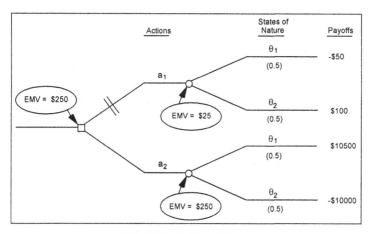

Fig. 6.35. A Tree Diagram

Note in Figure 6.35 that we have indicated our initial decision node (deciding between a_1 and a_2) as having an optimal EMV of $250. That is, we have compared the two actions and carried the EMV of the better one back to our original decision node. The process of moving progressively, from the end points of the tree toward the initial decision point by carrying EMV's backwards, is called *backward induction*.

Multiple Attribute Decision Making

In formulating a manufacturing strategy, trade-offs among cost, quality, time, and flexibility must be explicitly considered. A company may place a high weight on the cost criterion and gear its manufacturing organization and plant to low cost production. Another company may emphasize on customization and thus choose a flexible process and emphasize on flexibility in its production planning methods. The relative weights placed on different criteria will determine many other intermediate manufacturing positions.

Problem Statement

Consider a simple situation in which there are N decision alternatives denoted $a_1, a_2, ..., a_N$. The desirability of each alternative is measured on m criteria. The performance of the alternative a_j on the ith criterion is denoted as f_i^j. Table 6.11 [16] depicts this notation. To illustrate the notation, consider a dispatching problem with three resources (R1, R2, R3) and three tasks (T1, T2, T3). Each possible allocation of tasks to resources is denoted by a_j; for example, $a_1 = $ (R1T1, R2T2, R3T3), $a_2 = $ (R1T2, R2T3, R3T1), $a_3 = $ (R1T3, R2T2, R3T1) and so on. The criteria that are relevant to selecting the resource-task allocation may be mean flowtime (the average amount of time that the tasks spend in the system), mean tardiness (the average amount of time of the actual completion time exceeding the due date), capacity utilization, and quality (surface finish and deviations from dimensional specifications). Further, suppose that the mean flowtime and mean tardiness are measured in minutes, capacity utilization is measured as the percentage of available capacity, and quality is measured on a subjective scale as *excellent, very good, good, fair and poor*.

	Criteria			
Alternatives	1	2	...	m
a_1	f_1^1	f_2^1	f_3^1	f_m^1
a_2	f_1^2	f_2^2	f_3^2	f_m^2
$\vdots$	$\vdots$	$\vdots$	$\vdots$	$\vdots$
$\vdots$	$\vdots$	$\vdots$	$\vdots$	$\vdots$
a_N	f_1^N	f_2^N	f_3^N	f_m^N

Table 6.11. Notation for Multicriteria Decision Problem

Note that multicriteria decision methods do not preclude the use of subjectively measured criteria. However, the subjective scale must be precisely defined.

In this example:

Criterion 1 = mean flowtime (minutes)
Criterion 2 = mean tardiness (minutes)
Criterion 3 = capacity utilization (percentage of available capacity)
Criterion 4 = quality (excellent, very good, good, fair and poor)

If the mean flowtime for alternative a_1 is 10 minutes, then in our notation, f_1^1 = 10. Similarly, if quality for alternative a_2 is rated as excellent then f_4^2 = excellent.

Two questions now arise. How should one identify criteria for a given decision problem and what should the desirable properties be in defining a set of criteria? A literature survey or an analysis of the inputs and outputs of the system under consideration, will often suggest criteria. For important strategic decisions, a hierarchical approach for structuring the objectives and criteria may be useful. To illustrate this approach, consider a manufacturing firm's decision to add capacity. The decision may include where the capacity is to be added (location), the size of the addition, and the technology to be employed (manual versus automatic).

The set of criteria should be complete so that no concern of the decision maker is excluded. Using the capacity addition decision as an example, if the employment of domestic labor is a relevant concern, then it must be included as a criterion. Further, the set of criteria must not be redundant in order for double counting of outcomes to be avoided. Finally, the number of criteria should be kept as small as possible while maintaining understandability and measurability. Too broad an objective (e.g., a consumer's view of quality) must be broken down further; however, a large number of criteria for measuring a consumer's view of quality may not be appropriate as the subsequent analysis might become too cumbersome if the size of the criteria set is not kept at a manageable level.

Classification of Multiple Criteria Decision Methods

The key objective of all multicriteria methods is to formalize and capture the trade-offs among criteria of interest. Table 6.12 [16] summarizes the different classes of multiple criteria methods.

Multiple criteria decision methods can be distinguished along several dimensions. From a decision maker's perspective, a classification scheme based on the information required by him or her to implement the method is useful. In such a classification, the first class of methods is one to which the decision maker supplies no information during the analysis phase. Since the essential information about the decision maker's trade-offs among criteria is not conveyed, these methods provide only a trade-off curve (also called efficient frontier, non-dominated solutions, or pareto optimal solutions) among criteria. The decision maker can then choose a preferred point, reflecting his or her inherent preferences, from the trade-off curve.

Class	Information Obtained from Decision Maker	Situations where useful	Examples of methods and References
I	None	a. User is not known or multiple users b. User cannot provide information until some solutions are presented c. Problem is simple, or it is convenient to involve user only at the end of the process	Generation of efficient frontier or trade-off curves [20,21,22]
II	Trade-offs among criteria or choice between two options sought as needed	a. Preference function too complex to be stated explicitly b. User can provide only local information	Interactive mathematical programming [23,24]
III	Complete specification of trade-offs and preference function	a. Explicit preference function can be obtained b. Problem is of strategic nature and, therefore, thinking is focused on values (preferences) desirable before the decisions are evaluated	Multiattribute preference function theory [19,25]

Table 6.12. Classification of Multicriteria Methods

Alternatively, formal methods can be employed to elicit the decision maker's preference function, which can in turn be used to evaluate the points on the trade-off curve; the optimal decision is identified as the point that maximizes the utility or value specified by the decision maker.

In the second class of methods, the decision maker is an integral part of the analytical process. Information is solicited from the decision maker about his or her preferences for various levels of attainment on the criteria of interest. Based on this information, the method progresses so as to achieve better solutions, with more information, sought by the decision maker as it is needed. This sequential approach terminates when an acceptable or the most preferred decision is identified. These methods are interactive in nature, as information from the decision maker is essential for continuing the progress of the method toward the preferred decision.

Finally, in the third class of methods, the decision maker is asked to supply complete information about his or her preferences and trade-offs among criteria. A preference function (utility or value function) is then constructed based on the elicited information. Once the preference function is completely known, the selection procedure substitutes the preference function for an objective function.

Hybrid procedures, which combine the strategies employed in the three classes of methods just described, have also been developed. Other classification schemes can also be developed, based on whether the outcomes of the decisions are known or unknown, whether the time horizon over which the outcomes occur is modeled as single or multiple time periods, whether the decision alternatives are finite or infinite, whether there is a single decision maker or a group of decision makers, whether the competitive reaction is implicitly considered or explicitly modeled, and similar questions [32].

Efficient Frontier or Trade-off Method

Suppose $f_1(x)$ and $f_2(x)$ denote the performance or achievement levels (also referred to as the outcomes or consequences) on two criteria when a decision x is chosen. A decision x *dominates* a decision y whenever both of the following occur:

    1. $f_i(x)$ is at least as good as $f_i(y)$ for all i
    2. $f_i(x)$ is better than $f_i(y)$ for at least one i

Since decision x is at least as good as decision y for all criteria and is better for at least one criterion, it is reasonable to exclude decision y from consideration. Now suppose that we plot all the possible values of f_1 and f_2 for decision x as shown in Figure 6.36 [16]. We assume that higher levels of f_1 and f_2 are more desirable. In Figure 6.36a, decision x is a continuous variable and there are an infinite number of possible outcomes. In Figure 6.36b, decision x has a finite number of alternatives (e.g., 15 possible sites) and there are only a finite number of possible consequences. The set of darkened points in Figure 6.36 is called the efficient frontier (the pareto optimal set or the trade-off curve). The key characteristic of the efficient frontier is that every point on it is non-dominated. In other words, any point that is not on the efficient frontier or the trade-off curve, is dominated by some point on the efficient frontier. No matter what the underlying preference function (objective function or cost function) is, the most preferred decision (the optimal decision) must correspond to one of the points on the efficient frontier.

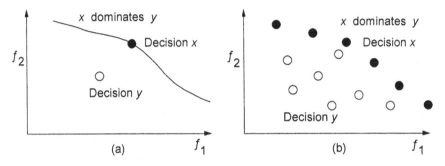

Fig. 6.36. The Trade-off Curves for Two-Attribute Cases (a) Infinite Consequences and (b) Finite Consequences

Suppose that in an inventory planning decision there are two criteria of interest:

f_1 = The average investment in inventory

f_2 = The work load on the system measured by the number of orders placed per year

Let R be the annual demand, p be the purchase cost per unit, and Q be the order quantity. The demand is known and is constant per unit of time period. Thus,

$$f_1 = \frac{Q}{2} \times p$$

$$f_2 = \frac{R}{Q}$$

The decision variable is Q. The efficient frontier can be constructed directly by simply varying the levels of Q. In Table 6.13 [16], the levels of f_1 and f_2 for several values of Q are shown by assuming $R = 12,000$ and $p = \$10/\text{unit}$. In Figure 6.37 [16], the efficient frontier is plotted. The decision maker can select a point from the efficient frontier. Suppose the decision maker chooses a point shown in Figure 6.37 that corresponds to 10 orders per year and an average investment of $6000. The value of the preferred decision Q can be derived by substituting the chosen point in the equation for f_1 and f_2.

$$f_1 = \frac{Q}{2} \times 10 = 6000$$
$$Q = 1200 \text{ units}$$

Order Quantity Q	Average Investment in Inventory, $f_1 = (Q/2)p$	Work Load or Annual Number of Orders, $f_2 = R/Q$
500	2,500	24
1,000	5,000	12
2,000	10,000	6
4,000	20,000	3
6,000	30,000	2
12,000	60,000	1

Table 6.13. Computation of f_1 and f_2 for the Inventory Problem ($R = 12,000$ and $p = $10/Unit$)

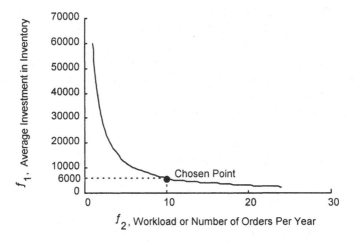

Fig. 6.37. Efficient Frontier for the Inventory Problem

Note that if the opportunity cost of capital is 10 percent per annum and the order cost is \$60 per order, then $Q = 1200$ will be optimal. A given point will be optimal for all combinations of the order cost c_p and the opportunity cost of capital I such that $c_p/I = $ constant. In our illustration, $Q = 1200$ is optimal for all combinations of c_p and I such that $c_p/I = 600$ (e.g., $c_p = \$120$ and $I = 20\%$, or $c_p = \$30$ and $I = 5\%$).

Mathematical Programming Approach

Consider a project planning example where the two criteria of interest are

$$f_1 = \text{Project completion time}$$
$$f_2 = \text{Cost of crashing activities}$$

This problem may be formulated as a linear programming problem. To construct the efficient frontier or trade-off curve between f_1 and f_2, we simply solve the problem:

Minimize f_2
Subject to: precedence constraints and $f_1 \leq T$

Now, we systematically vary T, the completion time, to obtain the efficient frontier shown in Figure 6.38 [16]. The efficient frontier is piecewise linear because f_1 and f_2 as well as all the constraints are linear in this problem. The key idea in this approach is that one criterion is contained in the constraint set and is parametrically varied to minimize the other criterion. The approach is also applicable when there are more than two criteria. In this more general case, all but one criterion must be brought into the constraint set and the remaining criterion is successfully optimized by varying the right hand sides of the criteria in the constraint set.

Another approach of obtaining the efficient frontier is to use a linear weighted average method. In this approach, we optimize

$$w_1 f_1 + w_2 f_2 + w_3 f_3 + \dots + w_m f_m$$

subject to specified constraints, where

$$w_i \geq 0, i = 1 \text{ to } m$$

$$\sum_{i=1}^{m} w_i = 1$$

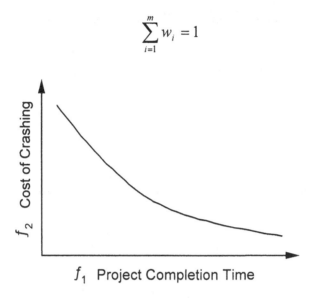

Fig. 6.38. Efficient Frontier for Project Planning Example

The solution of this problem, which is a simple linear program if all the f_i's and the constraints are linear, provides a point on the efficient frontier. By choosing different values of the weights, w_i, we can trace the entire efficient frontier.

This approach can be applied to product mix decisions, aggregate planning decisions, and location and distribution decisions when the appropriate criteria functions and the constraints are linear. In nonlinear cases, more complex procedures will be required to generate the efficient frontier, especially if the problem is nonconvex (i.e., if the efficient frontier contains local dips and valleys).

Many production problems, such as scheduling, may not be easily solved by mathematical programming approaches. This is because, even for a single criterion, the size of the problem is too large to be solved in reasonable computational time. Specialized procedures that exploit the problem characteristics and heuristics may be used to solve this type of problems.

Preference Function

In some production and operations situations, it may be desirable to ask the decision maker to quantify his or her preference function for the criteria of interest. The preference function can then be used to evaluate decision alternatives. A procedure for assessing the preference function of the decision maker is described in the following steps.

Step 1 - Determining the Form for the Preference Function

The preference function P can be decomposed into some simple forms if the decision maker's preferences for various combinations of the levels of criteria satisfy certain conditions. The simplest and the most widely used form is the weighted additive form:

$$P(f_1, f_2, f_3) = w_1 v_1(f_1) + w_2 v_2(f_2) + w_3 v_3(f_3) \qquad (6\text{-}11)$$

where w_i is the importance weight and v_i is the value function for each criterion. The weights and the value functions are scaled so that

$$\Sigma w_i = 1, \quad 0 \le w_i \le 1$$

and

$$v_i(\text{best level}) = 1$$
$$v_i(\text{worst level}) = 0, \text{ for } i = 1 \text{ to } m$$

In order to verify the legitimacy of this form, we need to check a condition called *difference independence* of each criterion. Essentially, this condition means that the magnitude of the difference in the strength of preference between two levels of the criterion i does not change when fixed levels of the other criteria are changed.

If difference independence fails to hold, then we have two options. One option is to use a more complicated model that will allow interaction among the criteria. The second is to redefine the criteria so that difference independence does really hold. Fortunately, the additive form is quite robust and, in most situations, will produce low errors even when there are moderate interactions among the criteria.

Step 2 - Constructing Single Dimensional Value Functions

In Equation 6-11, we need to assess $v_i(f_i)$, the value function for criterion i. A simple method for assessing these value functions is to use a 100 point rating scale on which 0 indicates the worst level and 100 indicates the best level (Fig. 6.39). The different levels of an attribute are rated on this scale to represent the strength of preference of the decision maker. The value of level of a criterion is simply the rating/100. Thus, the best level a criterion can take on, always has a value of 1 and the worst level has a value of 0.

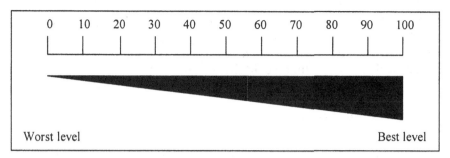

Fig. 6.39. Rating Scale

Step 3 - Determining Importance Weights for the Criteria

The most important criterion is first identified. This and one of the other criteria are handled two at a time. For example, let case A denote an alternative where the best criterion is at the best level and one of the other criteria is at the worst. Let case B denote a situation where the other criterion is at its best level. We then pose a hypothetical question to the decision maker and ask him the level of the best criterion that would make him indifferent between cases A and B. By equating the equations that will result from using Equation 6-11, for cases A and B, one would have an equation relating the weight of the other criterion with that of the most important criterion. If the above procedure is repeated for the remaining criteria, equations relating the weight of the most important criterion with that of the other criteria, will also be obtained. Finally, since all the weights should add up to one, one could solve for the value of the weight of the most important criterion. This value is then backsubstituted into the previous equations to determine the weight associated with the other criteria.

Step 4 - Computing the Overall Values

The computations of the overall weighted values for each of the alternatives are performed using Equation 6-11.

Step 5 - Sensitivity Analysis

Since both ratings of the different levels of the criteria and the importance weights are subjectively derived, it is useful to analyze the impact of changing some of these values. Alternatively, if several decision makers are involved in the process, it is quite possible that they differ in their assignments of ratings or importance weights. Sensitivity analysis clarifies whether such differences are relevant to selecting the most preferred alternative. In many situations, disagreements on ratings or on importance weights do not change the ranking. If, however, the differences in opinion about the ratings or the importance weights do indeed have an influence on the final ranking, then an open discussion or additional information must be used for reconciling the differing viewpoints.

6.3.3 Dispatching Heuristics

Traditionally, much of the research done in dispatching has been concentrated on heuristics called dispatch rules. In general, a heuristic for a problem is a solution procedure, which is based on intuition. Its application is based on the empirical evidence of its success, with 'success' generally considered as being the ability to find, by some measure, a good solution that is within a certain percentage of the optimum. A dispatch rule is a method for ranking a set of tasks, which are waiting to be processed on a machine. The task with the best rank is selected in order to be processed on the machine. A schedule results from the repeated application of the rule whenever a machine is available for assignment. When more than one machine is available at a given time, a machine selection rule is also required to specify the order in which the machines will be receiving assignments.

Many dispatch rules have been proposed [39-43] and a number of these rules are listed below [44]:

No.	Rule	Description
1	RANDOM	The job selected has the smallest value of a random priority assigned at the time of arrival at the queue (note that a job receives a new number for each of its operations).
2	FCFS	The job selected has arrived at queue first.
3	LCFS	The job selected has arrived at queue last.
4	FASFS	The job selected has reached the shop first.
5	SPT	The job selected has the shortest operation processing time.
6	LPT	The job selected has the longest operation processing time.
7	FOPNR	The job selected has the fewest operations remaining to be performed.
8	MOPNR	The job selected has the most operations remaining to be performed.

No.	Rule	Description
9	LWKR	The job selected has the least work remaining to be performed.
10	MWKR	The job selected has the most work remaining to be performed.
11	TWORK	The job selected has the greatest total work (all operations on the routing.
12	NINQ	The job selected will go on for its next operation to the shortest queue.
13	WINQ	The job selected will go on for its next operation to the queue with the least work.
14	XWINQ	The job selected will go on for its next operation to the queue with the least work. The queue is considered to include jobs now on other machines that will arrive before the subject job.
15	DDATE	The job selected has the earliest due date.
16	SLACK	The job selected has the least slack time determined by due date minus all remaining processing time for the job.
17	OPNDD	The job selected has the earliest operation due date and equally spaced due dates are assigned to each operation at the time the job enters shop.
18	S/OPN	The job selected has the least ratio of slack time divided by the remaining number of operations.
19	S/PT	The job selected has the least ratio of slack time divided by the remaining processing time.
20	DS	The job selected has the least slack time determined by due date, less the remaining expected flow time, minus the current date (dynamic slack).
21	DS/PT	The job selected has the least ratio of dynamic slack time divided by remaining processing time.
22	DS/OPN	The job selected has the least ratio of dynamic slack time divided by remaining number of operations.

No.	Rule	Description
23	P+WKR(a)	The job selected has the smallest weighted sum of the next processing time and work remaining. 'a' is a weighting constant, which is greater than 0.
24	P/WKR(a)	The job selected has the smallest weighted ratio of the next processing time to work remaining. 'a' is a weighting constant, which is greater than 0.
25	P/TWK	The job selected has the smallest ratio of the next processing time to total work.
26	P+WQ(a)	The job selected has the smallest weighted sum of the next processing time and work in the following queue. 'a' is a weighting constant, which is greater than 0.
27	P+XWQ(a)	The job selected has the smallest weighted sum of the next processing time and work (including expected work) in the following queue. 'a' is a weighting constant which is greater than 0.
28	P+S/OPN (a,b)	The job selected has the smallest weighted sum of next processing time and slack time per operation remaining. Both 'a' and 'b' are weighting constants. Both are greater than or equal to 0. Each particular set of the parameters (a,b) represents a different priority rule.
29	SPT-T(a)	This is the truncated version of SPT. As long as no job in the queue from which selection is made, has waited for more than 'a' time units in this queue, normal SPT selection is made. When a job has waited too long, it is given dominating priority.
30	FCFS(a)	This is the variation of the FCFS in which SPT selection is invoked for a particular queue whenever that queue becomes too long. If there are fewer than 'a' jobs in queue at the time of selection, the earliest arrival (to the queue) is chosen. If 'a' or more jobs exist, then the job with the shortest processing time is chosen.

No.	Rule	Description
31	SPT-2C(a)	This is a two-class variation of the SPT. A fraction of the jobs, denoted by 'a' is identified as preferred job with the SPT being selected. If there are no preferred jobs in this particular queue, then the regular job with the SPT is selected.
32	FCFS(pr)	Priority depends on the (dollar) value of the job. Jobs are divided into two classes - a high value class and a low value class. All high value jobs are assigned greater priorities than all low value jobs. Within the class, priority is assigned in arrival order (FCFS). There is actually a family of rules of this type which can be parameterized by 'pr', the proportion of jobs assigned to the low value class.
33	VALUE	Priority is directly related to the (dollar) value of the job. The priority is taken to be equal to the value of the job.
34	NINQ(q)	Priority is related to the subsequent move. Maximum priority is given to that job, which on leaving this machine will go on to the next one that has the shortest (in the sense of least processing time) critical queue. If no queue is critical, the selection is in arrival order (FCFS). A queue is considered critical when it has less than a specified number of time units of processing time waiting. There is actually a family of rules of this type which can be parameterized by 'q', the value below which a queue becomes critical.
35	ESDATE(a)	The job selected has the earliest start date determined by multiplying the number of operations (through which a job must pass) by an arbitrary allowance 'a' (a>0).

No.	Rule	Description
36	SEQ	This rule takes the flow allowance (delay) [43] into consideration, as compared with the elapsed waiting time and the number of remaining operations. The flow allowance is determined empirically on the basis of value of a job. The higher the value is, the lower the flow allowance; a job is selected for on-time completion on the basis of the number of remaining operations.
37	PRIORITY INDEX	The job selected has the smallest value of P(ab) [42]; P(ab)=(due date) - (time required for future moves) - a*(expected time required for future machining) + b*(expected time for present machining operation).
38	FIXSEQ	The job selected first is contained in a given class. After processing all jobs in the class, another one is processed in a fixed class sequence.
39	MINSEQ	The job selected has the minimum setup time.
40	FIXSEQ-S	The job selected has the shortest service time after choosing a particular class of jobs in the fixed sequence.
41	MIXSEQ-S	The job selected has the shortest service time after choosing a particular class for which setup time is minimum.
42	SSS	The job selected has the minimum sum of setup time and service time.

Dispatch rules may be task-based, such as EDD or SPT, or be job related, such as Least Remaining Work (LRM). Here we note that the distinction between a job and a task is that a job in general, corresponds to a part, while a task in general, corresponds to an individual operation to be performed on a part. Dispatch rules may also include definitions, which vary over time, such as Critical Ratio (CR), which is a function of task processing time and the slack time remaining in the job. Another method of classifying dispatch rules, divides them into four classes: rules involving processing time, rules involving due dates, simple rules involving neither due dates nor processing time, and rules involving two or more of the other classes [45].

As dispatch rules are heuristics, there is no dispatch rule that has been found to perform best in all situations. Factors which affect the suitability of a dispatch rule include:

- The distribution of the processing times of the tasks to be dispatched
- The distribution of the job due dates
- The distribution of job arrivals over time
- The performance measure to be optimized (e.g., mean tardiness, production rate, etc.)
- The nature of the jobs' process plans (presence of assembly tasks, number of tasks per job, etc.)

Extensive research has been performed to determine which dispatch rules perform best in which situations [46-51]. However, because of the multitude of factors involved, it has not been possible to provide a comprehensive answer to this question. Trial and error experimentation, either on the actual system or on a simulation model of the actual system, is still necessary before deciding which dispatch rule to apply [52].

Dispatch rules have many limitations; for example, they only prescribe which task to assign, not which resource the task should be assigned to. Therefore, when several resources become available at the same time, they must be arbitrarily assigned to the tasks, selected by the dispatch rule or another rule for the machine selection to be applied to. Dispatching rules thus, partition the dispatching problem by first considering the resources and then considering the tasks. In addition, dispatch rules do not "look ahead" to the consequence of a particular assignment on the future performance of the system, and their application is based on extensive trial and error. Finally, each dispatch rule is usually "hard-wired" to optimize a single criterion. This is a serious drawback because in most manufacturing environments, a number of criteria need to be considered simultaneously. However, dispatch rules are computationally efficient and provide a quick solution, which is often required in a constantly changing manufacturing system.

6.3.4 Evolution Programs

"Metaheuristics" – mainly evolution programs, genetic algorithms, simulated annealing and tabu search – have been efficiently used for addressing combinatorial optimization problems. Evolutionary algorithms and evolution programs, in particular, comprise a kind of stochastic heuristic search method, based on a simplified natural evolution model (Figure 6.40).

```
procedure evolution program
begin
    t ← 0
    initialize P(t)
    evaluate P(t)
    while (not termination - condition) do
    begin
        t ← t + 1
        select P(t) from P(t - 1)
        alter P(t)
        evaluate P(t)
    end
end
```

Fig. 6.40. An evolution program structure

The term evolution programs was introduced [53] to describe variations of genetic algorithms (GA), using problem-specific data structures along with appropriate genetic operators, apart from the fixed-length binary strings used in GA. The evolution program handles a population of randomly generated alternative solutions called individuals $P(t) = \{x_1^t, ..., x_n^t\}$ for each iteration t. Every alternative solution x_i^t, also called a chromosome, is made of units (genes), which define a specific data structure (which may be complex) S. Each chromosome is evaluated in order to calculate its value (fitness). A new population of alternative solutions is then generated (iteration $t + 1$), usually by selecting the best solutions (chromosomes).

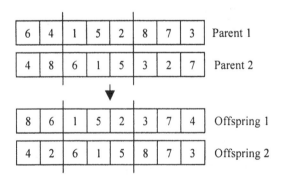

Fig. 6.41. The order crossover operator

Genetic operators are used to control the way new populations are generated. The order crossover operator, for instance, is used for merging the information contained in two different chromosomes (parents), while the displacement mutation operation is used for modifying the information contained in a chromosome. Each offspring produced by the order crossover operator chooses a sub-sequence of a tour from one parent and preserves the relative order of the remaining genes from the other parent (Figure 6.41). The displacement mutation operator selects a sub-tour composed of no more than three genes and inserts it in a random place (Figure 6.42).

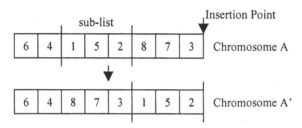

Fig. 6.42. The displacement mutation operator

In this example, each chromosome may represent an alternative solution, identifying the sequence of tasks to be dispatched in a single machine.

In each iteration, a number of chromosomes are selected for the next population through an evaluation and selection mechanism. After a number of iterations, it is *hoped* that the best chromosome of the final generation should be a near optimum solution.

Genetic algorithms and evolution programs have been used for the solution of job shop [53], flow shop [54] and project scheduling problems [55]. In some cases, evolution programs are used in combination with other methods, such as dispatch rules [56].

Sometimes, it is not very easy to model complex problems or constraints by using evolution programs. In such cases, encoding mechanisms are used for modeling the problem with specific data structures; then, decoding mechanisms are used for transforming these data structures into alternative solutions. In other cases, constraint satisfaction is enforced by implementing a penalty function, activated when one or more constraints are violated. In all of these cases, genetic operators are quite harder to implement [57]. Furthermore, the values of the evolution programs' parameters (crossover ratio, mutation ratio, size of the population, number of generations, etc.) are not known in advance and usually a lot of experimentation is required.

6.3.5 Gantt Charts and Performance Measures

Gantt charts are useful tools for presenting scheduling results, in terms of the time duration of operations and the time interrelationships of those operations, having also been used widely in both industry and academia. A Gantt chart is simply a specialized form of a bar chart, (Fig. 6.43 [3]) where the horizontal axis represents time and the vertical axis represents resources, (which can be, for example machines or personnel.) Each bar represents the processing time of a task.

The Gantt chart is useful for defining performance measures of scheduling problems. For this purpose, the following notation (refer to Fig. 6.43) will be used:

J_i job i.

M_j machine j.

$t_{ij(k)}^{proc}$ *processing time* of the kth task of J_i on M_j.

r_i *ready time* of J_i. The arrival time of J_i into the system.

t_i^{due} *due date*, i.e., the promised delivery date of J_i. The time by which ideally, we would like to have completed J_i.

t_i^a *allowance* for J_i. The period allowed for processing between the ready time and the due date: $t_i^a = t_i^{due} - r_i$

t_{ik}^{wait} *waiting time* of J_i after the end of its $(k-1)^{th}$ operation and before the start of its kth operation. By k^{th} operation we do not mean the one performed on M_k (although it may be), but the one that comes k^{th} in order of processing.

t_i^{wait} *total waiting time* of J_i. Clearly $t_i^{wait} = \sum_{k=1}^{m} t_{ik}^{wait}$.

t_i^{comp} *completion time* of J_i. The time at which processing of J_i finishes.

We have the equality: $t_i^{comp} = r_i + \sum_{k=1}^{m} \left(t_{ik}^{wait} + t_{ij(k)}^{proc} \right)$

F_i *flow time* of J_i. The time that J_i spends in the workshop $F_i = t_i^{comp} - r_i$.

L_i *lateness* of J_i. This is simply the difference between its completion time and its due date: $L_i = t_i^{comp} - t_i^{due}$. Note that when a job is early, i.e. when it completes before its due date, L_i is negative. It is often more useful to have a variable which, unlike lateness, only takes non-zero values when a job is tardy, i.e. when it completes after its due date. Hence we also define the tardiness and the earliness of a job.

T_i tardiness of J_i: $T_i = \max\{L_i, 0\}$.

E_i earliness of J_i: $E_i = \max\{-L_i, 0\}$.

$N_w(t)$ number of jobs waiting between machines or not ready to be processed at time t.

$N_p(t)$ number of jobs actually being processed at time t.

$N_c(t)$ number of jobs completed by t.

$N_u(t)$ number of jobs still to be completed by time t.

An overbar is used to indicate a time average of the quantities listed above. For example,

$\overline{N}_u$ is defined so that $\displaystyle \overline{N}_u = \frac{1}{t_{max}^{comp}} \int_0^{t_{max}^{comp}} N_u(t)\,dt$

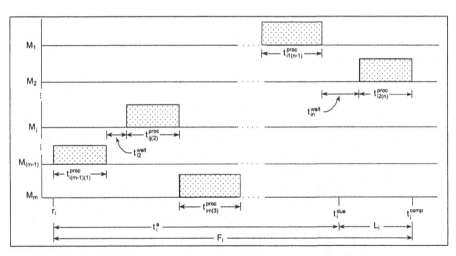

Fig. 6.43. Gantt Chart of a Typical Job J_i with n Tasks.

Referring to the Gantt chart in Figure 6.43, the processing order given by the technological constraints is (M_{m-1}, M_j, M_m, …, M_1, M_2). We can also see that the waiting times t_{i1}^{wait} and t_{i3}^{wait} are zero; t_{i2}^{wait} and t_{in}^{wait} are non-zero as shown. For this job, $T_i = L_i$ and $E_i = 0$, since the job is completed after its due date. Some of the common performance measures used in scheduling problems are:

Performance Measures Based on Completion Times

F_{max} *Maximum flow time.* Minimizing this performance measure makes the schedule cost directly related to the longest job.

t_{max}^{comp} *Maximum completion time* or *total production time* or *make-span.* Minimizing this causes the cost of the schedule to be dependent on how long the system is devoted to the entire set of jobs. t_{max}^{comp} and F_{max} are identical when all the ready times are zero.

$\overline{F}$ *Mean flow time.* Minimizing $\overline{F}$ implies that a schedule's cost is directly related to the average time it takes to process a single job.

$\overline{t}_{max}^{comp}$ *Mean completion time.* Minimizing this is the same as minimizing $\overline{F}$.

Performance Measures Based on Due Dates

$\overline{L}$ *Mean lateness.* Minimizing this performance measure is appropriate when there is a positive reward for completing a job early and the reward is larger, the earlier the job is completed.

L_{max} *Maximum lateness.* Minimizing this has the same effect as minimizing $\overline{L}$.

$\overline{T}$ *Mean tardiness.* Minimizing this performance measure is appropriate when early jobs do not bring any reward and when penalties are incurred for late jobs.

n_T *Number of tardy jobs.* Penalties incurred by a late job may sometimes be independent of how late the job is completed. That is, the penalty is the same irrespective of the time the late job was completed. In such instances, minimizing this performance measure is appropriate.

Performance Measures Based on Inventory and Utilization Costs

$\overline{N}_w$ *Mean number of jobs waiting for machines.* Minimizing this performance measure relates to the in-process inventory costs.

$\overline{N}_u$ *Mean number of unfinished jobs.* Minimizing this has the same effect as minimizing $\overline{N}_w$.

$\overline{N}_c$ *Mean number of completed jobs.* Minimizing this, results in the reduction of inventory costs of finished products.

$\overline{N}_p$ *Mean number of jobs actually being processed at any time.* Maximizing this performance measure ensures the most efficient use of the machines.

$\overline{I}$ *Mean machine idle time.* Minimizing this, results in the efficient use of machines.

I_{max} *Maximum machine idle time.* Minimizing this has the same effect as $\overline{I}$.

6.4 Applications

In this section, applications of the methods and tools, presented in both Chapters 5 and 6 to the operation of manufacturing systems, will be described. The intent is to provide an indicative discussion as to the way such methods and tools can be applied; no attempt at a comprehensive review of the existing applications will be made.

6.4.1 A Mathematical Programming Approach to Dispatching

A number of mathematical programming formulations of dispatching problems have been proposed. The necessity of deriving analytical expressions for both the objective function and the constraints usually dictates that a number of simplifying assumptions be made. It is usually assumed that a single performance measure is to be optimized and that all information about arriving jobs is known ahead of time. In one particular formulation [58], the following assumptions are made:

1. There are m machines, and n jobs which arrive at time 0.
2. Each job consists of exactly m tasks, with one task to be performed on each of the m machines. In addition, the order in which each job visits

the machines is the same, and the order in which each machine processes each job is also the same. This assumption makes the manufacturing system a *permutation shop*.
3. There are no alternative job processing routes.
4. Machines never break down.
5. There are no duplicate machines.
6. There is no randomness. The number of jobs, the number of machines, the processing times, the job arrival times, etc. are all known and fixed.
7. Each job, once it enters the manufacturing system, must be processed to completion.
8. An individual task, once started on the machine, may not be interrupted.

A solution to this problem is specified by a single permutation of the numbers 1, 2, ..., n, giving the sequence in which the jobs are processed on each and every machine. An example of a feasible dispatching solution for this permutation shop is shown in Figure 6.44. Notice that all three jobs A, B, and C visit the machines in the sequence (Machine 1, Machine 2, Machine 3). Furthermore, all three machines process the jobs in the order (Job C, Job A, Job B). The objective is to minimize the time it takes to complete the processing of all n jobs, or, equivalently, to minimize the completion time of the last (n^{th}) job on the last (m^{th}) machine. In Figure 6.44, this completion time is equal to 15.

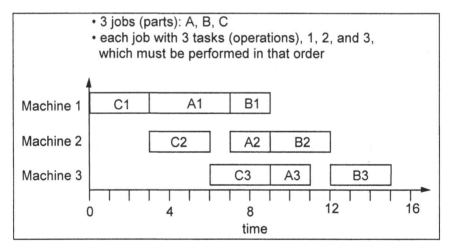

Fig. 6.44. A Sample Solution to the Permutation Shop Static Sequencing Problem

A description of the mixed integer programming formulation of this problem requires the introduction of some notation. Let the subscript i refer to the job J_i; the subscript j to the machine M_j, and the subscript (k), in parentheses, refer to the kth position in the processing sequence. Let $J_{(k)}$ be the kth job in the processing sequence. The distinction between J_i and $J_{(k)}$ is that J_i refers to a specific job, irrespective of where it appears in the processing sequence, whereas $J_{(k)}$ refers to whichever job $(J_1, J_2, ..., or J_n)$ is in the kth position of the processing sequence. Each job visits the machines in the sequence $(M_1, M_2, ..., M_m)$. The processing time of J_i on M_j is $t_{i,j}^{proc}$.

We require a set of n^2 variables constrained to take the values 0 or 1:

$$X_i = \begin{cases} 1, \text{ if } J_i = J_{(k)} \\ 0, \text{ otherwise} \end{cases} \tag{6-12}$$

Since exactly one job must be scheduled in each of the k positions,

$$\sum_{i=1}^{n} X_{i,(k)} = 1 \quad \text{for } k = 1,2,...,n \tag{6-13}$$

Furthermore, each job must be scheduled in one position only:

$$\sum_{k=1}^{n} X_{i,(k)} = 1 \quad \text{for } i = 1,2, ..., n. \tag{6-14}$$

Two other sets of decision variables are required to formulate the problem (Fig. 6.44). The variables $t_{(k),j}^{idle}$ represent the idle time of machine M_j between the completion of $J_{(k)}$ and the start of $J_{(k+1)}$. The variables $t_{(k),j}^{wait}$ represent the time that $J_{(k)}$ must spend between completion on M_j and the beginning of processing on M_{j+1}. It is convenient to define a set of temporary variables $t_{(k),j}^{proc}$ which represents the processing time of $J_{(k)}$ on M_j.

From Figure 6.45, it can be seen that a feasible solution satisfies the following constraint:

$$t_{(k),j}^{idle} + t_{(k+1),j}^{proc} + t_{(k+1),j}^{wait} = t_{(k),j}^{wait} + t_{(k),j+1}^{proc} + t_{(k),j+1}^{idle}$$

Or, equivalently,

$$t_{(k)j}^{idle} + \sum_{i=1}^{n} X_{i,(k+1)} t_{ij}^{proc} + t_{(k+1)j}^{wait} - t_{(k)j}^{wait} - \sum_{i=1}^{n} X_{i,(k)} t_{ij+1}^{proc} - t_{(k)j+1}^{idle} = 0 \quad (6\text{-}15)$$

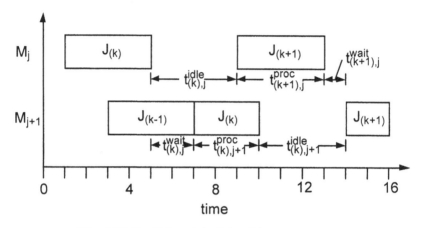

Fig. 6.45. Variables of the Mixed Integer Program

where the temporary variables $t_{(k+1)j}$proc and $t_{(k)j+1}$proc have been re-placed with expressions involving only the decision variables $X_{i,(k+1)}$ and $X_{i,(k)}$, and the known processing time constants $t_{i,j}$proc and $t_{i,j+1}$proc.

Constraints 6-13, 6-14, and 6-15 above, together with the requirements that $X_{i,(k)}$ are non-negative integers (Eq. 6-12) and that $t_{(k)j}$wait and $t_{(k)j}$idle are non-negative real numbers, form the entire constraint set.

The objective function may be defined as follows. Minimizing the maximum completion time is equivalent to minimizing the completion time of the last (n^{th}) job, which is in turn equivalent to minimizing the idle time on the last (m^{th}) machine. The total idle time on M_m is given by the sum of the inter-job idle times $t_{(k),m}$idle plus the idle time that occurs be-fore job $J_{(1)}$ begins processing on M_m. Thus, we seek to minimize

$$\sum_{k=1}^{n-1} t_{(k),m}^{idle} + \sum_{j=1}^{m-1} t_{(1)j}^{proc}$$

Expressing the temporary variables $t_{(1)j}$proc in terms of decision vari-ables and known constants, the objective function can be written:

$$\sum_{k=1}^{n-1} t_{(k),m}^{idle} + \sum_{j=1}^{m-1} \left(\sum_{i=1}^{n} X_{i,(1)} t_{ij}^{proc} \right)$$

This completes the mixed integer programming formulation.

At this point, let us summarize some of the essential features of the mathematical programming approach to dispatching. The translation of a dispatching problem into a mathematical program is only a part of the total solution procedure. Once the math program is formulated, the space of legal decision variable combinations (those combinations which satisfy all of the constraints) must be searched to find the optimum combination, via methods such as the branch and bound procedure described in Chapter 5. In general, such methods require extensive computation, partly because they rely only on the general properties of mathematical programs and do not consider particular features of the problem being solved. In practice, only small problems can be solved. Therefore, mathematical programming approaches have been applied mostly to problems of research interest only, where response time is not a critical factor.

A second feature to note about the math programming formulation is that it requires very restrictive assumptions regarding the nature of the process plans of the jobs (parts) to be dispatched, the structure of the manufacturing facility, and the performance measure to be optimized. A change in any of these elements requires the formulation of a different math program, a difficult to impossible task for more general dispatching situations.

6.4.2 A Dynamic Programming Approach to Dispatching

Similar to their mathematical programming counterparts, dynamic programming formulations of dispatching problems require restrictive assumptions, which only permit very rigidly structured problems to be handled. One example of a dispatching problem that can be formulated as a dynamic program is an n job, one-machine problem, in which the objective is to minimize the mean tardiness $\overline{T}$ of the jobs. Tardiness is defined to be the amount of time by which the completion time of a job exceeds its due date. Since there is only one machine, a solution to the problem is a sequence of the form $(J_{(1)}, J_{(2)}, ..., J_{(n)})$. It is assumed that all jobs arrive at time $t = 0$.

Some notation is required. Once again, J_i is a particular job i, and $J_{(k)}$ is the kth job in the solution sequence. The variable t_i^{comp} representing the completion time of J_i, t_i^{proc} is the processing time of J_i, and t_i^{due} is the due time of J_i. Likewise, the variable $t_{(k)}^{comp}$ represents the completion time of $J_{(k)}$, $t_{(k)}^{comp}$ which is the processing time of $J_{(k)}$, and $t_{(k)}^{due}$ is the due date of $J_{(k)}$. The notation $\{J_1, J_2, \ldots, J_n\}$ refers only to the *set* of jobs contained within the braces, without any implication of a particular processing order, while $(J_1, J_2, \ldots, J_n)$ is a *sequence* in which the jobs are processed in the order shown. If Q is any set of jobs containing J_i, then $Q-\{J_i\}$ is the set Q with J_i removed. Finally, we define t_Q^{comp} as the sum of the processing times of all the jobs in Q:

$$t_Q^{comp} = \sum_{J_i\,in\,Q} t_i^{proc}$$

If Q is the set of jobs to be scheduled, then t_Q^{comp} is also equal to the completion time of the last job in the solution sequence (since there is only one machine).

The dynamic programming formulation of this problem [59] is based on the observation that in an optimal schedule the first K jobs (any $K = 1, 2, \ldots, n$) must form an optimal schedule for the reduced problem, based on those K jobs alone. To perceive why this is true, we first note that minimizing the mean tardiness $\overline{T}$ is equivalent to minimizing the total tardiness $n\overline{T}$. Now suppose that $(J_{(1)}, J_{(2)}, \ldots, J_{(n)})$ is the optimal solution to the full problem. We observe that $n\overline{T}$ can be decomposed as follows:

$$n\overline{T} = \sum_{k=1}^{K} \max\left(t_{(k)}^{comp} - t_{(k)}^{due}, 0\right) + \sum_{k=K+1}^{n} \max\left(t_{(k)}^{comp} - t_{(k)}^{due}, 0\right) \qquad (6\text{-}16)$$

$$= \qquad A \qquad + \qquad B$$

Term A is the contribution to the total tardiness from the first K jobs in the solution sequence: $\{J_{(1)}, J_{(2)}, \ldots, J_{(K)}\}$. Term B is the contribution to the total tardiness from the remaining $(n-K)$ jobs: $\{J_{(K+1)}, J_{(K+2)}, \ldots, J_{(n)}\}$. Now consider the reduced problem involving only the K jobs $\{J_{(1)}, J_{(2)}, \ldots, J_{(K)}\}$. Term A is the total tardiness for this problem. If $(J_{(1)}, J_{(2)}, \ldots, J_{(K)})$ is not the optimal solution to this problem, then there is another sequence of the K jobs that yields a smaller total tardiness. However, this would mean that for the original problem, we could construct a solution sequence using the improved sequence for the first K jobs and the original sequence for the remaining $n-K$ jobs. The new solution would have a smaller Term A and the same Term B as the original solution. This would be in contradiction with our assumption that the original solution is optimum.

Now we can use the above observation to find an optimal solution sequence. Define $T_{opt}(Q)$ to be the minimum total tardiness obtained by sequencing the jobs in Q *optimally*. If Q contains a single job, say $Q = \{J_i\}$, then

$$T_{opt}(Q) = \max\left(t_i^{proc} - t_i^{due}, 0\right) \qquad (6\text{-}17)$$

since there is only one way to sequence one job, and that job must complete at $t_i^{comp} = t_i^{proc}$. If Q contains $K > 1$ jobs, then we can calculate $T_{opt}(Q)$ by using the facts that the last job must complete at t_Q^{comp}, and that in any optimal solution sequence, the first $K-1$ jobs are sequenced optimally for a reduced problem involving just those jobs. Therefore, we have

$$T_{opt}(Q) = \min_{J_i \ in \ Q}\left\{T_{opt}(Q-\{J_i\}) + \max\left(t_Q^{comp} - t_i^{due}, 0\right)\right\} \qquad (6\text{-}18)$$

In other words, to find the sequence of the jobs in Q that will minimize total tardiness, we consider placing each job in turn, at the last position in the sequence. We select the possibility with the minimum total tardiness.

Equation 6-17 defines the minimum total tardiness $T_{opt}(Q)$ for all Q containing a single job. Using these values and Equation 6-18, we can calculate $T_{opt}(Q)$ for all Q containing two jobs. Subsequent applications of Equation 6-18 yield $T_{opt}(Q)$ for all Q containing three jobs, then four jobs, and so on. In the end, we find $T_{opt}(\{J_1, J_2, \ldots, J_n\})$. In doing so, we find the optimum sequence. This process is demonstrated in the example below.

The example problem consists of four jobs. Their processing times and due dates are shown below.

J_i	J_1	J_2	J_3	J_4
$t_i$proc	8	6	10	7
$t_i$due	14	9	16	16

First, $T_{opt}(Q)$ for the four single-job sets can be calculated. For example, by Equation 6-17, we have

$$T_{opt}(\{J_1\}) = \max\{t_1{}^{proc} - t_1{}^{due}, 0\}$$
$$= \max\{8-14, 0\}$$
$$= 0$$

Similar calculations yield the results in Table 6.14.

J_i	$\{J_1\}$	$\{J_2\}$	$\{J_3\}$	$\{J_4\}$
$t_i{}^{proc} - t_i{}^{due}$	-6	-3	-6	-9
$T_{opt}(Q)$	0	0	0	0

Table 6.14. $T_{opt}(Q)$ for the Single-Job Sets

Next, we apply Equation 6-18 to calculate $T_{opt}(Q)$ for the six sets containing two jobs. For example, if $Q = \{J_1, J_2\}$, $t_Q{}^{comp} = t_1{}^{proc} + t_2{}^{proc} = 14$, and

$$T_{opt}(Q) = \min\{T_{opt}(Q-\{J_2\}) + \max\{t_Q{}^{comp} - t_1{}^{due}, 0\},$$
$$T_{opt}(Q-\{J_1\}) + \max\{ t_Q{}^{comp} - t_2{}^{due}, 0\}$$
$$= \min\{T_{opt}(\{J_1\}) + \max\{14-14, 0\},$$
$$T_{opt}(\{J_2\}) + \max\{14-9, 0\}\}$$
$$= \min\{0+0, 0+5\}$$
$$= 0$$

Results of the remaining calculations are shown in Table 6.15:

Q	$\{J_1,J_2\}$	$\{J_1,J_3\}$	$\{J_1,J_4\}$	$\{J_2,J_3\}$	$\{J_2,J_4\}$	$\{J_3,J_4\}$
$t_Q{}^{comp}$	14	18	15	16	13	17
'last' job J_i	J_1 J_2	J_1 J_3	J_1 J_4	J_2 J_3	J_2 J_4	J_3 J_4
$T_{opt}(Q-\{J_i\}) +$ $\max(t_Q{}^{comp} -$ $t_i{}^{due}, 0)$	0 5	4 2	1 0	7 0	4 0	4 4
minimum	*	*	*	*	*	*
$T_{opt}(Q)$	0	2	0	0	0	4

Table 6.15. $T_{opt}(Q)$ for the Two-Job Sets

For each set Q there are two possibilities for the last job in the sequence. For each of these possibilities, the resulting total tardiness $T_{opt}(Q -\{J_i\}) + \max(t_Q{}^{comp} - t_i{}^{due}, 0)$ is displayed in a separate column. The minimum total tardiness for each Q, $T_{opt}(Q)$, is marked with an asterisk.

Continuing in this fashion, $T_{opt}(Q)$ for all four sets Q containing three jobs may be calculated. For example, if $Q = \{J_1, J_2, J_3\}$, $t_Q{}^{\text{comp}} = t_1{}^{\text{proc}} + t_2{}^{\text{proc}} + t_3{}^{\text{proc}} = 24$, and

$$\begin{aligned}
T_{\text{opt}}(Q) &= \min\{\, T_{\text{opt}}(Q-\{J_1\}) + \max\{t_Q{}^{\text{comp}}{-}t_1{}^{\text{due}},\, 0\}, \\
&\qquad T_{\text{opt}}(Q-\{J_2\}) + \max\{t_Q{}^{\text{comp}}{-}t_2{}^{\text{due}},\, 0\}, \\
&\qquad T_{\text{opt}}(Q-\{J_3\}) + \max\{t_Q{}^{\text{comp}}{-}t_3{}^{\text{due}},\, 0\}\} \\
&= \min\{T_{\text{opt}}(\{J_2,J_3\}) + \max\{24{-}14,\, 0\}, \\
&\qquad T_{\text{opt}}(\{J_1,J_3\}) + \max\{24{-}9,\, 0\}, \\
&\qquad T_{\text{opt}}(\{J_1,J_2\}) + \max\{24{-}16,\, 0\}\} \\
&= \min\{0{+}10,\ 2{+}15,\ 0{+}8\}\ =\ 8
\end{aligned}$$

The most important characteristic of dynamic programming is the decomposition of the overall problem into nested sub-problems, with the solution of one problem being derived from that of the preceding problem. In the above calculations, we see that the solution for the three-job set has been derived from values obtained during the solution of the two-job set: $T_{\text{opt}}(\{J_2,J_3\})$, $T_{\text{opt}}(\{J_1,J_3\})$, and $T_{\text{opt}}(\{J_1,J_2\})$. Results of the remaining calculations are shown in Table 6.16:

Finally, $T_{\text{opt}}(Q)$ for the one set Q containing all four jobs may be calculated. Here, $t_Q{}^{\text{comp}} = t_1{}^{\text{proc}} + t_2{}^{\text{proc}} + t_3{}^{\text{proc}} + t_4{}^{\text{proc}} = 31$, and

$$\begin{aligned}
T_{\text{opt}}(Q) &= \min\{T_{\text{opt}}(Q-\{J_1\}) + \max\{t_Q{}^{\text{comp}}{-}t_1{}^{\text{due}},\, 0\}, \\
&\qquad T_{\text{opt}}(Q-\{J_2\}) + \max\{t_Q{}^{\text{comp}}{-}t_2{}^{\text{due}},\, 0\}, \\
&\qquad T_{\text{opt}}(Q-\{J_3\}) + \max\{t_Q{}^{\text{comp}}{-}t_3{}^{\text{due}},\, 0\}, \\
&\qquad T_{\text{opt}}(Q-\{J_4\}) + \max\{t_Q{}^{\text{comp}}{-}t_4{}^{\text{due}},\, 0\}\} \\
&= \min\{\, T_{\text{opt}}(\{J_2,J_3,J_4\}) + \max\{31{-}14,\, 0\}, \\
&\qquad T_{\text{opt}}(\{J_1,J_3,J_4\}) + \max\{31{-}9,\, 0\}, \\
&\qquad T_{\text{opt}}(\{J_1,J_2,J_4\}) + \max\{31{-}16,\, 0\}, \\
&\qquad T_{\text{opt}}(\{J_1,J_2,J_3\}) + \max\{31{-}16,\, 0\}\} \\
&= \min\{7{+}17,\ 9{+}22,\ 5{+}15,\ 8{+}15\} \\
&= 20
\end{aligned}$$

Q	$\{J_1, J_2, J_3\}$	$\{J_1, J_2, J_4\}$	$\{J_1, J_3, J_4\}$	$\{J_2, J_3, J_4\}$
$t_Q$comp	14	18	15	16
'last' job J_i	$J_1\ J_2\ J_3$	$J_1\ J_2\ J_4$	$J_1\ J_3\ J_4$	$J_2\ J_3\ J_4$
$T_{opt}(Q-\{J_i\})$ + $\max(t_Q$comp$-t_i$due$, 0)$	10 17 8	7 12 5	12 9 11	15 7 7
minimum	*	*	*	*
$T_{opt}(Q)$	8	5	9	7

Table 6.16. $T_{opt}(Q)$ for the Three-Job Sets

This calculation is summarized in Table 6.17:

Q	$\{J_1, J_2, J_3, J_4\}$
$t_Q$comp	31
'last' job J_i	$J_1\ \ J_2\ \ J_3\ \ J_4$
$T_{opt}(Q-\{J_i\})$ + $\max(t_Q$comp$-t_i$due$, 0)$	24 31 20 23
minimum	*
$T_{opt}(Q)$	20

Table 6.17. $T_{opt}(Q)$ for the Four-Job Set

By solving a sequence of nested problems, we have drawn the conclusion that the optimal solution sequence yields a total tardiness of 20, or a mean tardiness of 5. The asterisk in Table 6.17 tells us that the minimum mean tardiness is obtained when J_3 is in the last position. This means that $\{J_1, J_2, J_4\}$ occupy the first three positions in some order. Looking at Table 6.16, we see that out of these three remaining jobs, J_4 should come last. This leaves $\{J_1, J_2\}$ for the first two positions. From Table 6.15, we see that J_1 should come last out of these two jobs, leaving J_2 as the first job to be processed. Thus, the optimal schedule is (J_2, J_1, J_4, J_3).

Several points about the dynamic programming approach are worth emphasizing. The objective function to be optimized must be decomposable as is in Equation 6-19, which is repeated below for reference:

$$n\overline{T} = \sum_{k=1}^{K} \max\left(t_{(k)}^{comp} - t_{(k)}^{due}, 0\right) + \sum_{k=K+1}^{n} \max\left(t_{(k)}^{comp} - t_{(k)}^{due}, 0\right)$$
$$= \qquad\qquad A \qquad\qquad + \qquad\qquad B \qquad\qquad\qquad (6\text{-}19)$$

Specifically, it must be possible to minimize Term A independently of

Term B. In our example this condition was satisfied because Term A is strictly a function of the sequence $(J_{(1)}, J_{(2)}, \ldots, J_{(K)})$, while Term B is a function of the sequence $(J_{(K+1)}, J_{(K+2)}, \ldots, J_{(n)})$ and the sum of the processing times in the *set* $\{ J_{(1)}, J_{(2)}, \ldots, J_{(K)} \}$. Thus $\{ J_{(1)}, J_{(2)}, \ldots, J_{(K)} \}$ can be re-ordered to minimize Term A without affecting the value of Term B. The dynamic programming approach breaks down if this independence criterion is not fulfilled.

We also note that the solution of each one of the nested sub-problems provides many possible partial solutions. The selection of the correct partial solution depends on the solution of the *following* sub-problem, so that only after the last sub-problem is solved will any part of the overall solution be determined. At this point, in fact, the overall solution is completely determined.

6.4.3 An Application of Network Analysis for Progress Plots

The *progress plot* is an application of standard PERT network data, which provides a clear, simple, visual comparison of current progress versus expected progress for the control of a one-time manufacturing project such as the construction of a manufacturing facility. Furthermore a progress plot provides data about tolerable deviations from the plan and developments that may call for remedial action. Progress plotting is useful for comparing different projects with respect to planning accuracy. Once a project is underway, however, the probability of finishing on schedule is not easily read from traditional PERT or Gantt charts. The progress plot however, shows these probabilities at a glance.

Figure 6.46 [60] shows a progress plot for a completed project. The horizontal axis represents the time. The vertical axis represents the project progress, as a percentage of the critical path (CP). The original plan called for the project start at t_0 and for the completion at planned completion time t_{PC}. Time t_{PC} equals the critical path from a task/event network, such as a PERT chart.

Had this project gone as planned, the actual progress line would have lain on the original plan line, reaching 100-percent completion at t_{PC}. Instead, the project ended at actual completion time t_{AC}. Filled circles are planned CP events, open circles are actual event occurrences. Solid lines connecting these circles reveal a continually increasing discrepancy between the actual progress and the planned progress.

The control lines of Figure 6.46 [60] were warned of a probable late finish early in the project. At t_1 the probability of finishing by t_{PC} was 0.25, (i.e. the actual progress line crossed the $p = 0.25$ line at t_1) and falling. By t_2, the project only had a 0.01 chance of finishing on time.

Note that all segments of the actual progress line slope were fewer than those of the original plan line. This suggests that the task times were consistently underestimated or that inadequate resources were used. The smoothness of the progress line, however, showed that the progress was relatively continuous and that the expected critical path was in fact, the actual critical path.

Figure 6.47 [60] is a top level PERT network for a new product development project. Numbered circles are events and arrows are tasks that lead from event to event. Figure 6.47 also gives an estimated completion time and variance for each task, in weeks (variances are inside parentheses). These come from standard PERT methods: if $t_{i,j}$ is the task time needed to go from event i to event j, then the expected task time $E(t_{i,j})$ and the estimated variance of the task time probability distribution $\sigma^2_{i,j}$ are given by Equations 6-20 and 6-21, that is,

$$E(t_{i,j}) = \frac{a + 4m + b}{6} \tag{6-20}$$

and

$$\sigma^2_{i,j} = \left(\frac{b-a}{6}\right)^2 \tag{6-21}$$

where   a = estimated shortest possible task time
          b = estimated longest possible task time
          m = estimated most likely task time

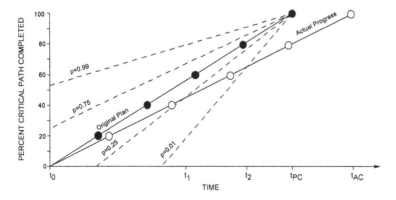

Fig. 6.46. A Progress Plot

For instance, in task 1,2 (the task between events 1 and 2, which is represented by the arc between nodes 1 and 2), a = 2, b = 6.6 and m = 4 weeks. Analysis of the estimated task times reveals a 32.5 week CP linking events 1,4,7,11,14, and 15. Figure 6.48 [60] shows the CP by itself, E ($t_{i,j}$) for each CP task, each task's contribution to the CP, expressed as a percentage of the total CP. The progress plot framework is built from this data. Horizontal segment $t_0 - t_{PC}$ of Figure 6.46 [60] equals the CP length of 32.5 weeks. The vertical axis is scaled from 0 to 100 percent. The original plan lines run from point (t_0, 0 percent) to point (t_{PC}, 100 percent). CP events appear on this line at the event's estimated occurrence (horizontal axis) and critical path location (vertical axis).

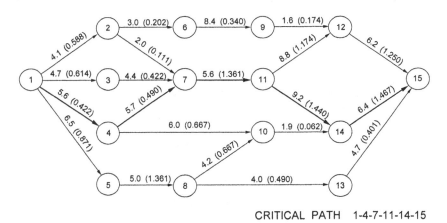

CRITICAL PATH   1-4-7-11-14-15

Fig. 6.47. The PERT Activity/Event Network for the Example Calculations

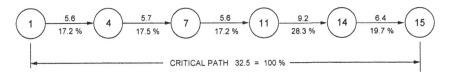

Fig. 6.48. The Critical Path (CP) for the Progress Plot Examples

These are the filled circles in Figure 6.46 [60]. Actual progress is tracked by plotting the actual completion times of the CP events, as total time after t_0 (horizontal axis) and original CP percentages for those events (vertical axis). These are the open circles in Figure 6.46.

Figure 6.49 [60] shows how the progress plot reveals the planning mistakes. All segments of the actual progress line A slope more steeply than those of the original plan line. All tasks have therefore been completed under the planned (estimated) time. By contrast, Line B illustrates consistently underestimated task times. Line C shows the effects of overestimating some tasks and underestimating others. The extent to which the actual progress line either deviates from or "hugs" the original plan line represents its planning accuracy. Line D shows what happens when tasks, not on the original CP, took longer than planned and became CP tasks. This kind of planning error creates a horizontal line segment in the progress line (there is a real time gap between the end of one CP task and the start of the next).

The original plan and actual progress lines are only descriptive statistics. A progress plot's control value is enhanced if task time variances are used to make probabilistic inferences about the actual project completion time.

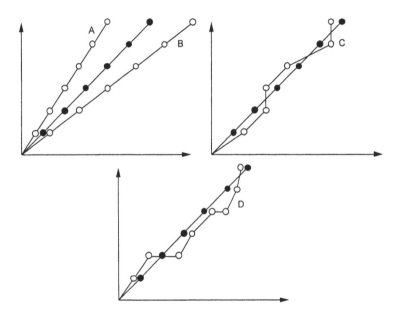

Fig. 6.49. Four Types of Progress Lines

As illustrated in Figure 6.46 [60], these inferences follow when the actual progress crosses a control line. Each control line is associated with a probability – the probability that the project reaches completion time on or before t_{PC}. Control lines for any number of p values between 0 and 1.0 may be calculated, but the plot is clearer if only a few are plotted, e.g. $p = 0.01$, $p = 0.25$, $p = 0.75$, and $p = 0.99$. The original plan line is also the $p = 0.50$ line.

The form and location of control lines depend on the assumptions that one makes about the project. The following assumptions are made:

- Task times are independent (actual time used for any task has no effect on estimated task time or variance of any other task).
- Probability distributions for task times have means and variances given by the Equations 6-20 and 6-21.
- Probability distributions for the sums of individual task times are approximately normal with means and variances as follows:

$$\mu_T = \sum E\!\left(t_{i,j}\right)$$

$$\sigma_T^2 = \sum \sigma_{ij}^2$$

for all tasks i, j included in the total.

- Estimated completion time for the project equals the sum of the estimated critical path task times.
- During the project, estimates a, b, and m for each task do not change.
- The likelihood of non-CP tasks intruding on the CP is so small that only CP tasks need to be considered when constructing control lines.
- Variances of individual task time probability distributions are proportional to task length.

When the actual progress line crosses a control line, e.g., the $p = 0.25$ line, the probability of completing the project by t_{PC} or sooner is 0.25, as long as the original a, b, and m estimates still apply. However, if a deviation from the original plan occurs, and more or less resources than originally planned for a specific task are now utilized (for instance) then the original control line probabilities will not be valid during the task.

New control lines may have to be drawn when changes occur in the quantity of project resources or in the estimates of a, b, and m. If the changes represent corrective actions for one CP task, to get the project back on schedule, the original control lines should remain. They will apply again when the "unusual" task has been completed. If however, a, b, and m estimates change for several CP tasks, new control lines should be drawn, since the old ones will not be valid again. And if changes in a, b, and m during the project alter the CP itself, the original plan line may stay in place, as far as it represents completed tasks, and a revised original plan line can be drawn to the new point (t_{PC}, 100 percent).

Based on the assumptions, one can use several methods for plotting the control lines [60]. The simplest method is that called "Proportional CP Variances". Individual task time variance estimates are used only to obtain an estimated variance for total project completion time. The variance is then "redistributed" among individual tasks. Figure 6.50 [60] shows the way that control lines are drawn. Under this method, each control line is a straight line from a point on the horizontal (time) axis, point t_x, 0 percent, to the project end, (t_{PC}, 100 percent). For the control line at $p = x$, the value of t_x is computed as:

$$t_x = z_x \sqrt{\sigma_T^2}$$

where z_x is the value under a unit-normal distribution above which a proportion x of the distribution lies, and

$$\sigma_T^2 = \sum \sigma_{ij}^2 \qquad \text{for all tasks } i, j \text{ on the CP}$$

z_x values come from a unit-normal table. For $x = 0.01$, for instance, $z_x = 2.326$ and for $x = 0.25$, $z_x = 0.675$. For the CP and other data shown in Figure 6.48 [60]:

$$\sigma_T^2 = 0.422 + 0.490 + 1.361 + 1.440 + 1.467 = 5.180$$

and, for $x = 0.01$:

$$t_x = (2.326)\sqrt{5.180} = 5.29 \text{ weeks}$$

Similarly, t_x for $x = 0.25$ is $(0.675)(5.180)^{0.5} = 1.54$ weeks.

Control lines for probabilities greater than 0.50 lie above the original plan line, intersecting the vertical axis at some value greater than 0 percent. The horizontal axis coordinate of these is found, as Figure 6.50 [60] suggests, by extending the axis to the left of t_0, where negative values represent an actual start before the planned start at t_0.

From the unit-normal distribution, all z values for the p values, greater than 0.50, will be negative. From the unit-normal distribution's symmetry note that

$$t_{1-x} = -t_x$$

The function of progress plots is similar to that of the statistical process-control charts used for the monitoring and controlling of manufacturing processes. Progress plots and control charts put the present performance immediately into historical context; they show, at a glance, the succeeding or failing of actions at keeping to the overall plan, and they show the cumulative effects of many small deviations from the plan. On the other hand, progress plots and control charts do not automatically expose the root cause behind deviations from the plan. As indicated, a progress plot can imply certain reasons for project slippage (consistently underestimated task durations or inadequate application of resources for example), but it will not explain this slippage. Nor will the progress plot predict unforeseen problems at the time the original project plans were made. The predictive power of the progress plot is limited to probability statements about on-time project completion - given that all of the original project planning assumptions still hold.

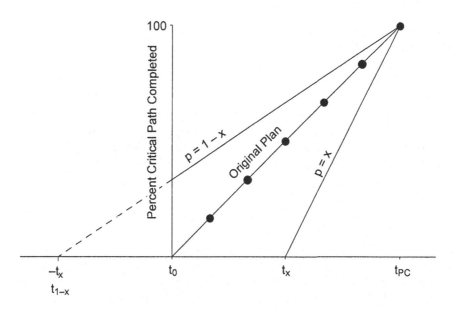

Fig. 6.50. Elements of Progress Plot Construction

6.4.4 A Simulation and Rule-Based System Approach to the On-line Control of Manufacturing Systems

A consensus among researchers in scheduling is that a combination of dispatch rules in many cases works better than an individual dispatch rule does. Discrete event simulation can be used in combination with dispatch rules for controlling and scheduling a manufacturing system [60-64]. In this approach, discrete simulation of a manufacturing system model is used to *evaluate* the performance of a set of plausible dispatching rules over a short planning horizon, Δt. The rule with the best simulated performance in the planning horizon is then *applied* to the physical system. At the end of Δt, the state of the physical system is incorporated into the system model. The evaluation/application process is carried on repeatedly.

The selection of a dispatch rule can be performed by a rule-based system [65,66], and can be based, for example, on the question as to which of the following performance measures are most relevant to the manufacturing system.

- maximum completion time
- mean flowtime
- maximum flowtime
- number of tardy jobs
- mean tardiness
- maximum tardiness

An overview of a possible scheduling mechanism is shown in Figure 6.51 [64]. The major components of the scheduling mechanism are the controller and the simulator. A subset of dispatch rules from a global set are selected based on the system objectives (performance measures) and on the current system status (a description of the on-going operations of an FMS).

The simulator then configures a model of the FMS with the current system status and performs a series of simulation runs, each for a short planning horizon Δt, for each of the dispatch rules. The best dispatch rule is then passed on to the controller, which generates a series of execution commands as prescribed by the best dispatch rule. Parts in the shop are dispatched based on these execution commands until the next planning horizon Δt. If no significant changes, either in the status of the system or in the objectives are detected, then the simulation/execution cycle is repeated using the existing subset of dispatch rules. If significant changes in the system status or in the objectives are encountered, the process begins anew with the selection of another, more appropriate subset of dispatch rules.

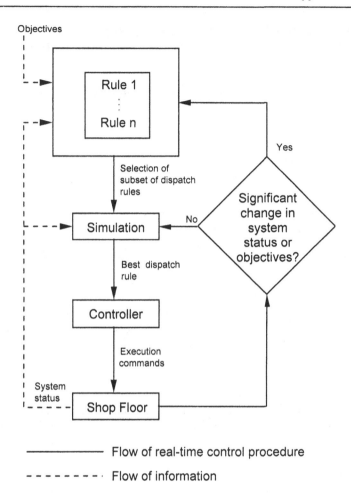

Fig. 6.51. Simulation/Rule-Based Approach to On-Line Control

The major drawback of this proposed scheduling methodology is that the selection of a subset of dispatch rules from a global set is a fundamentally arbitrary process. This applies even if a rule-based system is used to perform this function. Although much research has been devoted to determining the conditions under which particular dispatch rules perform well [45-51], the results apply only to the particular manufacturing systems and conditions studied, and are difficult to be generalized further. Unfortunately, since simulation is a computationally expensive process, it is not possible to evaluate a large subset of dispatch rules under the time constraints of real-time control.

6.4.5 An Intelligent Manufacturing Planning and ConTrol (IMPACT) Approach to the Operation of Manufacturing Systems

In actual manufacturing systems, dispatching (the assignment of resources, such as machines to production tasks at specific times) is typically performed ad hoc, or via the application dispatch rules (Section 6.3.3). Decision-making procedures may provide a comprehensive, fundamentally sound alternative to empirically stated dispatch rules. In making resource-task assignment decisions, resources and tasks can be considered simultaneously and each task can be assigned to a specific resource, in contrast to dispatch rules, which only select the next task to be performed. Decision-making procedures can also consider multiple criteria simultaneously.

The Intelligent Manufacturing Planning and ConTrol (IMPACT) approach, which makes use of decision-making procedures [67, 68], and which has been applied to actual industrial facilities [69, 70], is described below. In the IMPACT approach, a manufacturing system can be viewed generically as a hierarchy (Fig. 6.52). The highest level of the hierarchy, the factory, corresponds to the system as a whole. A factory can be divided into job shops, which are sets of work centers which produce a family of products or sub-assemblies. A work center consists of resources capable of performing similar manufacturing processes. For example, a turning work center may include some or all of the lathes of a job shop. There is no need for all individual resources to be at the same location in the factory, since only in a work center can the logical grouping of resources take place. A *resource* is an individual production unit, such as a machine, a human worker, or a manufacturing cell (a group of machines and auxiliary devices (e.g. robots) which work together to perform an operation).

The IMPACT approach performs dispatching dynamically during the operation of the manufacturing system. Within each work center, whenever one or more resources become free after completing their tasks, a dispatching decision takes place which assigns one pending task (a task which has been released to the work center but which has not yet been processed) to each free resource. Simulating this dispatching function over time, results in a list of assignments (indicating which resources process which tasks, and when) to each work center. *The schedule for the entire manufacturing system is built up by combining the assignment lists for the individual work centers.* Dispatching must be performed dynamically because task arrivals and resource breakdowns, which cannot be predicted with certainty ahead of time, constantly change the tasks and the resources involved. If dispatching decisions were performed ahead of time, then unforeseen interruptions, such as breakdowns would cause many of the resulting assignments to become infeasible. Since this dispatching must result in feasible assignments, it is a *finite-capacity* process.

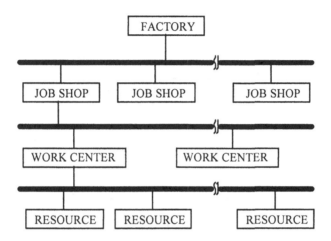

Fig. 6.52. The Manufacturing System Hierarchy

The IMPACT procedure is invoked at every change in the status of a work center (when a resource becomes available upon completion of its task, etc.). At these *decision points,* the procedure assigns one pending task to each of the available resources. The decision making process consists of the following four steps:

1. Form a set of alternatives.
   An alternative is defined as an assignment of tasks to available manufacturing resources. Only one task is assigned to each available resource at a time, to provide the most flexibility for adapting to unforeseen future events. For example, if resources R1 and R2 are available, and three tasks T1, T2 and T3 are pending, two possible alternatives are as shown in Fig. 6.52.
2. Determine a set of relevant decision making criteria.
   A criterion is an estimate of some relevant aspect of the performance of a schedule, given that a particular alternative is implemented. It is calculated for each alternative formed at a decision point in order to quantify the alternative's "goodness." Different manufacturing systems will have different criteria. Also, a manufacturing system will usually have more than one criterion. Examples of criteria are mean tardiness and

mean flowtime:

$$TARD(\text{alt}_q) = \frac{\sum_{i=1}^{L} \max\left[0; \left(T_i^{comp}(\text{alt}_q) - T_i^{dd}\right)\right]}{L} \qquad (6\text{-}22)$$

$$FLOW(\text{alt}_q) = \frac{\sum_{i=1}^{L} \left(\left(T_i^{comp}(\text{alt}_q) - T_i^{arr}\right)\right)}{L} \qquad (6\text{-}23)$$

where:
- alt_q: the qth alternative formed at the decision point
- L: the number of pending tasks in the work center at the decision point
- $T_i^{comp}(\text{alt}_q)$: the completion time of the ith pending task if alt_q is implemented
- T_i^{dd}: the due date of the ith pending task
- T_i^{arr}: the time at which the ith pending task arrived at the work center
  At a decision point, the aspect of future schedule performance, estimated by the mean tardiness criterion is called the mean tardiness performance measure, and it is the average amount of time beyond the due date that a job is completed. Similarly, the aspect of future schedule performance estimated by the mean flowtime criterion is called the mean flowtime performance measure, and it is the average time that a job (part) spends on the system.

3. Determine the consequences of the different alternatives with respect to the different criteria.
   The consequence of an alternative, with respect to a criterion, is the estimated value of that criterion should the alternative be implemented. For example, the consequence of Alternative 1 (Fig. 6.53) with respect to the mean flowtime criterion may be determined as follows (Fig. 6.54).

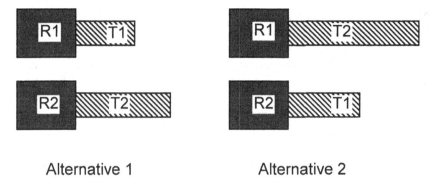

Alternative 1   Alternative 2

Fig. 6.53. Examples of Alternatives

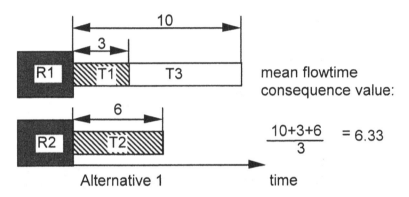

mean flowtime
consequence value:

$$\frac{10+3+6}{3} = 6.33$$

Alternative 1   time

Fig. 6.54. Example of Determining the Consequence of an Alternative With Respect to a Criterion (Mean Flowtime)

Here the mean flowtime criterion has been estimated by considering (but not actually implementing) the assignment of the unassigned task T3 to resource R1, and then applying the mean flowtime criterion definition (Eq. 6-23). We could just as well have considered assigning T3 to R2 to perform the estimate. The true criterion value is the average of the flowtimes of the two possible assignments. Although this is a small calculation for this simple example and an estimate is unnecessary, in a realistic dispatching situation, it is impossible to consider all the ways of assigning the unassigned tasks, and we must estimate the true criterion value by considering a sample of the possible ways of assigning the unassigned tasks. The more assignments we consider, the better our estimate will be.

4. Select the best alternative

The last step is to select the best alternative using a utility value, which measures the goodness of the consequence values. The alternative with the highest utility is defined as the best alternative (Fig. 6.55). The utility of an alternative is calculated by normalizing its consequence values. In Figure 6.55, the mean flowtime consequence values can be normalized by dividing the mean flowtime by the sum of the mean flowtimes, and then by subtracting the result from one. This normalizes the consequence values so that they will sum up to one, moreover the lower (better) values will result in higher utility values. The utility for Alternative 1 can then be calculated as 1–6.33/(5.43+6.33) = 0.46; similarly, the utility for Alternative 2 can be calculated as 1–5.43/(5.43+6.33) = 0.54. In this case Alternative 2 would be selected. The normalization of the consequence value c_{ij} (of an alternative i with respect to a criterion j) is defined formally below. For benefit criteria (e.g. quality), which should be maximized, the normalization formula is shown in Equation 6-24. For cost criteria (e.g. mean flowtime, mean tardiness), which should be minimized, the normalization formula is shown in Equation 6-25.

Alt #	Mean Flowtime	Utility	Selected Alternative
1	6.33	0.46	
2	5.43	0.54	←

Fig. 6.55. Selecting the Best Alternative When There is a Single Criterion

$$\bar{c}_{ij} = \frac{c_{ij}}{\sum\limits_{k=1}^{m} c_{kj}} \qquad (6\text{-}24)$$

$$\bar{c}_{ij} = \frac{1}{m-1}\left[1 - \frac{c_{ij}}{\sum\limits_{k=1}^{m} c_{kj}}\right] \qquad (6\text{-}25)$$

where:
- c_{ij}: the consequence value of alternative i with respect to criterion j
- $\bar{c}_{ij}$: the normalized value of cij

- m: the number of alternatives

When multiple criteria are considered, each alternative has multiple consequence values. The alternative's utility is then calculated by taking a weighted sum of the normalized consequence values. For example, in Figure 6.55, there are two criteria, mean flowtime and mean tardiness. Weighting the two criteria equally, the utility of Alternative 1 can be calculated as $0.5 \cdot [1-6.33/(5.43+6.33)] + 0.5 \cdot [1-1.25/(1.25+3.88)] = 0.61$, and the utility of Alternative 2 can be calculated as $0.5 \cdot [1-5.43/(5.43+6.33)] + 0.5 \cdot [1-3.88/(1.25+3.88)] = 0.39$. In this case, Alternative 1 would be selected.

The decision making steps can be formalized in a decision matrix, where the rows represent the alternatives and the columns represent the criteria. Each matrix element represents the consequence of an alternative with respect to a criterion (Fig. 6.57).

Alt #	Mean Flowtime	Mean Tardiness	Utility	Selected Alternative
1	6.33	1.25	0.61	←
2	5.43	3.88	0.39	

Fig. 6.56. Selecting the Best Alternative When There Are Multiple Criteria

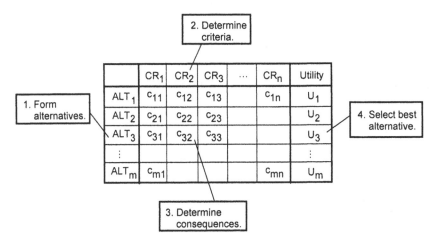

Fig. 6.57. A Decision Matrix

The quality of a schedule can be judged by a set of *performance meas-*

ures. It is useful to re-emphasize the distinction between performance measures and criteria at this point. Performance measures are defined with reference to the entire schedule, after it has been produced. Decision making criteria, on the other hand, are estimates, rather than measures, of performance. They are calculated during the production of the schedule, based both on the information about the tasks and on the resources available in a work center, at a decision point. They serve to estimate the effect of local assignment decisions within a work center on the schedule of an entire job shop or of a system. Examples of performance measures are *mean tardiness* and *mean flowtime*:

$$\text{TARD}_j = \max\left[0; \left(T_j^{comp} - T_j^{dd}\right)\right] \tag{6-26}$$

$$\text{FLOW}_j = T_j^{comp} - T_j^{arr} \tag{6-27}$$

$$\overline{\text{TARD}}(T) = \frac{\sum_{j=1}^{C} \text{TARD}_j}{C} \tag{6-28}$$

$$\overline{\text{FLOW}}(T) = \frac{\sum_{j=1}^{C} \text{FLOW}_j}{C} \tag{6-29}$$

where:
- T: the point in time at which the performance measure is calculated
- C: the number of tasks which were completed on or before time T
- T_j^{comp}: the time at which task j was completed
- T_j^{dd}: the due date of task j
- T_j^{arr}: the time at which task j arrived

The determination of the mean flowtime performance measure for a completed schedule is shown in Figure 6.58.

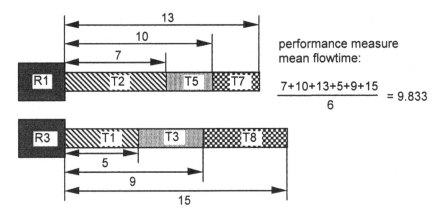

performance measure
mean flowtime:

$$\frac{7+10+13+5+9+15}{6} = 9.833$$

Fig. 6.58. Calculation of the Mean Flowtime Performance Measure

The decision making procedure is embedded in software, which simulates the operation of a manufacturing system. To establish a schedule for a manufacturing system, a model of the system is simulated for the period of time for which a schedule is needed. During the simulation, the decision making procedure is invoked whenever a new event, such as a task completion occurs. The output of the simulation is a schedule, which can be viewed either as a printed list or in the form of a Gantt chart. It consists of a detailed sequence of resource-task assignments as they have been produced by the decision making process in each work center of the simulation model. This schedule can be transmitted to the actual resources on the factory floor either electronically, via computer terminals, or manually via a hard copy. The schedule can be generated periodically as necessary, and if the factory floor conditions change locally in a particular work center, the decision making procedure can produce a local assignment decision.

The combined nesting – scheduling problem

This paragraph addresses the scheduling of the carpet weaving process, which is a problem of nesting rectangular patterns under complex production constraints (Figure 6.59, [71]). A schedule, in cases where nesting is required, cannot be properly evaluated without first having solved the nesting problem, since its solution influences the evaluation of the schedule. The concept presented includes the use of the IMPACT approach in combination with a rule-based mechanism (Figure 6.60, [71]) constructed and expanded by accumulating problem-solving expertise from knowledge sources, such as human experts and factory log files [71]. The objective is to find a good - not necessarily the optimum - nesting schedule

(Figure 6.59), taking into consideration the overall production objectives, such as meeting due dates, minimizing of the cost, maximizing of the machines and the stock sheet utilization.

The tasks to be scheduled correspond to the product items of the customers' orders and they may be dispatched in parallel machines. Each task corresponds to a product item, which is characterized by a specific rectangular pattern (tasks T0 – T9 in Figure 6.59). Each product item is defined by a set of attributes, such as size and type, whilst on the other hand, each machine is characterized by the production attributes, namely, maximum width, production rate and sequence dependent times.

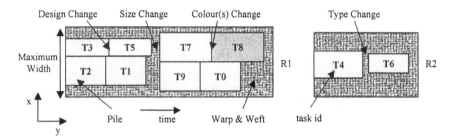

Fig. 6.59. A carpet nesting schedule

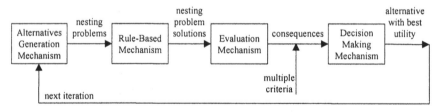

Fig. 6.60. IMPACT approach with rule-based mechanism

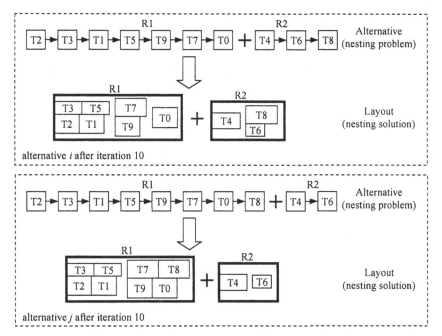

Fig. 6.61. Different alternatives (nesting problems) and their layouts (nesting problem solutions) in last iteration

The approach (Figure 6.60) involves the generation of scheduling alternatives, their transformation through a rule based mechanism into nesting solutions (Figure 6.61, [71]) and finally, their evaluation by using four different criteria that reflect the overall production objectives: tardiness minimization, cost minimization and maximization of the machines and stock sheet utilization.

A continuous process case: the short-term refinery scheduling problem

In a refinery, a number of operations take place for turning raw material (crude oils) into higher value end (petroleum) products. Demand drives the definition of the production targets (quantities of final products with specific properties within specified time periods) to a refinery. The first steps for achieving these targets involve the processing of the different crude oils on their own, or as blends, in the distillation units. All crudes are usually delivered to storage tanks by vessels. In a typical refinery, there is often an intermediate level between distillation units and storage tanks, consisting of charging tanks for storing and mixing the crudes. The distillation process takes place in the crude oil distillation unit (CDU), which separates the charged oil into fractions, such as gasoline, kerosene, gas oil and residual. Lighter fractions are produced in lower temperatures, while in turn, the heaviest fraction, called the residue fraction, is a collection of compounds with a very high boiling point. Every type of crude oil has different characteristics, i.e. when it is processed under the same operating conditions, there will be different yields and properties for each fraction [72].

This paragraph does not deal with the entire refinery process but focuses on the short-term scheduling of the crude oil unloading, blending and distillation activities of a typical refinery. In particular, the scheduling problem, under discussion, includes a) the assignment of a docking station to each vessel, b) the specification of the volume flow transfer among storage tanks, charging tanks and distillation units per time interval and c) the determination of the temperature cut points for each distillation unit per time interval.

The material flow related to the refinery processes examined in this paper can be modeled as a system consisting of vessels, docking stations, storage tanks, charging tanks, CDUs and streams (Figure 6.62, [72]).

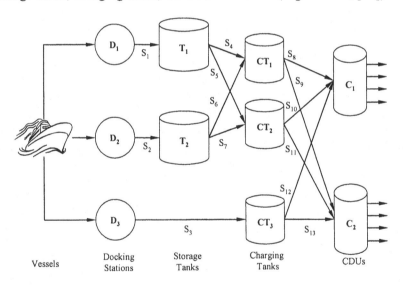

Fig. 6.62. Example of a refinery model

A number of vessels, carrying various types of crude oil, are scheduled to arrive. The arrival dates of the vessels along with the crude oil types and composition data are known in advance. The types of crude oil and blends may be categorized according to their composition: for instance, the crude oil types and the blends could be distinguished into Medium

Sulfur (MS) and Low Sulfur (LS) categories, according to their sulfur concentration. Crude oil and blends are transferred from the storage tanks to the charging tanks and from the latter to CDUs. Furthermore, the following major operation constraints must be met: a) equipment capacity limitations (minimum and maximum volume capacity for tanks and transfer rate for streams), b) minimum and maximum acceptable component (such as sulfur in the mixed crude oil stream) concentration for the blends feeding CDUs, c) minimum and maximum acceptable quantity and density for each distillation product, produced in the CDUs for every time interval, according to the monthly demand forecast [72].

In this problem, the IMPACT approach is adapted to the continuous nature of the problem, through the discretization of time and decision variables and the implementation of a constraint satisfaction mechanism. Ii is assumed all operating decisions are made at the beginning of each decision (time) interval. The duration of the time interval (for instance a 24-hour day or an 8-hour shift) depends on how often scheduling and operating decisions are made in a specific refinery and is, therefore, user-defined. The decision variables are shown in Table 6.18 [72]: the stream feed rate S_s, ranging between the minimum and maximum capacity of the stream s, the temperature $W_{d,p}$ for the pth cut of the dth distillation column, ranging between the minimum ($W_{d,p,min}$) and the maximum ($W_{d,p,max}$) temperatures for the respective cut and the docking station TT_v to which the vth vessel is connected. All decision variables except for TT_v are continuous and they are discretized (discretization steps represented by the user-defined parameters $S_{s,step}$ and $W_{d,p,step}$ for variables S_s and $W_{d,p}$ respectively), taking into account the existing pumps and distillation units hardware and the corresponding control strategy.

Decision Variables	Range	Discretization step
S_s	$[S_{s,min}, S_{s,max}]$	$0 < S_{s,step} \leq S_{s,max} - S_{s,min}$
$W_{d,p}$	$[W_{d,p,min}, W_{d,p,max}]$	$0 < W_{d,p,step} \leq W_{d,p,max} - W_{d,p,min}$
TT_v	0 if no docking station is selected, else 1...NK (NK: number of docking stations)	-

Table 6.18. Decision Variables in the refinery scheduling problem

Two criteria are used for the evaluation of the alternatives: the cost criterion and the blend category changeover criterion, which should both be minimized. The cost of a single time period (interval) is equal to the sum of the vessels' waiting cost, tank inventory cost and oil transfer cost.

Blend category changeover (e.g. from MS to LS and vice versa) occurs every time the charged blend category, reaching the distillation unit, is changed.

6.4.6 An Agent-Based Framework for Manufacturing Decision Making

During the last decades, a new generation of decentralized, agent-based factory control algorithms has appeared in literature. A software agent [73] a) is a self-directed object, b) has its own value systems and a means of communicating with other such objects, c) continuously acts on its own initiative. A system of this kind of agents, called a multi-agent system, consists of a group of identical or complementary agents that act together.

This paragraph describes a flexible agent-based system named RIDER (Real tIme DEcision-making in manufactuRing) [74], which encompasses both real-time and decentralized manufacturing decision-making capabilities.

In RIDER, each agent, as a software application instance, is responsible for monitoring a specific set of resources, namely, machines, buffers, or labor that belong to a production system. When an event, such as a machine breakdown, is identified, the agent supervising the corresponding domain of resources, initiates the decision making process and generates a set of local alternatives (Figure 6.63). Each local alternative proposes different values for the decision variables, such as the production rate or the task dispatching sequence, thus suggesting a series of actions to be taken.

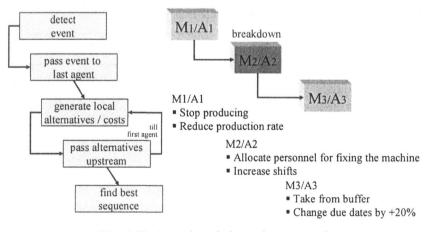

Fig. 6.63. Agent-based alternatives generation

When the proposed actions affect other resource domains supervised by neighbor agents, a message exchange communication mechanism is employed. This mechanism is used for notifying the neighbor agents to commence generating new alternatives that will be based on the ones that have already been created.

The local alternatives, originating from the involved agents, form a tree consisting of alternative decision paths. User defined objective functions are utilized for estimating the consequences of each decision path. The built-in evaluation mechanism is flexible enough so as to take into account the varied importance of different business aspects, such as production cost, as they are affected by the proposed actions of each alternative.

The real-time information required for monitoring the system status and for generating valid alternatives is obtained through a special data exchange mechanism, incorporated in the RIDER framework, in order to communicate with other installed manufacturing information technology systems (Figure 6.64).

The overall schedule is finally generated by a backward scheduling algorithm. Although, not too many applications have been reported in the literature, agent-based and decentralized decision-making systems are expected to be applied extensively in the future to the engineering and manufacturing worlds.

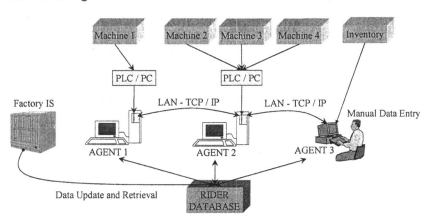

Fig. 6.64. RIDER architecture

Further Reading

The operation of manufacturing systems is a multi-disciplinary field, in which the methods and tools developed in operations research, artificial intelligence, manufacturing and other related fields are blended. The treatment of the operation of manufacturing systems, in a single chapter, can only provide the reader with an overview, which is the primary intention of this chapter. For a detailed analysis, we refer the reader to the existing literature from which the various sections in this chapter have been derived.

The introductory discussion on academic and industrial perspectives was primarily drawn from "A Review of Production Scheduling" by Graves [2], *Sequencing and Scheduling* by French [2], "A Recent Survey of Production Scheduling" by Rodammer [3], "Deterministic job-shop scheduling: Past, present and future" by Jain and Meeran [26] and "Resource-Constrained Project Scheduling: A Survey of Recent Developments"by Herroelen *et al.* [29]. These works provide a categorization of scheduling methods and perspectives from which scheduling can be approached, such as control theory and artificial intelligence. Descriptions of the various levels of manufacturing planning and the associated planning paradigms, such as MPS, MRP, capacity planning, shop floor control and JIT were adapted from *Shop Floor Control* by Melnyk *et al* [12], *Modern Production/Operations Management* by Buffa and Sarin [13] and *Operations Management* by Bennet *et al* [14]. These books provide very readable accounts of the various planning concepts and supplement them with relevant examples. The book entitled "*Designing and evaluating Value Added Services in manufacturing e-marketplace*" by Perrone *et al.* [75], is an excellent textbook, discussing the concepts related to manufacturing e-marketplaces, the extended enterprise as well as the agent-based systems.

The sections on network analysis and decision making were adapted from *Modern Production/Operations Management* by Buffa and Sarin [16], where a thorough treatment of PERT and CPM, and multiple attribute decision making techniques, complete with examples, are found. The section on decision making was also complemented with material drawn from *An Introduction to Management Science* by Cabot and Harnett [31].

The section on Gantt charts and performance measures and on the application of mathematical programming and dynamic programming techniques to dispatching, was adapted from *Sequencing and Scheduling* by French [3]. This book provides a survey of operations research techniques as applied to the operation of manufacturing systems. The use of network analysis for project management was drawn from "Schedule Monitoring for Engineering Projects" by Schmidt [60]. This paper provides an innovative approach to predicting the probability of a project being completed in time.

References

1. Wiers, V.C.S., "A Review of the Applicability of OR and AI Scheduling Techniques in Practice, Omega", *International Journal of Management Science*, (Vol. 25, No. 2, 1997), pp. 145-153.

2. Graves, S.C., "A Review of Production Scheduling", *Operations Research* (Vol. 29, 1981), pp. 646-675.

3. French, S., *Sequencing and Scheduling: An Introduction to the Mathematics of the Job Shop*, Ellis Horwood Limited, Chichester, West Sussex, England, 1982.

4. Rodammer, F.A., "A Recent Survey of Production Scheduling," *IEEE Transactions on Systems, Man and Cybernetics*, (Vol. 18, No. 6, 1988), pp. 841-851.

5. Akella, R., Y. Choong and S.B. Gershwin, "Performance of a Hierarchical Production Scheduling Policy", *IEEE Transactions on Components Hybrid Manufacturing Technology*, (Vol. CHMT 7, No. 3, 1984), pp. 225-238.

6. Gershwin, S.B., R.R. Hildebrant, R. Suri and S.K. Mitter, "A Control Perspective on Recent Trends in Manufacturing Systems", *IEEE Control Systems* (1986), pp. 3-15.

7. Abraham, C., B. Dietrich, S. Graves and W. Maxwell, "A Research Agenda for Models to Plan and Schedule Manufacturing Systems", *IBM Watson Research Center, working paper*, 1985.

8. Ho, Y.C., "A Short Tutorial on Perturbation Analysis of Discrete Event Dynamic Models to Plan and Schedule Manufacturing Systems", *Proceedings of the First ORSA/TIMS Special Interest Conference on Flexible Manufacturing Systems* (1984), pp. 372-378.

9. Suri, R. and J.W. Dille, "On-Line Optimization of Flexible Manufacturing Systems Using Perturbation Analysis", *Proceedings of the First ORSA/TIMS Special Interest Conference on Flexible Manufacturing Systems* (1984), pp. 379-384.

10. Bensana, E., G. Bel and D. Dubois, "OPAL: A Multi-Knowledge Based System for Industrial Job-Shop Scheduling", *International Journal of Production Research*, (Vol. 26, No. 5, 1988), pp. 795-819.

11. Anthony, R.N., *Planning and Control Systems: A Framework for Analysis*, Harvard University Graduate School of Business Administration, Boston, MA, 1965.

12. Berry, W.L., T.E. Vollman and D.C. Whybark, *Master Production Scheduling: Principles and Practice*, American Production and Inventory Control Society, Washington, D.C., 1979.

13. Melnyk, S.A., P.L. Carter, D.M. Dilts and D.M. Lyth, *Shop Floor Control*, Dow Jones-Irwin, Homewood, IL, 1985.

14. Keung, K.W., W.H. Ip and C. Y. Chan, "An enhanced MPS solution for FMS using GAs", *Integrated Manufacturing Systems*, (Vol. 12, No. 5, 2001), pp. 351-359.

15. Venkataraman, R. and M.P. D'Itri, "Rolling horizon master production schedule performance: a policy analysis", *Production Planning & Control*, (Vol. 12, No. 7, 2001), pp. 669-679.

16. Buffa, E.S. and R.K. Sarin, *Modern Production / Operations Management*, 8th edition, John Wiley and Sons, Singapore, 2003.

17. Bennet, D., C. Lewis and M. Oaklley, *Operations Management*, Philip Allan, Oxford, 1988.

18. Benton, W.C. and H. Shin, "Manufacturing planning and control: The evolution of MRP and JIT integration", *European Journal of Operational Research*, (Vol. 110, 1998), pp. 411-440.

19. Nicholson, T.A.J. and R.D. Pullen, "A Practical Control System for Optimizing Production Schedule," *International Journal of Production Research*, (Vol. 9, No.2, 1971), pp. 219-227.

20. Irastorza, J.C. and R.H. Deane, "Starve the Shop - Reduce Work-in-Process", *Production and Inventory Management* (Vol. 17, No. 2, 1976), pp. 20-25.

21. Jacobs F.R. and E. Bendoly, "Enterprise resource planning: Developments and directions for operations management research," *European Journal of Operational Research*, 146, 233-240, 2003.

22. Umble E. J., R.R. Haft and M.M. Umble, "Enterprise resource planning: Implementation procedures and critical success factors," *European Journal of Operational Research*, 146, 241-257, 2003.

23. Missbach M. and U. Hoffmann, *SAP Hardware Solutions: Servers, Storage, and Networks for mySAP.com*, HP Professional Books, 2001.

24. http://www.standishgroup.com, 1998.

25. Chryssolouris, G., S. Makris, N. Papakostas and V. Xanthakis, "A co-operative software environment for planning and controlling ship-repair contracts," *Proceedings of the 4th International Conference on e-ENGDET*, Leeds Metropolitan University U.K, (September 2004), pp. 321-330.

26. Jain A.S. and S. Meeran, "Deterministic job-shop scheduling: Past, present and future", *European Journal of Operational Research*, (Vol. 113, 1999), pp. 390-434

27. Chryssolouris G., N. Papakostas, S. Makris and D. Mourtzis, "Planning and Scheduling of Shipyard Processes", *Application of Information Technologies to the Maritime Industries*, C. Guedes Soares, J. Brodda (Eds.), 255-274, 1999.

28. Kolisch R. and R. Padman, "An Integrated Survey of Project scheduling - Models, algorithms, Problems, and Applications", *Technical Report,* August 1997.

29. Herroelen W.B. de Reyck and E. Demeulemeester, "Resource-Constrained Project Scheduling: A Survey of Recent Developments", *Computers and Operations Research*, (Vol. 25, No. 4, 1998), pp. 279-302.

30. Davis, E.W., and J.H. Patterson, "A Comparison of Heuristic and Optimum Solutions in Resource Constrained Project Scheduling", *Management Science*, (Vol. 21, 1975), pp. 944-955.

31. Cabot, A.V and D.L. Harnett, *An Introduction to Management Science*, Addison-Wesley, Reading, MA, 1977.

32. Keeney, R.L. and H. Raiffa, *Decisions With Multiple Objectives: Preferences and Value Tradeoffs*, Wiley, New York, NY, 1976.

33. Philip, J., "Algorithms for the Vector Maximization Problem", *Mathematical Programming*, (Vol. 2, 1972), pp. 207-229.

34. Zeleny, M., *Linear Multiobjective Programming*, Springer Verlag, Berlin, 1980.

35. Evans, J.P. and R.E. Steuer, "A Revised Simplex Method for Linear Multiple Objective Programs", *Mathematical Programming*, (Vol. 5, 1973), pp. 54-72.

36. Geoffrion, A., J.S. Dyer and A. Feinberg, "An Interactive Approach for Multi-Criterion Optimization, with an Application to the Operation of an Academic Department", *Management Science*, (Vol. 19, 1972), pp. 367-368.

37. Zoints, S. and J. Wallenius, "An Interactive Programming Method for Solving the Multiple Criteria Problem", *Management Science* (Vol. 22, 1976), pp. 652-663.

38. Sarin, R.K., "Ranking of Multiattribute Alternatives and an Application to Coal Power Plant Siting," in *Multiple Criteria Decision Making: Theory and Applications,* G. Fandel and T.Gal (eds.), Springer Verlag, Berlin, 1980.

39. Baker, K.R., "Priority Dispatching in the Single Channel Queue with Sequence-Dependent Set-ups", *Journal of Industrial Engineering* (Vol. 19, No. 4, 1968).

40. Buffa, E.S. and J.G. Miller, *Production Inventory Systems-Planning and Control*, Third edition, Richard D. Irwin Inc., Homewood, IL, 1979.

41. Conway, R.W., W.L. Maxwell and L.W. Miller, *Theory of Scheduling*, Addison-Wesley Publishing Company, Reading, MA, 1967.

42. Jackson, J.R., "Simulation Research on Job-Shop Production", *Naval Research Logistics Quarterly* (Vol. 4, No. 4, 1957).

43. Rowe, A.J., "Toward a Theory of Scheduling", *Journal of Industrial Engineering*, (Vol. 11, No. 2, March 1960).

44. Malstrom, E.M., "A Literature Review and Analysis Methodology for Traditional Scheduling Rules in a Flexible Manufacturing System", *Final Technical Report performed under CAM-I Contract LA-83-FM-01*, (1983).

45. Blackstone, J.H., D.T. Philips and G.L. Hogg, "A State of the Art Survey of Dispatching Rules for Manufacturing Job Shop Operations", *International Journal of Production Research*, (Vol. 20, No. 1, 1982), pp. 27-45.

46. Conway, J.H., B.M. Johnson and W.L. Maxwell, "An Experimental Investigation of Priority Dispatching," Journal of Industrial Engineering (Vol. 11, No. 3, 1960).

47. Moodie, C.L. and S.D. Roberts, "Experiments with Priority Dispatching Rules in a Parallel Processor Shop", *International Journal of Production Research*, (Vol. 6, No. 4, 1968).

48. Elvers, D.E., "The Sensitivity of the Relative Effectiveness of Job Shop Dispatching Rules with Various Arrival Distributions," *Transactions of the American Institute of Industrial Engineers*, (Vol. 6, 1974), pg. 41.

49. Nanot, Y.R., "An Experimental Investigation and Comparative Evaluation of Priority Disciplines in Job-shop-like Queuing Networks", *Management Sciences Research Project, UCLA, Research Report No. 87*, 1963.

50. Panwalker, Y.R. and W. Iskander, "A Survey of Scheduling Rules", *Operations Research*, (Vol. 25, 1977), pg. 48.

51. Rochette, R. and R.P. Sadowski, "A Statistical Comparison of the Performance of Simple Dispatching Rules for a Particular set of Job Shops", *International Journal of Production Research*, (Vol. 14, 1976), pg. 63.

52. Chryssolouris, G., N. Giannelos, N. Papakostas and D. Mourtzis, "Chaos Theory in Production Scheduling", *Annals of the CIRP*, (Vol. 53, No. 1, 2004), pp. 381-383.

53. Michalewicz, Z., *Genetic Algorithms + Data Structures = Evolution Programs*, Springer Verlag, 3rd Edition, 1996.

54. Fleury G. and M. Gourgand, "Genetic algorithms applied to workshop problems", *International Journal of Computer Integrated Manufacturing*, (Vol. 11, No. 2, 1998), pp.183-192.

55. Cheng R. and M. Gen, "An evolution program for the resource-constrained project scheduling problem", *International Journal of Computer Integrated Manufacturing*, (Vol. 11, No. 3, 1998), pp. 274-287.

56. Cavalieri S. and P. Gaiardelli, "Hybrid genetic algorithms for a multiple-objective scheduling problem", *Journal of Intelligent Manufacturing*, (Vol. 9, 1998), pp. 361-367.

57. Mori M. and C.C. Tseng, "A genetic algorithm for multi-mode resource constrained project scheduling problem", *European Journal of Operational Research*, (Vol. 100, 1997), pp. 134-141.

58. Wagner, H.M., "An Integer Programming Model for Machine Scheduling," *Naval Logistics Research Quarterly*, (Vol. 6, 1959), pp. 131-140.

59. Held, M. and R.M. Karp, "A Dynamic Programming Approach to Sequencing Problems", *Journal of the Society for Industrial and Applied Mathematics,* (Vol. 10, 1962), pp. 196-210.

60. Schmidt, M.J., "Schedule Monitoring for Engineering Projects", *IEEE Transactions on Engineering Management,* (Vol. 35, No. 2, 1988), pp. 108-114.

61. Fisher, H. and G.L. Thomson, "Probabilistic Learning Combinations of Local Job Shop Scheduling Rules", in *Industrial Scheduling*, edited by J.F. Muth and G.L. Thomson, Prentice-Hall, (1963), pp. 225-251.

62. Gere, W.S. "Heuristics in Job Shop Scheduling", *Management Science*, (Vol. 13, 1966), pp. 167-190.

63. Trybula, W.J. and R.G. Ingalls, *Simulation of Hybrid Automation*, General Electric Company, Electronics Automation Application Center, Charlottesville, Virginia, 1985.

64. Wu, S.Y. and R.A. Wysk, "An Application of Discrete-event Simulation to On-line Control and Scheduling in Flexible Manufacturing", *International Journal of Production Research*, (Vol. 27, No. 9, 1989), pp. 1603-1623.

65. Mentink, W., B. van der Pluym, B. Goedhart, H. de Swann Arons, and H. Steinen, "Realtime Expert Systems in CIM", *Proceedings of Artificial Intelligence Applications*, (1988), pp. 577-581.

66. Van der Pluym, B. "Knowledge-based Decision-Making for Job Shop Scheduling", *International Journal of Computer Integrated Manufacturing*, (Vol. 3, No. 6, 1990), pp. 354-363.

67. Chryssolouris, G., "MADEMA: An Approach to Intelligent Manufacturing Systems", *CIM Review* (Spring 1987), pp. 11-17.

68. Chryssolouris, G., K. Wright, J. Pierce and W. Cobb, "Manufacturing Systems Operation: Dispatch Rules Versus Intelligent Control", *Robotics and Computer Integrated Manufacturing*, (Vol. 4, No. 3/4, 1988), pp. 531-544.

69. Chryssolouris, G., J. Pierce and K. Dicke, "A Decision-Making Approach to the Operation of Flexible Manufacturing Systems", *International Journal of Flexible Manufacturing Systems* (Vol. 4, 1991).

70. Chryssolouris, G., K. Dicke, M. Lee, "An Approach to Short Interval Scheduling for Discrete Parts Manufacturing", *International Journal of Computer-Integrated Manufacturing*, (Vol. 4, No. 3, 1991), pp. 157-168.

71. Chryssolouris, G., N. Papakostas and D. Mourtzis, "A Decision-Making Approach for Nesting Scheduling: A Textile Case", *International Journal of Production Research*, (Vol. 38, No. 17, 2000), pp. 4555-4564.

72. Chryssolouris, G., N. Papakostas and D. Mourtzis, "Refinery short-term scheduling with tank farm, inventory and distillation management: an integrated simulation-based approach", *European Journal of Operational Research*, (Vol.166, No.3, 2005), pp. 812-827.

73. Baker, A., "A Survey of Factory Control Algorithms That Can Be Implemented in a Multi-Agent Hetararchy: Dispatching, Scheduling, and Pull", *Journal of Manufacturing Systems*, (Vol. 17, No. 4, 1998), pp. 297-320.

74. Papakostas, N., D. Mourtzis, K. Bechrakis, G. Chryssolouris et al., "A Flexible Agent Based Framework for Manufacturing Decision Mak-

ing", *Proceedings of the FAIM99 Conference*, Begell House Inc., Tilburg, Netherlands, (June 1999), pp. 789-800.

Perrone G., M. Bruccoleri and P. Renna, "Designing and evaluating Value Added Services in manufacturing e-marketplace" *to be published*, Kluwer editions.

Review Questions

1. What is a manufacturing system?

2. What is scheduling?

3. How can production scheduling problems be classified?

4. What are the major levels in manufacturing planning, and what are the characteristics of these levels?

5. Discuss the role of master production scheduling in a manufacturing system.

6. Compare and contrast the single stage and multistage planning decision system models.

7. Describe some of the tools that could be used in conjunction with master production scheduling.

8. Discuss the types of information that are required by the material requirements planning system?

9. Describe some of the statistical techniques of establishing safety stocks in a MRP system.

10. What is Manufacturing Resource Planning (MRP II)?

11. What is the role of capacity planning and how is it related to material requirements planning?

12. Describe some of the major resources that are managed by the shop floor control system.

13. What are the major activities of the shop floor control system?

14. State the characteristics of the shop floor control system and how do they vary with different types of manufacturing processes?

15. What are the goals of just in time manufacturing?

16. Describe the implications of just in time manufacturing to the shop floor control system.

17. What tools may be used for the efficient operation of manufacturing systems?

18. What is the critical path?

19. What is the significance of slack in network analysis?

20. How do probabilistic network methods differ from deterministic network methods?

21. What is meant by resource smoothing or load levelling?

22. Differentiate decision making under certainty with decision making under uncertainty.

23. What is the expected monetary value criterion?

24. Discuss some of the methods that may be used with multicriteria decision making problems.

25. When will one decision dominate another?

26. What is the efficient frontier and how can it be constructed?

27. What is the preference function and describe the procedure for assessing the preference function of the decision maker?

28. State some of the common dispatching rules.

Index

Mechanical Engineering Series *(continued from page ii)*